Nitric Oxide Signaling in Plants

Nitric Oxide Signaling in Plants

Editors

John T. Hancock
Steven J. Neill

MDPI • Basel • Beijing • Wuhan • Barcelona • Belgrade • Manchester • Tokyo • Cluj • Tianjin

Editors
John T. Hancock
University of the West of England
UK

Steven J. Neill
University of the West of England
UK

Editorial Office
MDPI
St. Alban-Anlage 66
4052 Basel, Switzerland

This is a reprint of articles from the Special Issue published online in the open access journal *Plants* (ISSN 2223-7747) (available at: https://www.mdpi.com/journal/plants/special_issues/nitric_oxide_sig).

For citation purposes, cite each article independently as indicated on the article page online and as indicated below:

LastName, A.A.; LastName, B.B.; LastName, C.C. Article Title. *Journal Name* **Year**, *Volume Number*, Page Range.

ISBN 978-3-0365-0006-5 (Hbk)
ISBN 978-3-0365-0007-2 (PDF)

© 2020 by the authors. Articles in this book are Open Access and distributed under the Creative Commons Attribution (CC BY) license, which allows users to download, copy and build upon published articles, as long as the author and publisher are properly credited, which ensures maximum dissemination and a wider impact of our publications.

The book as a whole is distributed by MDPI under the terms and conditions of the Creative Commons license CC BY-NC-ND.

Contents

About the Editors vii

Preface to "Nitric Oxide Signaling in Plants" ix

John T. Hancock
Nitric Oxide Signaling in Plants
Reprinted from: Plants **2020**, 9, 1550, doi:10.3390/plants9111550 1

Joana R. Expósito, Sara Martín San Román, Eva Barreno, José Reig-Armiñana, Francisco José García-Breijo and Myriam Catalá
Inhibition of NO Biosynthetic Activities during Rehydration of Ramalina farinacea Lichen Thalli Provokes Increases in Lipid Peroxidation
Reprinted from: Plants **2019**, 8, 189, doi:10.3390/plants8070189 7

John T. Hancock and Steven J. Neill
Nitric Oxide: Its Generation and Interactions with Other Reactive Signaling Compounds
Reprinted from: Plants **2019**, 8, 41, doi:10.3390/plants8020041 23

Francisco J. Corpas, Luis A. del Río and José M. Palma
Impact of Nitric Oxide (NO) on the ROS Metabolism of Peroxisomes
Reprinted from: Plants **2019**, 8, 37, doi:10.3390/plants8020037 37

Marie Agatha Mohn, Besarta Thaqi and Katrin Fischer-Schrader
Isoform-Specific NO Synthesis by Arabidopsis thaliana Nitrate Reductase
Reprinted from: Plants **2019**, 8, 67, doi:10.3390/plants8030067 47

Manuel Tejada-Jimenez, Angel Llamas, Aurora Galván and Emilio Fernández
Role of Nitrate Reductase in NO Production in Photosynthetic Eukaryotes
Reprinted from: Plants **2019**, 8, 56, doi:10.3390/plants8030056 63

Tamara Lechón, Luis Sanz, Inmaculada Sánchez-Vicente and Oscar Lorenzo
Nitric Oxide Overproduction by cue1 Mutants Differs on Developmental Stages and Growth Conditions
Reprinted from: Plants **2020**, 9, 1484, doi:10.3390/plants9111484 77

Lorena Aranda-Caño, Beatriz Sánchez-Calvo, Juan C. Begara-Morales, Mounira Chaki, Capilla Mata-Pérez, María N. Padilla, Raquel Valderrama and Juan B. Barroso
Post-Translational Modification of Proteins Mediated by Nitro-Fatty Acids in Plants: Nitroalkylation
Reprinted from: Plants **2019**, 8, 82, doi:10.3390/plants8040082 99

Misa Takahashi and Hiromichi Morikawa
Nitrogen Dioxide at Ambient Concentrations Induces Nitration and Degradation of PYR/PYL/RCAR Receptors to Stimulate Plant Growth: A Hypothetical Model
Reprinted from: Plants **2019**, 8, 198, doi:10.3390/plants8070198 121

Jana Jahnová, Lenka Luhová and Marek Petřivalský
S-Nitrosoglutathione Reductase—The Master Regulator of Protein S-Nitrosation in Plant NO Signaling
Reprinted from: Plants **2019**, 8, 48, doi:10.3390/plants8020048 139

About the Editors

John T. Hancock (Prof.) has had a longstanding interest in how small reactive compounds control cellular events, including reactive oxygen species, nitric oxide, hydrogen sulfide and, more recently, hydrogen gas.

Steve J. Neill (Prof.) has worked for many years on the way plants manage drought stress, including work on the roles of ABA and nitric oxide.

Preface to "Nitric Oxide Signaling in Plants"

With nitric oxide being recognized as an important signaling molecule in plants as well as in animals, it has been a pleasure to collect together a series of research articles focused on this topic. It is even more poignant that this Special Issue comes just after the forty-year celebration of the work on nitric oxide in plants. There have been many milestones and controversies along the way. Much of the research has looked to the work being reported in animals; while this can have some advantages, especially in the development of methods and technologies, it also has some serious drawbacks. It is tempting to think that the biochemistry in animals and plants is the same, or at least very similar, but there are major differences, too. A good example is the failure to find a true nitric oxide synthase in higher plants. Assays and inhibitor use hint at its existence, but outside of algae it seems elusive. The downstream signaling from nitric oxide production is not clear cut either. Therefore, the continued publication of papers in the field of nitric oxide and plants, and the continued enhancement of the profile of this area of science by the use of Special Issues, is crucial if a holistic understanding of the production and role of nitric oxide in plants is to be obtained. It only remains for me to thank all those involved in this Special Issue, from the editorial team to the authors. Thank you for all your support—John T. Hancock, November 2020.

John T. Hancock, Steven J. Neill
Editors

Editorial

Nitric Oxide Signaling in Plants

John T. Hancock

Department of Applied Sciences, University of the West of England, Bristol BS16 1QY, UK; john.hancock@uwe.ac.uk; Tel.: +44-(0)117-328-2475

Received: 3 November 2020; Accepted: 10 November 2020; Published: 12 November 2020

Abstract: Nitric oxide (NO) is an integral part of cell signaling mechanisms in animals and plants. In plants, its enzymatic generation is still controversial. Evidence points to nitrate reductase being important, but the presence of a nitric oxide synthase-like enzyme is still contested. Regardless, NO has been shown to mediate many developmental stages in plants, and to be involved in a range of physiological responses, from stress management to stomatal aperture closure. Downstream from its generation are alterations of the actions of many cell signaling components, with post-translational modifications of proteins often being key. Here, a collection of papers embraces the differing aspects of NO metabolism in plants.

Keywords: nitrate reductase; nitration; nitric oxide; reactive oxygen species; stress responses; S-nitrosation; S-nitrosylation; SNO-reductase; thiol modification

1. Introduction

Nitric oxide (NO) is now well acknowledged as an instrumental signaling molecule in both plants and animals [1]. First recognized as important as a signal in the control of vascular tone [2], its role in plants came to prominence in the late 1990s [3–5]. The forty years of research into NO in plants has just been highlighted by a review by Kolbert et al. [6].

In plants, NO has been found to be involved in a wide range of developmental stages and physiological responses. For example, NO has been found to be generated during pollination and pollen tube growth [7–9], seed germination [10], root development [11,12], and stomatal aperture control [13,14]. It is also instrumental in the orchestration of responses to stress in plants [15], including to heavy metals such as cadmium [16], salt [17], temperature [18], light [19] and pathogens [20].

NO in animals is known to be generated by several sources, but primarily nitric oxide synthase (NOS) is the enzyme which has a dominant role in NO accumulation in cells [21]. However, there is some controversy over whether NOS-like enzymes exist in higher plants [22]. There are homologues which have been found in algal species [23,24], but if higher plants have such an enzyme, its protein and gene(s) are being very elusive [25]. An enzyme which is known to make NO in plants is nitrate reductase, and this enzyme has been the focus of attention for several research groups [26–29].

Downstream of NO generation is also not without controversy. In animals, the classical pathway involves the generation of cGMP through the action of soluble guanylyl cyclase [30]. However, such pathways have recently been questioned in plants [31]. What is clear is that NO can lead to post-translational modification of proteins. Most commonly studied is the modification of thiol groups, so called S-nitrosation (otherwise called S-nitrosylation: for a recent overview of terminology to be used in plant NO research see [32]) [33,34]. Other modifications include tyrosine nitration [35]. Such modifications are often reversible and can be thought of as being akin to phosphorylation, where proteins can be toggled between functional states.

It can be seen, therefore, that NO is a crucial signaling molecule in plants. It can be generated endogenously, be seen to interact with many signal transduction components, and has numerous physiological responses. In this Special Issue, authors were invited to contribute papers encompassing this field of biochemistry.

2. Aspects of NO Metabolism

Life evolved in the presence of reactive compounds and many of these have been adopted as signaling molecules [36]. Looking at an ancient species, i.e., the lichen *Ramalina farinacea*, Expósito et al. [37] showed that NO production was likely to be dependent on NR. An inhibitor of NOS did not reduce NO levels in the lichen, whereas they reported the activity of NR to be 91 µU/mg protein, comparable with other systems.

The synthesis of NO in plants remains controversial [22], with the terminology to be used around NOS-like enzymes in plants recently being discussed [32]. In this Special Issue, Hancock and Neill [38] used a bioinformatic approach but failed to find evidence of an obvious NOS protein in plant databases. They also discussed how NO needs to interact with other reactive signaling molecules, a theme also picked up by Corpas et al. [39]. They, in a mini-review, discussed how NO is produced by peroxisome and that the NO produced interacts with glutathione and reactive oxygen species metabolism.

Two papers returned to the theme of NO production by discussing the enzyme nitrate reductase. Mohn et al. [40] reported on a comparative study between two NR isoforms, NIA1 and NIA2, and suggested that the different isoform have specialist functions. Tejada-Jimenez et al. [41] took a critical look at NR function and how interacting proteins may be involved. Lechón et al. [42] continued this theme by investigating the overproduction of NO in *cue1* mutants, and found that NO accumulation only occurs once seedlings are established.

Downstream events in NO-mediated signaling are embraced in the remaining papers. Post-translational modification (PTM) of proteins via *S*-nitrosation was discussed by Corpas et al. [39] as part of their discussions on peroxisome, but PTMs were also discussed by Aranda-Caño et al. [43]. Here, the role of nitrate fatty acids (NO_2-FAs) was discussed as signaling molecules and also how they may affect the modification of proteins, and hence function and activity. A second PTM, nitration, is a subject discussed by Takahashi and Morikawa [44]. In particular, they discussed the possible tyrosine nitration of PYR/PYL/RCAR receptors in leaves of *Arabidopsis thaliana*. Stimulated plant growth is the result of the signaling of PYR/PYL/RCAR receptors in *Arabidopsis thaliana*, and the authors studied how this may help to mediate the stimulation of plant growth.

The interaction of NO with glutathione and the formation of *S*-nitrosoglutathione is an immensely important aspect of NO biology [45]. Jahnová et al. [46] summarized the current thoughts on *S*-nitrosoglutathione reductase and how it has a crucial role in NO-based signaling.

3. Conclusions on *S*-Nitrosoglutathione Reductase (GSNOR) and How This Alters the Metabolism of *S*-Nitrosoglutathione and Hence *S*-Nitrosation of Proteins in Plant Cells

I hope that this Special Issue is a useful collection of papers which gives the reader an insight into the exciting area of NO biology in plants, and also hope that it inspires researchers to continue to work in this area, or indeed, to start investigations on plant NO metabolism. Such work would lead to the use of the manipulation of NO in plants as a way to enhance plant health and crop production, especially under stressful conditions [47].

Funding: This research received no external funding.

Acknowledgments: I would like to thank the unwavering support from the editorial office, and Prof Steve Neill, who helped edit this Special Issue. I would also like to thank all those who contributed papers and UWE, Bristol for supporting me to edit this Special Issue.

Conflicts of Interest: The author declares no conflict of interest.

References

1. Shapiro, A.D. Nitric oxide signaling in plants. *Vitam. Horm.* **2005**, *72*, 339–398. [CrossRef] [PubMed]
2. Palmer, R.M.J.; Ferrige, A.G.; Moncada, S. Nitric oxide release accounts for the biological activity of endothelium-derived relaxing factor. *Nature* **1987**, *327*, 524–526. [CrossRef] [PubMed]
3. Laxalt, A.M.; Beligni, M.V.; Lamattina, L. Nitric oxide preserves the level of chlorophyll in potato leaves infected by *Phytophthora infestans*. *Eur. J. Plant Pathol.* **1997**, *103*, 643–651. [CrossRef]
4. Delledonne, M.; Xia, Y.; Dixon, R.A.; Lamb, C. Nitric oxide functions as a signal in plant disease resistance. *Nature* **1998**, *394*, 585–588. [CrossRef] [PubMed]
5. Durner, J.; Wendehenne, D.; Klessig, D.F. Defense gene induction in tobacco by nitric oxide, cyclic GMP, and cyclic ADP-ribose. *Proc. Natl. Acad. Sci. USA* **1998**, *95*, 10328–10333. [CrossRef] [PubMed]
6. Kolbert, Z.; Barroso, J.B.; Brouquisse, R.; Corpas, F.J.; Gupta, K.J.; Lindermayr, C.; Loake, G.J.; Palma, J.M.; Petřivalský, M.; Wendehenne, D.; et al. A forty year journey: The generation and roles of NO in plants. *Nitric Oxide* **2019**, in press. [CrossRef]
7. Hiscock, S.J.; Bright, J.; McInnis, S.M.; Desikan, R.; Hancock, J.T. Signaling on the stigma: Potential new roles for ROS and NO in plant cell signaling. *Plant Signal Behav.* **2007**, *2*, 23–24. [CrossRef]
8. Reichler, S.A.; Torres, J.; Rivera, A.L.; Cintolesi, V.A.; Clark, G.; Roux, S.J. Intersection of two signalling pathways: Extracellular nucleotides regulate pollen germination and pollen tube growth via nitric oxide. *J. Exp. Bot.* **2009**, *60*, 2129–2138. [CrossRef]
9. Šírová, J.; Sedlářová, M.; Piterková, J.; Luhová, L.; Petřivalský, M. The role of nitric oxide in the germination of plant seeds and pollen. *Plant Sci.* **2011**, *181*, 560–572. [CrossRef]
10. Arc, E.; Galland, M.; Godin, B.; Cueff, G.; Rajjou, L. Nitric oxide implication in the control of seed dormancy and germination. *Front Plant Sci.* **2013**, *4*, 346. [CrossRef]
11. Correa-Aragunde, N.; Graziano, M.; Lamattina, L. Nitric oxide plays a central role in determining lateral root development in tomato. *Planta* **2004**, *218*, 900–905. [CrossRef] [PubMed]
12. Lombardo, M.C.; Graziano, M.; Polacco, J.C.; Lamattina, L. Nitric oxide functions as a positive regulator of root hair development. *Plant Signal. Behav.* **2006**, *1*, 28–33. [CrossRef] [PubMed]
13. Neill, S.; Barros, R.; Bright, J.; Desikan, R.; Hancock, J.; Harrison, J.; Morris, P.; Ribeiro, D.; Wilson, I. Nitric oxide, stomatal closure, and abiotic stress. *J. Exp. Bot.* **2008**, *59*, 165–176. [CrossRef] [PubMed]
14. Gayatri, G.; Agurla, S.; Raghavendra, A.S. Nitric oxide in guard cells as an important secondary messenger during stomatal closure. *Front Plant Sci.* **2013**, *4*, 425. [CrossRef] [PubMed]
15. Fancy, N.N.; Bahlmann, A.K.; Loake, G.J. Nitric oxide function in plant abiotic stress. *Plant Cell Environ.* **2017**, *40*, 462–472. [CrossRef]
16. Gill, S.S.; Hasanuzzaman, M.; Nahar, K.; Macovei, A.; Tuteja, N. Importance of nitric oxide in cadmium stress tolerance in crop plants. *Plant Physiol. Biochem.* **2013**, *63*, 254–261. [CrossRef]
17. Tailor, A.; Tandon, R.; Bhatla, S.C. Nitric oxide modulates polyamine homeostasis in sunflower seedling cotyledons under salt stress. *Plant Signal Behav.* **2019**, *17*, 1667730. [CrossRef]
18. Rai, K.K.; Pandey, N.; Rai, S.P. Salicylic acid and nitric oxide signaling in plant heat stress. *Physiol. Plant.* **2019**, *7*. [CrossRef]
19. Lytvyn, D.I.; Raynaud, C.; Yemets, A.I.; Bergounioux, C.; Blume, Y.B. Involvement of inositol biosynthesis and nitric oxide in the mediation of UV-B induced oxidative stress. *Front Plant Sci.* **2016**, *7*, 430. [CrossRef]
20. Mur, L.A.J.; Simpson, C.; Kumari, A.; Gupta, A.K.; Gupta, K.J. Moving nitrogen to the centre of plant defence against pathogens. *Ann. Bot.* **2017**, *119*, 703–709. [CrossRef]
21. Nathan, C.; Xie, Q.W. Nitric oxide synthases: Roles, tolls, and controls. *Cell* **1994**, *78*, 915–918. [CrossRef]
22. Jeandroz, S.; Wipf, D.; Stuehr, D.J.; Lamattina, L.; Melkonian, M.; Tian, Z.; Zhu, Y.; Carpenter, E.J.; Wong, G.K.-S.; Wendehenne, D. Occurrence, structure, and evolution of nitric oxide synthase-like proteins in the plant kingdom. *Sci. Signal.* **2016**, *9*, re2. [CrossRef] [PubMed]
23. Foresi, N.; Correa-Aragunde, N.; Parisi, G.; Caló, G.; Salerno, G.; Lamattina, L. Characterization of a nitric oxide synthase from the plant kingdom: NO generation from the green alga *Ostreococcus tauri* is light irradiance and growth phase dependent. *Plant Cell* **2010**, *22*, 3816–3830. [CrossRef] [PubMed]
24. Astier, J.; Jeandroz, S.; Wendehenne, D. Nitric oxide synthase in plants: The surprise from algae. *Plant Sci.* **2018**, *268*, 64–66. [CrossRef] [PubMed]

25. Santolini, J.; André, F.; Jeandroz, S.; Wendehenne, D. Nitric oxide synthase in plants: Where do we stand? *Nitric Oxide* **2017**, *63*, 30–38. [CrossRef] [PubMed]
26. Rockel, P.; Strube, F.; Rockel, A.; Wildt, J.; Kaiser, W.M. Regulation of nitric oxide (NO) production by plant nitrate reductase *in vivo* and *in vitro*. *J. Exp. Bot.* **2002**, *53*, 103–110. [CrossRef]
27. Chamizo-Ampudia, A.; Sanz-Luque, E.; Llamas, A.; Galvan, A.; Fernandez, E. Nitrate reductase regulates plant nitric oxide homeostasis. *Trends Plant Sci.* **2017**, *22*, 163–174. [CrossRef]
28. Desikan, R.; Grifitths, R.; Hancock, J.; Neill, S. A new role for an old enzyme: Nitrate reductase-mediated nitric oxide generation is required for abscisic acid-induced stomatal closure in *Arabidopsis thaliana*. *Proc. Natl. Acad. Sci. USA* **2002**, *99*, 16314–16318. [CrossRef]
29. Hao, F.; Zhao, S.; Dong, H.; Zhang, H.; Sun, L.; Miao, C. *Nia1* and *Nia2* are involved in exogenous salicylic acid-induced nitric oxide generation and stomatal closure in Arabidopsis. *J. Integr. Plant Biol.* **2010**, *52*, 298–307. [CrossRef]
30. Montfort, W.R.; Wales, J.A.; Weichsel, A. Structure and activation of soluble guanylyl cyclase, the nitric oxide sensor. *Antioxid. Redox Signal.* **2017**, *26*, 107–121. [CrossRef]
31. Astier, J.; Mounier, A.; Santolini, J.; Jeandroz, S.; Wendehenne, D. The evolution of nitric oxide signalling diverges between animal and green lineages. *J. Exp. Bot.* **2019**, *70*, 4355–4364. [CrossRef] [PubMed]
32. Gupta, K.J.; Hancock, J.T.; Petrivalsky, M.; Kolbert, Z.S.; Lindermayr, C.; Durner, J.; Barroso, J.B.; Palma, J.M.; Brouquisse, R.; Wendehenne, D.; et al. Recommendations, golden rules on terminology and practices used in plant nitric oxide research. *New Phytol.* **2019**, in press.
33. Lindermayr, C.; Saalbach, G.; Durner, J. Proteomic identification of S-nitrosylated proteins. *Plant Physiol.* **2005**, *137*, 921–930. [CrossRef] [PubMed]
34. Hancock, J.T.; Craig, T.; Whiteman, M. Competition of reactive signals and thiol modifications of proteins. *J. Cell Signal.* **2017**, *2*, 170. [CrossRef]
35. Kolbert, Z.; Feigl, G.; Bordé, Á.; Molnár, Á.; Erdei, L. Protein tyrosine nitration in plants: Present knowledge, computational prediction and future perspectives. *Plant Physiol. Biochem.* **2017**, *113*, 56–63. [CrossRef]
36. Hancock, J.T. Harnessing evolutionary toxins for signaling: Reactive oxygen species, nitric oxide and hydrogen sulfide in plant cell regulation. *Front. Plant Sci.* **2017**, *8*, 189. [CrossRef]
37. Expósito, J.R.; San Román, S.M.; Barreno, E.; Reig-Armiñana, J.; García-Breijo, F.J.; Catalá, M. Inhibition of NO biosynthetic activities during rehydration of *Ramalina farinacea* lichen thalli provokes increases in lipid peroxidation. *Plants* **2019**, *8*, 189. [CrossRef]
38. Hancock, J.T.; Neill, S.J. Nitric oxide: Its generation and interactions with other reactive signaling compounds. *Plants* **2019**, *8*, 41. [CrossRef]
39. Corpas, F.J.; del Río, L.A.; Palma, J.M. Impact of nitric oxide (NO) on the ROS metabolism of peroxisomes. *Plants* **2019**, *8*, 37. [CrossRef]
40. Mohn, M.A.; Thaqi, B.; Fischer-Schrader, K. Isoform-specific NO synthesis by *Arabidopsis thaliana* nitrate reductase. *Plants* **2019**, *8*, 67. [CrossRef]
41. Tejada-Jimenez, M.; Llamas, A.; Galván, A.; Fernández, E. Role of nitrate reductase in NO production in photosynthetic eukaryotes. *Plants* **2019**, *8*, 56. [CrossRef] [PubMed]
42. Lechón, T.; Sanz, L.; Sánchez-Vicente, I.; Lorenzo, O. Nitric oxide overproduction by *cue1* mutants differs on developmental stages and growth conditions. *Plants* **2020**, in press.
43. Aranda-Caño, L.; Sánchez-Calvo, B.; Begara-Morales, J.C.; Chaki, M.; Mata-Pérez, C.; Padilla, M.N.; Valderrama, R.; Barroso, J.B. Post-translational modification of proteins mediated by nitro-fatty acids in plants: Nitroalkylation. *Plants* **2019**, *8*, 82. [CrossRef] [PubMed]
44. Takahashi, M.; Morikawa, H. Nitrogen dioxide at ambient concentrations induces nitration and degradation of PYR/PYL/RCAR receptors to stimulate plant growth: A hypothetical model. *Plants* **2019**, *8*, 198. [CrossRef] [PubMed]
45. Yun, B.W.; Skelly, M.J.; Yin, M.; Yu, M.; Mun, B.G.; Lee, S.U.; Hussain, A.; Spoel, S.H.; Loake, G.J. Nitric oxide and S-nitrosoglutathione function additively during plant immunity. *New Phytol.* **2016**, *211*, 516–526. [CrossRef]

46. Jahnová, J.; Luhová, L.; Petřivalský, M. *S*-nitrosoglutathione reductase—The master regulator of protein *S*-nitrosation in plant NO signaling. *Plants* **2019**, *8*, 48. [CrossRef] [PubMed]
47. Simontacchi, M.; Galatro, A.; Ramos-Artuso, F.; Santa-María, G.E. Plant survival in a changing environment: The role of nitric oxide in plant responses to abiotic stress. *Front Plant Sci.* **2015**, *6*, 977. [CrossRef]

Publisher's Note: MDPI stays neutral with regard to jurisdictional claims in published maps and institutional affiliations.

© 2020 by the author. Licensee MDPI, Basel, Switzerland. This article is an open access article distributed under the terms and conditions of the Creative Commons Attribution (CC BY) license (http://creativecommons.org/licenses/by/4.0/).

Article

Inhibition of NO Biosynthetic Activities during Rehydration of *Ramalina farinacea* Lichen Thalli Provokes Increases in Lipid Peroxidation

Joana R. Expósito [1,*], Sara Martín San Román [1], Eva Barreno [2], José Reig-Armiñana [2], Francisco José García-Breijo [3] and Myriam Catalá [1]

1. Department of Biology and Geology, Physics and Inorganic Chemistry, ESCET—Campus de Móstoles, C/Tulipán s/n, E-28933 Móstoles (Madrid), Spain
2. Universitat de València, Botánica & ICBIBE—Jardí Botànic, Fac. CC. Biológicas, C/Dr. Moliner 50, 46100 Burjassot, Valencia, Spain
3. U. Politècnica de València, Dpto. Ecosistemas Agroforestales, Camino de Vera s/n, 46020 Valencia, Spain
* Correspondence: joana.exposito@urjc.es

Received: 31 March 2019; Accepted: 17 June 2019; Published: 26 June 2019

Abstract: Lichens are poikilohydrous symbiotic associations between a fungus, photosynthetic partners, and bacteria. They are tolerant to repeated desiccation/rehydration cycles and adapted to anhydrobiosis. Nitric oxide (NO) is a keystone for stress tolerance of lichens; during lichen rehydration, NO limits free radicals and lipid peroxidation but no data on the mechanisms of its synthesis exist. The aim of this work is to characterize the synthesis of NO in the lichen *Ramalina farinacea* using inhibitors of nitrate reductase (NR) and nitric oxide synthase (NOS), tungstate, and NG-nitro-L-arginine methyl ester (L-NAME), respectively. Tungstate suppressed the NO level in the lichen and caused an increase in malondialdehyde during rehydration in the hyphae of cortex and in phycobionts, suggesting that a plant-like NR is involved in the NO production. Specific activity of NR in *R. farinacea* was 91 µU/mg protein, a level comparable to those in the bryophyte *Physcomitrella patens* and *Arabidopsis thaliana*. L-NAME treatment did not suppress the NO level in the lichens. On the other hand, NADPH-diaphorase activity cytochemistry showed a possible presence of a NOS-like activity in the microalgae where it is associated with cytoplasmatic vesicles. These data provide initial evidence that NO synthesis in *R. farinacea* involves NR.

Keywords: *Trebouxia*; microalgae; lipid peroxidation; diaphorase activity; lichens; nitric oxide; nitrate reductase; nitric oxide synthase

1. Introduction

Lichens are symbiogenetic organisms composed of fungi (mycobionts) and their photosynthetic partners (photobionts), which may be unicellular green algae (phycobionts, microalgae) or cyanobacteria [1,2] and bacterial communities. Lichens are nowadays in the focus of understanding multi-microbial symbioses evolutionary processes. They are poikilohydrous, subjected to repeated desiccation/rehydration cycles, and able to survive in extreme, frequently very dry environments, such as deserts or the arctic and Antarctic habitats. They can remain long periods with inactive metabolism and restart it again in the presence of water (reviewed by Kranner et al. [3]). Rehydration of lichens is a stressful situation that results in the massive release of reactive oxygen species (ROS). ROS are produced in the oxidative phosphorilation (respiratory) and photosynthetic electron chains, but their production increase during stress such as nutrient limitation and exposure to xenobiotics, and are a major cause of damage during desiccation-rehydration events, especially in photosynthetic organisms [4]. When desiccated, carbon fixation is limited by water deficiency, but electron flow continues, and excitation energy can be passed from photo-excited chlorophyll pigments to ground state oxygen, forming singlet oxygen ($^1O_2$). In addition, superoxides ($O_2^{\bullet-}$), hydrogen peroxides (H_2O_2), and the hydroxyl radicals ($^\bullet OH$) can be produced at photosystem (PS) II [5]. If antioxidant defenses are overcome by ROS production, the uncontrolled free radicals cause widespread cellular damage by provoking protein alterations, lipid peroxidation, and the formation of DNA adducts [6]. The lichen symbiosis is intricately linked to desiccation tolerance, for which potent ROS scavenging machinery is essential [4].

Nitric oxide (NO) is an intra- and intercellular signaling molecule involved in the regulation of diverse biochemical and physiological processes. These functions include signal transduction, cell communication, stress signaling, and metabolism of free radicals (reviewed by Wilson et al. [7], Mur et al. [8]). NO has been postulated as one of the first protective mechanisms against ROS in eukaryotic cells [9]. It's a dual functional molecule. While low levels of NO modulate the ROS such as superoxide anion [10,11], high concentrations of NO enhance superoxide production in mitochondria by inhibiting electron flow cytochrome *c* oxidase [12], producing peroxynitrite and causing lipid peroxidation and protein nitration. In the first case, modulation of superoxide formation and inhibition of lipid peroxidation by NO illustrates its less known potent antioxidant role [13,14]. Research on the role of NO in biological systems has increased since it was suggested in the latter part of the 1980s that it was an important signaling molecule in animals [15]. This function has also been studied in plants, bacteria [16] (reviewed by Gupta et al. [17]), algae [18–20], and fungi [21–24].

We have recently reported evidence that NO released during lichen *Ramalina farinacea* rehydration plays a fundamental role in the antioxidant defense and production appears to be regulated by ROS [25]. Regarding the phycobionts we have shown that they also generate significant quantities of NO, in contrast to the findings of Weissman and co-workers [26]. Moreover, our group has also demonstrated that NO is involved in the regulation of oxidative stress caused by exposure to the prooxidant air pollutant cumene hydroperoxide [27]. Although all these studies confirm the production of NO in *R. farinacea* and provide insight into its roles, no experimental designs have addressed the synthesis of NO in lichens or their symbionts.

In animal cells, biosynthesis of NO is primarily catalyzed by the enzyme NOS (reviewed by Wendehenne et al. [28]), that catalyzes the conversion of L-arginine to L-citrulline and NO using NADPH as electron donor, molecular oxygen as co-substrate, and FAD, FMN, tetrahydrobiopterin (BH_4), and calmodulin (CaM) as cofactors [29]. Regarding plants, they are not only affected by the atmospheric pollutant NO, but they also possess the ability to produce NO by enzymatic and non-enzymatic pathways. Non-enzymatic NO formation can be the result of chemical reactions between nitrogen oxides and plant metabolites, nitrous oxide decomposition, or chemical reduction of nitrite (NO_2^-) at acidic pH (reviewed by Wendehenne et al. [28]). The first enzymatic source of NO to be identified in plants was the nitrate reductase (NR) [30]. This enzyme not only reduce nitrate to nitrite, it also catalyzes the reduction of nitrite to NO using molybdenum (Mo) as a cofactor and NADH or NADPH as an electron donor. Two isoforms of NR have been described in higher plants and eukaryotic algae: EC 1.6.6.1 is specific for NADH whereas EC 1.6.6.2 is able to use both NADH or NADPH [31]. Recently, ARC (Amidoxime Reducing Component) has been reported to catalyze NO production from nitrite taking electrons from NR in the microalga *Chlamydomonas*, allowing its synthesis in the presence of nitrate by means of a newly described NO-forming nitrite reductase activity [32]. In addition to NR as a possible source for NO, the existence of a mammalian-type NOS in plants has been under debate in recent years (reviewed by Wendehenne et al. [28,33]). Despite the intensive quest for NOS in vascular plants, the only NOS known in the Viridiplantae has recently been identified, cloned, purified, and characterized in the marine free living green microalga *Ostreococcus tauri* (Trebouxiophyceae) showing a 45% homology with human NOS [34]. The researchers have observed that *O. tauri* cultures in the exponential growth phase produce 3-fold more NOS-dependent NO than do those in the stationary phase and NO production increases in high intensity light irradiation.

In regard to the synthesis of NO in fungi there is little information, the evidence that there is a NOS associated to NO production are indirect and all rely on the use of inhibitors of this enzyme [23]. A specific fungal NR (EC 1.6.6.3) using NADPH as co-factor has been described [31].

NO is revealing itself as a keystone in stress tolerance of symbiotic associations such as *Symbiodinium*—cnidarian (corals), plant—*Rhizobium* or mycorrhizae, critical in global geomorphology and nitrogen ecology [35]. Thus, it is of the utmost interest to elucidate the mechanisms that mediate its production in lichens, symbiotic organisms inhabiting almost every terrestrial habitat. *R. farinacea* (L.) Ach is a fruticose lichen bearing in each thallus two predominant microalgae, *Trebouxia sp.* TR9 and *T. jamesii*, and a mycobiont belonging to the phylum Ascomycota [36]. We have previously demonstrated that NO limits intracellular free radical release and modulates lipid peroxidation during rehydration of these lichen thalli also protecting phycobiont chlorophyll from photooxidation [25,37].

The aim of this work is to gain insight into the synthesis of NO in the lichen model *R. farinacea*. To this end we have studied the effect of specific enzyme inhibitors on lipid peroxidation upon rehydration and a preliminary quantification of plant-like NR specific activity has been obtained.

2. Results

2.1. Effects of NR Inhibition on Lipid Peroxidation during Lichen Rehydration

Our group previously reported that NO is involved in intracellular free radical modulation and lipid peroxidation prevention during *R. farinacea* thalli rehydration [25]. In order to test whether NR is involved in the production of this NO, the inhibitor tungstate was added during thalli rehydration. The results of lipid peroxidation when lichen thalli were rehydrated with tungstate inhibitor are shown in Table 1. In the case of the controls, MDA concentration was between a minimum value at 5 min of 81.47 ± 8.14 nEq MDA/g lichen and a maximum of 131.41 ± 18.80 nEq MDA/g lichen at 120 min. In thalli rehydrated with tungstate 100 µM, MDA concentration was between a minimum value at 5 min of 83.98 ± 6.28 nEq MDA/g lichen and a maximum of 191.88 ± 11.06 nEq MDA/g lichen at 120 min. At all test times, treatment MDA levels were higher than controls with statistically significance at 120 min.

Table 1. Effect of tungstate on the lipid peroxidation level in differently rehydrated *R. farinacea* thalli. * $p < 0.05$.

Time of Rehydration (min)	Lipid Peroxidation Level (nEq MDA/g Dry Weight)		p Value (Student's t-Test)
	Control	100 µM Tungstate	
5	81.47 ± 8.14	83.98 ± 6.28	0.809
30	102.21 ± 12.43	115.16 ± 7.42	0.381
60	113.70 ± 13.73	144.82 ± 18.42	0.189
120	131.41 ± 18.80	191.88 ± 11.06	0.011*
240	87.69 ± 7.61	108.60 ± 7.36	0.061

Morphological distribution of lipid peroxidation in pink and brown tones is shown in Figure 1B where only one representative picture from replicated experiments has been selected. Despite microscopy is not a quantitative technique, at all time points, the coloration in the controls was less intense than in thalli rehydrated with tungstate. However, visual differences were only perceived at 5 (B1) and 30 (B2) minutes. There were not remarkable visual differences at 60 (B3), 120 (B4), and 240 (B5) minutes. In both cases, controls and thalli rehydrated with tungstate, lipid peroxidation was primarily located in the hyphae of the cortex and chondroid area and in the microalgae. In the hyphae of medulla, lipid peroxidation was lower.

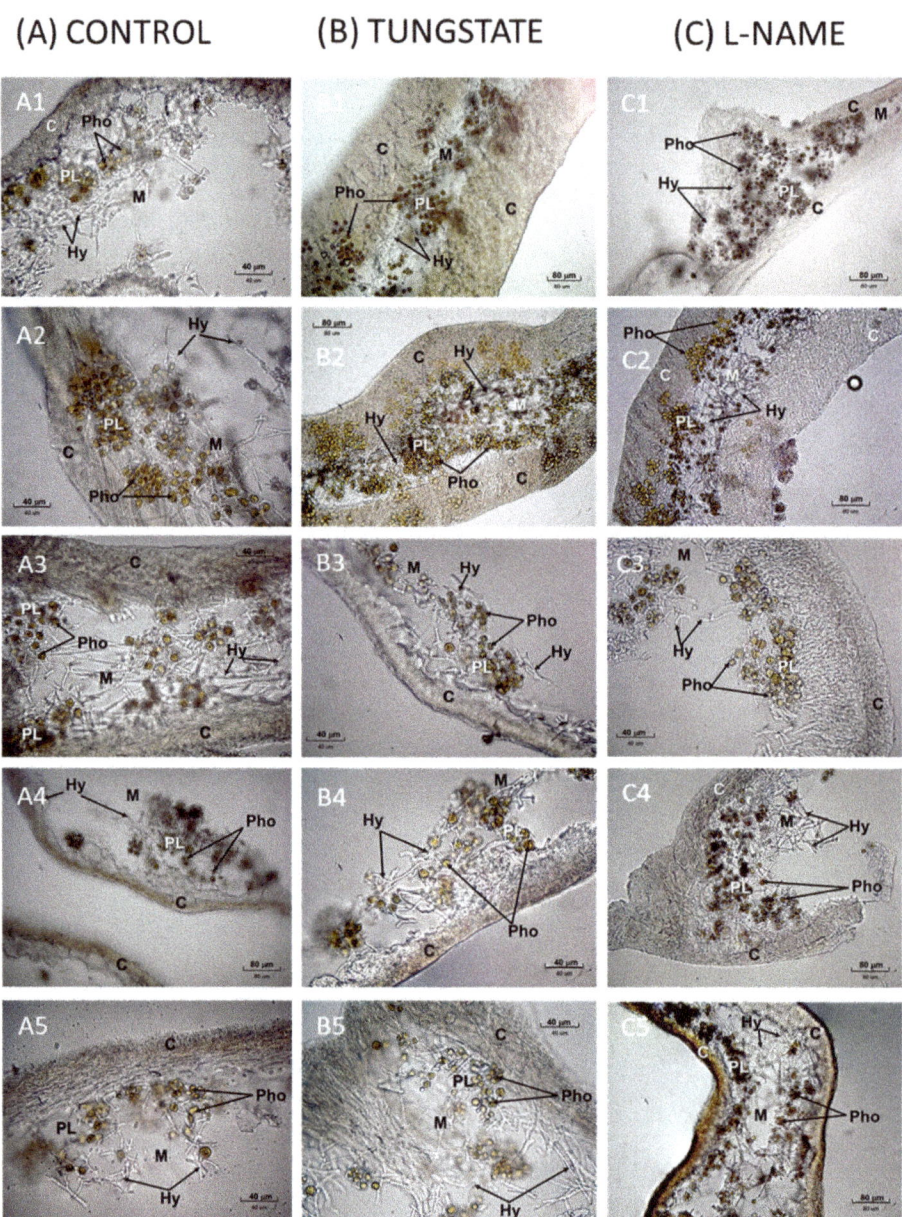

Figure 1. Bright field microscopic images of pink-brown TBARS in thalli of *R. farinacea* rehydrated with tungstate 100 μM (**B**), L-NAME 300 μM (**C**) vs. thalli rehydrated with deionized water (**A**). One representative image of different independent experiments has been selected for each condition. The number by the letter identifying the picture indicates the time post-rehydration when TBA reaction was revealed: (**1**) 5 min, (**2**) 30 min, (**3**) 60 min, (**4**) 120 min and (**5**) 240 min. Magnitude bars in the microphotographs correspond to 40 or 80 μm. C cortex with chondroid tissue, PL phycobiont layer, Pho microalgae, M medulla, Hy fungal hyphae.

2.2. Effects of Nitric Oxide Synthase (NOS) Inhibition on Lipid Peroxidation during Lichen Rehydration

The results of lipid peroxidation when lichen thalli were rehydrated with L-NAME are shown in Table 2. In the case of the control, a maximum of 110.51 ± 12.17 nEq MDA/g lichen at 30 min was observed and a minimum value of 44.74 ± 4.66 nEq MDA/g lichen at 240 min. In the rehydration with L-NAME a maximum of 137.51 ± 11.77 nEq MDA/g lichen at 30 min was found and a minimal value of 74.56 ± 6.29 nEq MDA/g lichen at 240 min. MDA concentration in the treated thalli was always higher than in the controls. The differences are statistically significant at 120 and 240 min.

Table 2. Effect of L-NAME on the lipid peroxidation level in differently rehydrated *R. farinacea* thalli. * $p < 0.05$.

Time of Rehydration (min)	Lipid Peroxidation Level (nEq MDA/g Dry Weight)		p Value (Student's t-Test)
	Control	300 µM L-NAME	
5	81.47 ± 8.14	110.86 ± 14.90	0.09741
30	110.51 ± 12.17	137.51 ± 11.77	0.12500
60	104.32 ± 11.76	121.60 ± 10.29	0.28040
120	72.77 ± 5.46	89.45 ± 4.69	0.03021 *
240	44.74 ± 4.66	74.56 ± 6.29	0.00096 *

Morphological distribution of lipid peroxidation in pink and brown tones is shown in Figure 1C. Only one representative picture from the experimental replicates is shown. Lipid peroxidation in the hyphae of the chondroid cortical area and medulla was lower than in controls. At 5 (C1), 30 (C2), and 120 (C4) minutes, the microalgae of thalli rehydrated with L-NAME were more affected by lipid peroxidation than controls (see brown color in phycobionts). However, these thalli show lower lipid peroxidation in the hyphae of the cortical zones and medulla than controls. At 60 min (C3) lipid peroxidation appeared to be higher in the hyphae of the cortex and in the phycobionts of the controls (dark brown areas). Finally, at 240 (C5) minutes lipid peroxidation was greater in the thalli treated with the inhibitor than in controls and it was localized in the peripheral areas and in the microalgae (very dark areas). This was the time when the greatest visual differences were observed.

2.3. NO Endproducts

At all times, NOx levels of thalli rehydrated with tungstate were lower than controls (Table 3). NOx production in controls was between a minimum absolute value of 0.05 ± 0.01 µmol NOx/g lichen (DW) and a maximum of 0.26 ± 0.03 µmol NOx/g lichen (DW). NOx production in thalli rehydrated with tungstate was between a minimum absolute value of 0.03 ± 0.01 µmol NOx/g lichen (DW) and a maximum of 0.16 ± 0.03 µmol NOx/g lichen (DW). At 30 and 120 min, statistically significant differences were found.

Table 3. Effect of tungstate and L-NAME on NO endproducts levels in differently rehydrated *R. farinacea* thalli. *p* value was calculated by Student's *t*-test. * $p < 0.05$.

Time of Rehydration (min)	NO Endproducts Levels (% Relative to Controls)			
	100 µM Tungstate	p Value	300 µM L-NAME	p Value
5	72.70 % ± 19.23 %	0.3496	88.77 % ± 16.45 %	0.4554
30	28.11 % ± 4.20 %	0.0018 *	188.56 % ± 24.96 %	0.0038 *
60	67.22 % ± 14.34 %	0.2946	128.40 % ± 18.73 %	0.2369
120	55.65 % ± 11.11 %	0.0077 *	235.78 % ± 41.25 %	0.0079 *
240	73.62 % ± 22.93 %	0.4206	143.39 % ± 11.59 %	0.0439 *

NOx levels of lichen thalli rehydrated with L-NAME were greater than controls at all times, except for 5 min (Table 3). NOx production in controls was between a minimum absolute value of 0.008 ± 0.001 µmol NOx/g lichen (DW) and a maximum of 0.034 ± 0.005 µmol NOx/g lichen (DW). NOx production in thalli rehydrated with L-NAME was between a minimum absolute value of 0.012 ± 0.001 µmol NOx/g lichen (DW) and a maximum of 0.030 ± 0.006 µmol NOx/g lichen (DW). Statistically significant differences were found at 30, 120, and 240 min.

2.4. Diaphorase Activity

Histochemical detection of NADPH-diaphorase activity has been related with NOS in animal and plant tissues [38]. At 2 h (Figure 2A–D) blue precipitates were observed in the hyphae, both in the cortex and chondroid area, but especially in the latter. Small vesicles with blue precipitate were seen inside phycobionts (Figure 2(C1,D1)). In the peripheral zone of microalgae, blue precipitates were also found (Figure 2(A1)). Assuming that the NADPH-diaphorase activity represent the NOS-like activity, the results here suggest the occurrence of NOS-like enzymes in *R. farinacea*.

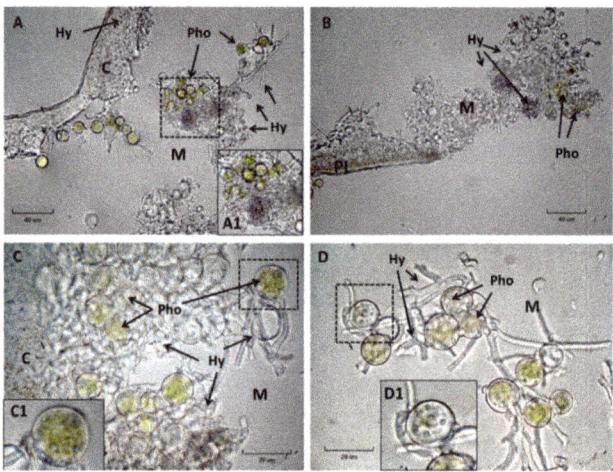

Figure 2. Diaphorase activity assayed with nitrotetrazolium blue chloride (NBT, blue precipitates) in *R. farinacea* thalli (**A–D**). Bright-field microscopy of slices cut in a freezing microtome (magnification 1000×). The areas framed with discontinuous lines have been digitally magnified in the corresponding insets (**A1**, **C1** and **D1**). C with chondroid tissue, PL phycobiont layer, Pho microalgae, M medulla, Hy fungal hyphae.

2.5. Specific Activity of NR

As other Chlorophyta, *Trebouxia* phycobionts of *R. farinacea* are likely to possess NADH-NR activity. Despite no method to assess NR activity has been reported for lichens to our knowledge, we used a general method for plants [39] in whole thalli in order to obtain a value of the specific activity of NADH-NR in *R. farinacea*. The value obtained for NADH-NR specific activity in fresh *R. farinacea* thalli was $(0.91 ± 0.13) \times 10^{-4}$ U/mg protein (U = µmoles nitrite/min). In order to check if tungstate was capable of inhibiting this measured activity, we used increasing concentrations of this inhibitor. The activity of NADH-NR decreased as the concentration of tungstate increased in a dose dependent manner and was not measurable above 50 µM of the inhibitor (Supplementary Figure S1).

3. Discussion

NO is revealing itself as a key molecule in the tolerance of abiotic stress of symbiotic organisms as mycorrhizae, *Rhizobium*, and lichens and the elucidation of the mechanisms and regulation of its synthesis will provide very valuable information both for conservation of biodiversity as well as for biotechnological purposes. The increase in lipid peroxidation upon the inhibition of NR and NOS-like activities in the model lichen *R. farinacea* described in the present work suggests the participation of both enzymes in the synthesis of NO under rehydration stress.

In lichen thalli rehydrated with the NR inhibitor tungstate, lipid peroxidation increased compared to control, while NO release decreased as expected. Lipid peroxidation indirectly indicates that the antioxidant defenses have been overcome by the formation of reactive oxygen species (ROS) [6]. Although NO donors have been shown to reduce antioxidants, inhibit or inactivate antioxidant enzymes and increase MDA through H_2O_2 accumulation in stressed plants (reviewed by Groß et al. [40]) it has also been shown to decrease the generation of ROS and thus, lipid peroxidation in plant microsomes [13,14] and lichens [27,37]. As a matter of fact, it is able to directly terminate free radical chain reactions [41]. The use of tungstate as NR inhibitor has to be considered with caution because of side effects due to the affection of other molybdenum-enzymes or heavy metal toxicity, especially at longer exposure times [42]. However, this important result correlates with c-PTIO lichen NO scavenging [25] and points to the existence of a NO related NR activity in *R. farinacea*. This is the first study providing evidence that NR may be implicated in the synthesis of NO during abiotic stress in lichens or lichen symbionts. NR has also been involved in the synthesis of NO in the green microalga *Chlamydomonas reinhardtii* [43]. As a matter of fact, Mallick et al. [44] and Medina-Andrés et al. [45] propose that the synthesis of NO is a common feature for algae as well as embryophytes and is strongly dependent on NR.

A study with *Pleurotus eryngii var. tuoliensis*, a basidiomycete fungus, showed that heat stress induced an increase in NO production in mycelial cells which was significantly blocked by NOS inhibition (L-NAME). In contrast, NR activities were not obviously altered during heat stress [23]. But NO levels required in the morphogenesis and reproduction of the ascomycetes fungus *Aspergillus* seem to be insufficient without a functional NR gene [46]. *R. farinacea* mycobiont is an ascomycete and the gene for NR has been reported to be part of a cluster of genes that were transferred horizontally from a basidiomycete to an ancestor of the ascomycetes [47]. These data support the existence of a functional NR enzyme also in ascomycetes as a plausible hypothesis.

In thalli rehydrated with NOS inhibitor L-NAME, lipid peroxidation slightly increased in both symbionts but, unexpectedly, NOx endproducts increased too. This suggests that NO levels are higher in thalli rehydrated with L-NAME than in thalli rehydrated with deionized water but yet not efficient in lipid peroxidation prevention. This L-arginine analogue is a reversible inhibitor whose paradoxical ability to induce NO increases by NOS activity upregulation was reviewed by Kopincová et al. [48]. NOS enzymes have been demonstrated to be finely regulated both at protein and expression levels depending on the physiological conditions of the organism. NO chemistry is complex and its sources, multiple, which could generate local effects linked to spatial and morphological constraints to NO bioavailability and activity [48]. Although thallus NOS-like activity was inhibited, NO could still be synthesized by NR activity or by non-enzymatic pathways (reviewed by Wendehenne et al. [28]) resulting in overproduction. Accumulation of NO in response to stress has been associated with increased ROS levels due to inhibition of antioxidant enzymes and formation of peroxinitrites (reviewed by Gross et al. [49]). Despite quantitatively lower, NOS-like activity inhibited by L-NAME seems to be especially critical since its inhibition triggered an upregulation of other NO sources which, in turn, seem not to be efficient in lipid peroxidation limitation. Although unknown side effects of L-NAME in lichens cannot be ruled out, the development of a method for total NR activity quantification in lichens to test a possible upregulation during NOS activity inhibition, together with morphological localization of NO release would shed light on this paradox.

In the same line, cytochemical NADPH-diaphorase activity demonstration points to a NOS-like activity in *R. farinacea* analogue to animals and plants [38]. Diaphorase activity has been detected in the hyphae, both the cortical plectenchyma and medulla, in vesicles inside and in the periphery of the microalgae. In the marine microalga *Chattonella marina*, the main source of NO production has been reported to be NOS activity [50]. Recently, the first NOS in Viridiplantae has been identified and characterized by Foresi et al. [34] in the marine green microalga *Ostreococcus tauri*. Valentovicová et al. [51] showed that L-NAME inhibited both NADPH-diaphorase activity and NO production in barley root tips. However, NADH-diaphorase activity has been reported for NADH-NR and this activity cannot be ruled out for fungal NADPH-NR or other enzymes, further experiments are necessary to confirm the presence of a NOS enzyme.

Our data show that when NR is inhibited, lipid peroxidation is primarily located in the hyphae of the cortex and chondroid plectenchymas and in phycobionts, while when NOS is inhibited, lipid peroxidation increases in microalgae. As pointed out above, this means that the mechanisms and kinetics of synthesis of NO determine, at least in part, its role: While NR has an important role in the protection of both mycobiont hyphae and phycobionts in the first hours, a fungal NOS, sensitive to animal NOS inhibitors and immunoreactive with animal NOS antibodies, has been described both in ascomycetes and yeasts although, alike plants, gene orthologues have not been found and responsible proteins have not been characterized [52]. NOS-like activity inhibited by L-NAME seems critical for microalgae from the very first minutes after rehydration. We don´t know how many NR and NOS-like enzymes there might be in this lichen and if so, which participates in the synthesis of NO in stress conditions. We must bear in mind that *R. farinacea* contains, at least, three different eukaryotic organisms (fungi, yeasts, microalgae) from two supergroups (Opisthokonta and Archaeplastida) and each could possess its own NR and NOS enzymes with specific peculiarities in expression regulation, suborganellar localization, kinetics, or allosteric modulation. This provides symbiotic organisms with a versatile set of tools to cope with abiotic stress.

The value found for plant-like NADH-dependent NR specific activity in *R. farinacea* (0.91×10^{-4} U/mg) is two orders of magnitude lower than NR specific activity reported for the Chlorophyceae *Ulva intestinalis* (Table 4) [53]. A much more similar value of 0.40×10^{-4} U/mg has been reported for the bryophyte *Physcomitrella patens* [45]. The specific activities reported for various marine macroalgae of the Rhodophyta show some divergences, whereas in *Kappaphycus alvarezii* (Gigartinales), specific activity is 0.16 U/mg [54], *Gracilaria tenuistipitata* (Gracilariales) specific activities of NR reported for crude extracts are ten times higher (3.0 ± 0.2 in apical part, 1.6 ± 0.1 U/mg for basal) [55]. However, a more recent study characterizing *Gracilaria chilensis* by Chow et al. [56] has reported 2.53×10^{-4} U/mg, and a value of 8.33×10^{-4} U/mg has been obtained for *Porphyra perforata* (Bangiales) [53] comparable to *Arabidopsis thaliana*'s 2.50×10^{-4} U/mg [57]. In the same order of magnitude, NR specific activity reported for *Gracilaria caudata* is 0.93×10^{-4} U/mg [39] and for *Gracilaria tikvahiae* is 0.43×10^{-4} U/mg [58].

Table 4. Specific NR activities referred to total soluble protein.

Species	Taxonomic Group	NR Specific Activity (U = μmol Nitrite/min)
Ramalina farinacea	Lichen (Chlorophyta – Trebouxiophyceae + Ascomycota)	$(0.91 \pm 0.13) \times 10^{-4}$ U/mg present work
Ulva intestinalis	Chlorophyta – Chlorophyceae	0.27×10^{-2} U/mg [53]
Physcomitrella patens	Bryophyta	0.40×10^{-4} U/mg [45]
Porphyra perforata	Rhodophyta, Bangiophyceae	8.33×10^{-4} U/mg [53]
Kappaphycus alvarezii	Rhodophyta, Gigartinales	0.16 U/mg [54]
Gracilaria tenuistipitata	Rhodophyta, Gracilariales	3.00 ± 0.20 (apical) U/mg [55] 1.60 ± 0.10 (basal) U/mg [55]
Gracilaria tikvahiae	Rhodophyta, Gracilariales	0.43×10^{-4} U/mg [58]
Gracilaria chilensis	Rhodophyta, Gracilariales	$(2.53 \pm 0.03) \times 10^{-4}$ U/mg [56]
Gracilaria caudata	Rhodophyta, Gracilariales	0.93×10^{-4} U/mg [39]
Arabidopsis thaliana	Magnoliophyta	2.50×10^{-4} U/mg [57]

As symbiotic organisms, lichens are composed of algae, fungi, and bacteria. Given that for this preliminary approach we have used a NADH-method designed for plants, we are only taking into account the plant/algae isoforms, namely EC 1.6.6.1 and EC 1.6.6.2. The possible existence of a fungal isoform (EC1.6.6.3) specific for NADPH as co-factor remains to be elucidated. On the other hand, most of photosynthetic organisms seem to possess NADH-specific EC 1.6.6.1 isoform, but some microalgae have shown a small nitrate reducing activity with NADPH (EC 1.6.6.2) [59]. NR specific activity studies with the isolated microalgae (Trebouxiophyceae) of *R. farinacea* are necessary to rule out whether one or both isoforms are present. In any case, in the absence of fungal biomass we can reasonably expect higher values, likely in the range of the Chlorophyceae *Ulva intestinalis*.

Phycobionts are probably the main source of the specific activity we report using a plant-designed method with *R. farinacea* whole extract. However, we report above the induction of lipid peroxidation in fungal hyphae upon NR inhibition with tungstate. Since the mycobiont accounts for the greater part of the biomass of the thallus, a remarkable NR specific activity could also be expected if NADPH were used as co-factor. We are currently working on optimizing a method specifically designed for lichens.

Our approach has allowed us to demonstrate NR and NOS-like enzymes activities in *R. farinacea*, but the evidence of these enzymes is indirect and the presence of the proteins themselves should be further verified. The quantification of NADH-NR, although preliminary, adds to the evidences. Nonetheless, in order to confirm the presence of NOS-like in each symbiont and to characterized NR isoforms, future studies are required about the biosynthesis of NO in the microalgae as well as in the isolated *R. farinacea* mycobiont. Studies to isolate the proteins and genetic studies will also be necessary.

4. Materials and Methods

4.1. Chemicals

2-Thiobarbituric acid (TBA), sodium tungstate dihydrate ($Na_2WO_4 \bullet 2H_2O$) and 1,1,3,3 tetraethoxypropane (TEP), nitrotetrazolium blue chloride (NTB), 2,6-di-tert-buthyl-4-methylphenol trichloroacetic acid (BHT), bovine serum albumin (BSA), L-Cysteine, sulfanilamide ($C_6H_8N_2O_2S$), N-1-(naphthyl) ethylenediamine dihydrochloride ($C_{12}H_{16}Cl_2N_2$), and NADPH were provided by Sigma Aldrich Química S.A (Tres Cantos, Spain); NG-Nitro-L-arginine methyl ester (L-NAME) was purchased from Sigma Aldrich (China); Ethylenediaminetetraacetic Acid (EDTA) and trichloroacetic acid (TCA) was from Merck (Germany); dithiothreitol (DTT) and NADH were from Roche Custom Biotech; inorganics and ethanol (etOH) were purchased from Panreac Quimica S.A.U (Spain); triton X-100 was from VWR Prolabo (Barcelona, Spain).

4.2. Lichen Material

R. farinacea (L.) Ach. lichen thalli were collected in the air-dry state from *Quercus pyrenaica* in San Lorenzo de El Escorial at 969 m altitude (Ermita Virgen de Gracia, Madrid, Spain; 40°34′25,6″ N, 4°9′146″ W). Samples were maintained in a silica gel atmosphere during 24 h and frozen at −20 °C until the experiment.

4.3. Treatments

Lichen thalli were cut and weighed between 20–30 mg. For each time point (0, 30, 60, 120, and 240 min) 12 replicates were processed with each inhibitor (100 µM sodium tungstate dihydrate or 300 µM L-NAME) and controls. The day of the experiment, fragments of lichen thalli were rehydrated during 5 min with deionized water for controls or one of the inhibitors. Then, they were maintained at room temperature for the times of study and subsequently frozen at −20 °C until lipid peroxidation analysis. Inhibitor concentrations were selected according to the literature regarding plant NR [42] and fungal NOS [23].

4.4. Measurement of Lipid Peroxidation: MDA

The most common method for measuring MDA in biological samples is the thiobarbituric acid reactive substances (TBARS), which is based on spectrophotometric quantification of the pink complex formed after reaction of MDA with two molecules of TBA [60] with maximum absorbance at 532 nm [61]. In our study, lipid peroxidation was evaluated as MDA by a variant of the method of Reilly and Aust [62]. As standards, 0–20 µM TEP were used as an MDA precursor. The reaction of TEP in acid medium generates the same complex TBA-MDA-TBA, allowing to relate the absorbance with the concentration of the complex. The presentation of the results of lipid peroxidation will expressed as nEq MDA/g of sample, as a measure of the amount of MDA in the sample.

Lichen thalli were homogenized with 1 mL of deionized water on ice and centrifuged at 16,060× g for 10 min to eliminate debris. Supernatants were frozen at −20°C for NO_x analysis and sediments were resuspended in 500 µL ethanol—BHT 2%. A volume of 900 µL of TBA ($2.57 \cdot 10^{-2}$ M), TCA ($9.18 \cdot 10^{-1}$ M) and HCl (3.20 M) working solution was added to each sample and standard. Then, samples and standards were vortexed in a Vortex Labnet × 100 for 5 min at 3000 rpm and placed in a water bath at 70°C for 30 min. Next, samples and standards were vortexed, cooled in ice and centrifuged 10 min at 16,060× g. Finally, absorbance of the supernatants from samples and standards was measured at 532 nm and 600 nm to eliminate the interferences of soluble sugars in samples, in a spectrophotometer Anthos 2010, model 17-550.

To analyze the morphological distribution of lipid peroxidation, fragments of treated lichen thalli were subjected directly to TBARS method (described above), but they were not homogenized. Then they were placed in a freezing microtome (CM 1325; Leica, Germany) and sliced into sections of 30 µm. The slices were washed with deionized water and mounted on slides prior to their observation by optical microscopy (OLYMPUS Provis AX 70 optical microscope) equipped with an infinity 2-3C Lumenera® digital camera and analysed with "Infinity Analyze" Software v.6.5.5 at the Plant Anatomy Laboratory "Julio Iranzo" in the Botanical Garden of the University of Valencia.

4.5. NO Endproducts Determination

The products formed by the oxidation of NO in an aqueous environment are mainly NO_2^-, which is further oxidized to NO_3^- [63]. In order to estimate NO generation, NO oxidation endproducts (nitrates and nitrites) were measured in the soluble fraction of different samples with an autoanalyzer Skalar, model SAN++. The automated determination of nitrates and nitrites is divided in two phases: first, the reduction of nitrates to nitrites by the cadmium reduction method, where the sample is passed through a column containing granulated copper-cadmium; second phase involves the reaction of nitrites with N-(1-naphthyl) ethylenediamine dihydrochloride in acid medium to form a highly coloured azo dye which is measured at 540 nm. This method is known as Griess reaction [64,65].

4.6. Diaphorase Activity

The basic protocol used to detect diaphorase activity in animal neurons [66,67] was used in a modified manner in this study. Diaphorase activity was observed after 2 h. During these times, lichen thalli were incubated in a solution of 0.5 mg/mL NADPH and 0.2 mg/mL NBT in phosphate buffered saline (PBS) with 0.25% of Triton X-100. Then, thalli were washed three times with deionized water and frozen at −20 °C. The samples were then placed in a freezing microtome (CM 1325; Leica, Germany) and cut in sections of 30 µm. The slices were washed with deionized water and mounted on slides prior to their observation by optical microscopy (OLYMPUS Provis AX 70 optical microscope) at the Jardí Botànic and SCSIE (UVEG, Valencia).

4.7. Specific Activity of NR

The enzymatic assay of NR was performed as described in [39] with minor changes. Samples of lichen thalli of *R. farinacea* (250 mg) were milled in a conical homogenizer and suspended in 5 mL of extraction buffer (1 mM DTT, 5 mM EDTA, 1 mM cysteine, 0.3% BSA (w/v), and 0.2 M phosphate buffer, pH = 7.5) to stabilize NR. Cell debris was removed by centrifugation at 17000 g for 15 min at 4 °C. An aliquot of the supernatant was taken for total soluble protein quantification. The supernatant (crude extract) was recovered and kept on ice until the activity of the enzyme was analyzed. To 100 µL of crude extract 20 µL of KNO_3 500 mM, 20 µL of $MgSO_4$ 9.5 mM, and 50 µL of NADH 380 µM were added to initiate the reaction. The reaction was interrupted after 10 min by adding 20 µL of $ZnSO_4$ 1.4 mM and 20 µL of cold ethanol 90% v/v. The precipitates were removed by centrifugation at 12,000 g

for 10 min at 20 °C and the Griess method [68] was used to analyze nitrite production as described in Chaki et al. [69] although some changes were applied. To 190 µL of sample 95 µL of 1% sulfanilamide (w/v) in 1.5 M HCl and then 95 µL of 0.2% n-naphthylethylenediamine (w/v) in 0.2 M phosphate buffer pH = 7.5 were added. A measurable azo dye at 540 nm was developed after 5 min. $NaNO_2$ was used as a standard in a range between 0–10,000 µg/L. Nitrate blanks were performed to account for non-enzymatic nitrite content of lichen samples. Bradford method [70] with some modification was used to quantify total soluble protein: 5 µL of sample were mixed with 250 µL of Bradford reagent, and absorbance was measured at 595 nm after 10 min. A standard curve was made with concentrations ranging from 0 to 1 mg BSA/ml extraction buffer. Blanks without the substrates were performed with each sample analyzed. The activity value obtained in the absence of these substrates informed about natural levels of nitrite in thalli. Nitrate blank was the highest and was subtracted from total activity to account for non-enzymatic nitrite. NR activity units (U) were defined as µmoles nitrite produced/min at 25 °C and pH 7.5.

4.8. Statistics

For each treatment and study times at least 12 replicates were prepared (n = 12). The results are expressed as means ± standard error. Software used for statistical analysis was "R-commander". We determined the significant differences between treatments in each time using a Student's *t*-test when the variances were equal, and the Welch's test when the variances were different. Comparison of variances was made with a statistical test based on the distribution F of Snedecor. Before statistical analysis, the normality of the data was verified by the application of Shapiro–Wilk test. in all cases was considered significant a *p*-value < 0.05. For the quantification assay of NR activity, 3 replicates were used. The results are expressed as means ± standard error.

5. Conclusions

Our results demonstrate the existence of NR activity correlated with NO generation in the lichen *R. farinacea* under stress conditions. NO role seems to be determined by its source: NO related to NR activity seems to have an important role in the hyphae of cortex and in phycobionts in the first hours, while NO correlated with NOS-like seems to be more important for microalgae. NADPH-diaphorase activity cytochemistry supports the existence of NOS-like activity in both the mycobiont and the phycobionts of *R. farinacea*, where it is associated with big cytoplasmatic vesicles. Preliminary quantification of NADH-NR specific activity has rendered 91.00 ± 13.23 µU/mg. Taken together these data indicate that NO regulation and synthesis in lichens is complex involving both NR and NOS-like activities.

Supplementary Materials: The following are available online at http://www.mdpi.com/2223-7747/8/7/189/s1, Figure S1: Tungstate inhibition of lichen *R. farinacea* nitrate reductase activity.

Author Contributions: S.M.S.R., J.R.E. and M.C. conceived objectives and designed the study and general design of the work with the critical review of E.B. which, also got the funds. J.R.E. and S.M.S.R. collected *R. farinacea* thalli and performed the biochemical and cytochemical studies. Microscopy and image handling were performed by F.J.G.-B. and J.R-.A. J.R.E., S.M.S.R., and M.C. prepared the draft of the paper and E.B. made final considerations. All authors read and approved the final manuscript.

Funding: This research was funded by Ministerio de Economía y Competitividad (MINECO - FEDER, Spain) (CGL2016-79158-P) and Generalitat Valenciana (GVA, Excellence in Research, Spain) (PROMETEOIII/2017/039).

Acknowledgments: The authors are indebted to the Jardí Botànic and the SCSIE of the University of Valencia as well as the Nutrilab of Rey Juan Carlos University. The authors wish to thank M. Feelisch (University of Southampton, United Kingdom) and F. J. Corpas (Estación Experimental El Zaidín, CSIC, Spain) for their expert comments and Rosa de las Heras (Rey Juan Carlos University, Spain) for organizing the manuscript.

Conflicts of Interest: The authors declare no conflict of interest.

Abbreviations

BHT	2,6-Di-tert-buthyl-4-methylphenol
BSA	Bovine serum albumin
DTT	Dithiothreitol
DW	Dry weight
etOH	Ethanol
EDTA	Ethylenediaminetetraacetic Acid
L-NAME	NG-nitro-L-arginine methyl ester
MDA	Malondialdehyde
NBT	Nitro blue tetrazolium
NOx	Nitric oxide oxidation end-products
NOS	Nitric oxide synthase
NR	Nitrate reductase
PBS	Phosphate buffered saline
ROS	Reactive oxygen species
TBA	2-Thiobarbituric acid
TBARS	Thiobarbituric Acid Reactive Substances
TCA	Trichloroacetic acid
TEP	1,1,3,3- Tetraexthoxypropane

References

1. Ahmadjian, V. *The Lichen Symbiosis*; John Wiley & Sons: New York, NY, USA, 1993; ISBN 0471578851.
2. Izco, J.; Barreno, E.; Brugués, M.; Costa, M.; Devesa, J.A.; Fernández-González, F.; Llimona, X.; Valdés, B. *Botánica*; McGraw Hill Interamericana: Madrid, Spain, 1997; ISBN 8448601823.
3. Kranner, I.; Beckett, R.; Hochman, A.; Nash, T.H. Desiccation-tolerance in lichens: A review. *Bryologist* **2008**, *111*, 576–593. [CrossRef]
4. Kranner, I.; Cram, W.J.; Zorn, M.; Wornik, S.; Yoshimura, I.; Stabentheiner, E.; Pfeifhofer, H.W. Antioxidants and photoprotection in a lichen as compared with its isolated symbiotic partners. *Proc. Natl. Acad. Sci. USA* **2005**, *102*, 3141–3146. [CrossRef] [PubMed]
5. McKersie, B.D.; Leshem, Y.Y. *Stress and Stress Coping in Cultivated Plants*; Kluwer Academic Publishers: Norwell, MA, USA, 1994.
6. Halliwell, B.; Cross, C.E. Oxygen-derived species: Their relation to human disease and environmental stress. *Environ. Health Perspect.* **1994**, *102*, 5–12. [PubMed]
7. Wilson, I.D.; Neill, S.J.; Hancock, J.T. Nitric oxide synthesis and signalling in plants. *Plant. Cell Environ.* **2008**, *31*, 622–631. [CrossRef] [PubMed]
8. Meilhoc, E.; Cam, Y.; Skapski, A.; Bruand, C. The Response to Nitric Oxide of the Nitrogen-Fixing Symbiont *Sinorhizobium meliloti*. *Mol. Plant Microbe Interact.* **2010**, *748*, 748–759. [CrossRef] [PubMed]
9. Feelisch, M.; Martin, J.F. The early role of nitric oxide in evolution. *Trends Ecol. Evol.* **1995**, *10*, 496–499. [CrossRef]
10. Mittler, R. Oxidative stress, antioxidants and stress tolerance. *Trends Plant Sci.* **2002**, *7*, 405–410. [CrossRef]
11. Vranová, E.; Inzé, D.; Van Breusegem, F. Signal transduction during oxidative stress. *J. Exp. Botany* **2002**, *53*, 1227–1236. [CrossRef]
12. Millar, A.; Day, D. Nitric oxide inhibits the cytochrome oxidase but not the alternative oxidase of plant mitochondria. *FEBS Lett.* **1996**, *398*, 155–158. [CrossRef]
13. Caro, A.; Puntarulo, S. Nitric oxide decreases superoxide anion generation by microsomes from soybean embryonic axes. *Physiol. Plant.* **1998**, *104*, 357–364. [CrossRef]
14. Boveris, A.D.; Galatro, A.; Puntarulo, S. Effect of nitric oxide and plant antioxidants on microsomal content of lipid radicals. *Biol. Res.* **2000**, *33*, 159–165. [CrossRef] [PubMed]
15. Wendehenne, D.; Hancock, J.T. New frontiers in nitric oxide biology in plant Preface. *Plant Sci.* **2011**, *181*, 507–508. [CrossRef] [PubMed]
16. Hayat, S.; Mori, M.; Pichtel, J.; Ahmad, A. *Nitric Oxide in Plant Physiology*; Whiley-Blackwell: Hoboken, NJ, USA, 2009.

17. Gupta, K.J.; Fernie, A.R.; Kaiser, W.M.; van Dongen, J.T. On the origins of nitric oxide. *Trends Plant Sci.* **2011**, *16*, 160–168. [CrossRef] [PubMed]
18. Lobysheva, I.; Vanin, A.; Sineshchekov, O.; Govorunova, E. Phototaxis in *Chlamydomonas reinhardtii* is modulated by nitric oxide. *Biofizika* **1996**, *41*, 540–541.
19. Mallick, N.; Mohn, F.H.; Soeder, C.J.; Grobbelaar, J.U. Ameliorative role of nitric oxide on H2O2 toxicity to a chlorophycean alga *Scenedesmus obliquus*. *J. Gen. Appl. Microbiol.* **2002**, *48*, 1–7. [CrossRef] [PubMed]
20. Chen, K.; Feng, H.; Zhang, M.; Wang, X. Nitric oxide alleviates oxidative damage in the green alga *Chlorella pyrenoidosa* caused by UV-B radiation. *Folia Microbiol. (Praha)* **2003**, *48*, 389–393. [CrossRef] [PubMed]
21. Wilken, M.; Huchzermeyer, B. Suppression of mycelia formation by NO produced endogenously in *Candida tropicalis*. *Eur. J. Cell Biol.* **1999**, *78*, 209–213. [CrossRef]
22. Maier, J.; Hecker, R.; Rockel, P.; Ninnemann, H. Role of nitric oxide synthase in the light-induced development of sporangiophores in *Phycomyces blakesleeanus*. *Plant Physiol.* **2001**, *126*, 1323–1330. [CrossRef] [PubMed]
23. Kong, W.; Huang, C.; Chen, Q.; Zou, Y.; Zhang, J. Nitric oxide alleviates heat stress-induced oxidative damage in *Pleurotus eryngii* var. *tuoliensis*. *Fungal Genet. Biol.* **2012**, *49*, 15–20. [CrossRef] [PubMed]
24. Song, N.; Jeong, C.-S.; Choi, H.-S. Identification of Nitric Oxide Synthase in *Flammulina velutipes*. *Mycologia* **2000**, *92*, 1027–1032. [CrossRef]
25. Catalá, M.; Gasulla, F.; Pradas del Real, A.E.; García-Breijo, F.; Reig-Armiñana, J.; Barreno, E. Fungal-associated NO is involved in the regulation of oxidative stress during rehydration in lichen symbiosis. *BMC Microbiol.* **2010**, *10*, 297. [CrossRef] [PubMed]
26. Weissman, L.; Garty, J.; Hochman, A. Rehydration of the lichen *Ramalina lacera* results in production of reactive oxygen species and nitric oxide and a decrease in antioxidants. *Appl. Environ. Microbiol.* **2005**, *71*, 2121–2129. [CrossRef] [PubMed]
27. Catalá, M.; Gasulla, F.; Pradas Del Real, A.E.; García-Breijo, F.; Reig-Armiñana, J.; Barreno, E. The organic air pollutant cumene hydroperoxide interferes with NO antioxidant role in rehydrating lichen. *Environ. Pollut.* **2013**, *179*, 277–284. [CrossRef] [PubMed]
28. Wendehenne, D.; Pugin, A.; Klessig, D.F.; Durner, J. Nitric oxide: Comparative synthesis and signaling in animal and plant cells. *Trends Plant Sci.* **2001**, *6*, 177–183. [CrossRef]
29. Bogdan, C. Nitric oxide and the regulation of gene expression. *Trends Cell Biol.* **2001**, *11*, 66–75. [CrossRef]
30. Yamasaki, H.; Sakihama, Y.; Takahashi, S. An alternative pathway for nitric oxide production in plants: New features of an old enzyme. *Trends Plant Sci.* **1999**, *4*, 128–129. [CrossRef]
31. Berges, J. Miniview: Algal nitrate reductases. *Eur. J. Phycol.* **1997**, *32*, 3–8. [CrossRef]
32. Chamizo-Ampudia, A.; Sanz-Luque, E.; Llamas, Á.; Ocaña-Calahorro, F.; Mariscal, V.; Carreras, A.; Barroso, J.B.; Galván, A.; Fernández, E. A dual system formed by the ARC and NR molybdoenzymes mediates nitrite-dependent NO production in *Chlamydomonas*. *Plant Cell Environ.* **2016**, *39*, 2097–2107. [CrossRef]
33. Corpas, F.J.; Barroso, J.B. Nitric oxide synthase-like activity in higher plants. *Nitric Oxide Biol. Chem.* **2017**, *68*, 5–6. [CrossRef]
34. Foresi, N.; Correa-Aragunde, N.; Parisi, G.; Caló, G.; Salerno, G.; Lamattina, L. Characterization of a nitric oxide synthase from the plant kingdom: NO generation from the green alga *Ostreococcus tauri* is light irradiance and growth phase dependent. *Plant Cell* **2010**, *22*, 3816–3830. [CrossRef]
35. Hichri, I.; Boscari, A.; Meilhoc, E.; Catalá, M.; Barreno, E.; Bruand, C.; Lanfranco, L.; Brouquisse, R. Nitric Oxide: A multitask Player in Plant–Microorganism Symbioses. In *Gasotransmitters in Plants. The Rise of a New Paradigm in Cell Signaling*; Lamattina, L., García-Mata, C., Eds.; Springer International Publishing: Berlin/Heidelberg, Germany, 2016; pp. 239–268.
36. Moya, P.; Molins, A.; Martinez-Alberola, F.; Muggia, L.; Barreno, E. Unexpected associated microalgal diversity in the lichen *Ramalina farinacea* is uncovered by pyrosequencing analyses. *PLoS ONE* **2017**, *12*, 1–21. [CrossRef] [PubMed]
37. Catalá, M.; Gasulla, F.; Pradas del Real, A.E.; García-Breijo, F.J.; Reig-Armiñana, J.; Barreno, E. Nitric oxide is involved in oxidative stress during rehydration of *Ramalina farinacea* (L.) Ach. in the presence of the oxidative air pollutant cumene hydroperoxide. *Biol. Lichens Symbiosis Ecol. Environ. Monit. Syst. Cyber Appl.* **2010**, *105*, 87–92.
38. Cueto, M.; Hernández-Perera, O.; Martín, R.; Bentura, M.L.; Rodrigo, J.; Lamas, S.; Golvano, M.P. Presence of nitric oxide synthase activity in roots and nodules of *Lupinus albus*. *FEBS Lett.* **1996**, *398*, 159–164. [CrossRef]

39. Chow, F.; Capociama, F.V.; Faria, R.; de Oliveira, M.C. Characterization of nitrate reductase activity in vitro in *Gracilaria caudata* J. Agardh (Rhodophyta, Gracilariales). *Rev. Bras. Bot.* **2007**, *30*, 123–129. [CrossRef]
40. Groß, F.; Durner, J.; Gaupels, F. Nitric oxide, antioxidants and prooxidants in plant defence responses. *Front. Plant Sci.* **2013**, *4*, 419. [CrossRef] [PubMed]
41. Darley-Usmar, V.M.; Pate, R.P.; O'Donnell, V.B.; Freeman, B.A. Antioxidant Actions of Nitric Oxide. In *Nitric Oxide: Biology and Pathology*; Ignarro, L., Ed.; Academic Press: Los Ángeles, CA, USA, 2000; pp. 265–276. ISBN 978-0-12-370420-7.
42. Xiong, J.; Fu, G.; Yang, Y.; Zhu, C.; Tao, L. Tungstate: Is it really a specific nitrate reductase inhibitor in plant nitric oxide research? *J. Exp. Bot.* **2012**, *63*, 33–41. [CrossRef] [PubMed]
43. Sakihama, Y.; Nakamura, S.; Yamasaki, H. Nitric oxide production mediated by nitrate reductase in the green alga *Chlamydomonas reinhardtii*: An alternative NO production pathway in photosynthetic organisms. *Plant Cell Physiol.* **2002**, *43*, 290–297. [CrossRef] [PubMed]
44. Mallick, N.; Rai, L.C.; Mohn, F.H.; Soeder, C.J. Studies on nitric oxide (NO) formation by the green alga *Scenedesmus obliquus* and the diazotrophic cyanobacterium *Anabaena doliolum*. *Chemosphere* **1999**, *39*, 1601–1610. [CrossRef]
45. Medina-Andres, R.; Solano-Peralta, A.; Saucedo-Vazquez, J.P.; Napsucialy-Mendivil, S.; Pimentel-Cabrera, J.A.; Sosa-Torres, M.E.; Dubrovsky, J.G.; Lira-Ruan, V. The nitric oxide production in the moss *Physcomitrella patens* is mediated by nitrate reductase. *PLoS ONE* **2015**, *10*, e0119400. [CrossRef] [PubMed]
46. Cánovas, D.; Marcos, J.F.; Marcos, A.T.; Strauss, J. Nitric oxide in fungi: Is there NO light at the end of the tunnel? *Curr. Genet.* **2016**, *62*, 513–518. [CrossRef] [PubMed]
47. Slot, J.C.; Hibbett, D.S. Horizontal transfer of a nitrate assimilation gene cluster and ecological transitions in fungi: A phylogenetic study. *PLoS ONE* **2007**, *2*, e1097. [CrossRef] [PubMed]
48. Kopincová, J.; Púzserová, A.; Bernátová, I. L-NAME in the cardiovascular system – nitric oxide synthase activator? *Pharmacol. Rep.* **2012**, *64*, 511–520. [CrossRef]
49. Gross, B.H.; Kreutz, K.J.; Osterberg, E.C.; McConnell, J.R.; Handley, M.; Wake, C.P.; Yalcin, K. Constraining recent lead pollution sources in the North Pacific using ice core stable lead isotopes. *J. Geophys. Res. Atmos.* **2012**, *117*, D16307. [CrossRef]
50. Kim, D.; Yamaguchi, K.; Oda, T. Nitric oxide synthase-like enzyme mediated nitric oxide generation by harmful red tide phytoplankton, Chattonella marina. *J. Plankton Res.* **2006**, *28*, 613–620. [CrossRef]
51. Valentovicová, K.; Halusková, L.; Huttová, J.; Mistrík, I.; Tamás, L. Effect of cadmium on diaphorase activity and nitric oxide production in barley root tips. *J. Plant Physiol.* **2010**, *167*, 10–14. [CrossRef]
52. Roszer, T. *The Biology of Subcellular Nitric Oxide*; Springer: Dordrecht, The Netherlands; Heidelberg, Germany; London, UK; New York, NY, USA, 2012; ISBN 9789400728189.
53. Thomas, T.E.; Harrison, P.J. A Comparison of Invitro and Invivo Nitrate Reductase Assays in 3 Intertidal Seaweeds. *Bot. Mar.* **1988**, *31*, 101–107. [CrossRef]
54. Granbom, M.; Chow, F.; Lopes, P.F.; De Oliveira, M.C.; Colepicolo, P.; De Paula, E.J.; Pedersén, M. Characterisation of nitrate reductase in the marine macroalga *Kappaphycus alvarezii* (Rhodophyta). *Aquat. Bot.* **2004**, *78*, 295–305. [CrossRef]
55. Lopes, P.F.; Oliveira, M.C.; Colepicolo, P. Diurnal Fluctuation of Nitrate Reductase Activity in the Marine Red Alga *Gracilaria Tenuistipitata* (Rhodophyta). *J. Phycol.* **1997**, *33*, 225–231. [CrossRef]
56. Chow, F.; De Oliveira, M.C.; Pedersén, M. In vitro assay and light regulation of nitrate reductase in red alga *Gracilaria chilensis*. *J. Plant Physiol.* **2004**, *161*, 769–776. [CrossRef]
57. Zhao, M.-G.; Chen, L.; Zhang, L.-L.; Zhang, W.-H. Nitric Reductase-Dependent Nitric Oxide Production Is Involved in Cold Acclimation and Freezing Tolerance in Arabidopsis. *Plant Physiol.* **2009**, *151*, 755–767. [CrossRef]
58. Hwang, S.; Williams, S.; Brinhuis, B. Changes in internal dissolved nitrogen pools as related to nitrate uptake and assimilaitonin *Gracilaria tikvahiae* McLachlan (Rhodophyta). *Bot. Mar.* **1987**, *30*, 11–19. [CrossRef]
59. Berges, J.A.; Harrison, P.J. Nitrate reductase activity quantitatively predicts the rate of nitrate incorporation under steady state light limitation. *Limnol. Oceanogr.* **1995**, *40*, 82–93. [CrossRef]
60. Botsoglou, N.A.; Fletouris, D.J.; Papageorgiou, G.E.; Vassilopoulos, V.N.; Mantis, A.J.; Trakatellis, A.G. Rapid, Sensitive, and Specific Thiobarbituric Acid Method for Measuring Lipid-Peroxidation in Animal Tissue, Food, and Feedstuff Samples. *J. Agric. Food Chem.* **1994**, *42*, 1931–1937. [CrossRef]

61. Du, Z.; Bramlage, W.J. Modified thiobarbituric acid assay for measuring lipid oxidation in sugar-rich plant tissue extracts. *J. Agric. Food Chem.* **1992**, *40*, 1566–1570. [CrossRef]
62. Reilly, A.; Aust, S.D. Measurement of lipid peroxidation. *Free Radic. Res.* **1999**, *28*, 659–671. [CrossRef] [PubMed]
63. Fukuto, J.M.; Cho, J.Y.; Switzer, C.H. The Chemical Properties of Nitric Oxide and Related Nitrogen Oxides. In *Nitric Oxide. Biology and Pathology*; Ignarro, L., Ed.; Academic Press: Los Ángeles, CA, USA, 2000; pp. 23–40. ISBN 978-0-12-370420-7.
64. Nagano, T. Practical methods for detection of nitric oxide, Luminescence. *J. Biol. Chem. Lumin.* **1999**, *14*, 283–290.
65. Nussler, A.K.; Glanemann, M.; Schirmeier, A.; Liu, L.; Nussler, N.C. Fluorometric measurement of nitrite/nitrate by 2,3-diaminonaphthalene. *Nat. Protoc.* **2006**, *1*, 2223–2226. [CrossRef] [PubMed]
66. Hope, B.T.; Vincent, S.R. Histochemical characterization of neuronal NADPH-diaphorase. *J. Histochem. Cytochem.* **1989**, *37*, 653–661. [CrossRef] [PubMed]
67. Hope, B.T.; Michael, G.J.; Knigge, K.M.; Vincent, S.R. Neuronal NADPH diaphorase is a nitric oxide synthase. *Proc. Natl. Acad. Sci. USA* **1991**, *88*, 2811–2814. [PubMed]
68. Griess, P. Bemerkungen zu der Abhandlung der HH. Weselsky und Benedikt "Ueber einige Azoverbindungen.". *Berichte Dtsch. Chem. Ges.* **1879**, *12*, 426–428. [CrossRef]
69. Chaki, M.; Valderrama, R.; Fernández-Ocaña, A.M.; Carreras, A.; Gómez-Rodríguez, M.V.; Pedrajas, J.R.; Begara-Morales, J.C.; Sánchez-Calvo, B.; Luque, F.; Leterrier, M.; et al. Mechanical wounding induces a nitrosative stress by down-regulation of GSNO reductase and an increase in S-nitrosothiols in sunflower (*Helianthus annuus*) seedlings. *J. Exp. Bot.* **2011**, *62*, 1803–1813. [CrossRef] [PubMed]
70. Noble, J.E.; Bailey, M.J.A. Quantitation of Protein. In *Methods in Enzymology*; Elsevier Inc.: Amsterdam, The Netherlands, 2009; Volume 463, pp. 73–95. ISBN 9780123745361.

© 2019 by the authors. Licensee MDPI, Basel, Switzerland. This article is an open access article distributed under the terms and conditions of the Creative Commons Attribution (CC BY) license (http://creativecommons.org/licenses/by/4.0/).

Review

Nitric Oxide: Its Generation and Interactions with Other Reactive Signaling Compounds

John T. Hancock [1],* and Steven J. Neill [2]

1. Department of Applied Sciences, University of the West of England, Bristol BS16 1QY, UK
2. Faculty of Health and Applied Sciences, University of the West of England, Bristol BS16 1QY, UK; steven.neill@uwe.ac.uk
* Correspondence: john.hancock@uwe.ac.uk; Tel.: +44-(0)117-328-2475

Received: 15 January 2019; Accepted: 10 February 2019; Published: 12 February 2019

Abstract: Nitric oxide (NO) is an immensely important signaling molecule in animals and plants. It is involved in plant reproduction, development, key physiological responses such as stomatal closure, and cell death. One of the controversies of NO metabolism in plants is the identification of enzymatic sources. Although there is little doubt that nitrate reductase (NR) is involved, the identification of a nitric oxide synthase (NOS)-like enzyme remains elusive, and it is becoming increasingly clear that such a protein does not exist in higher plants, even though homologues have been found in algae. Downstream from its production, NO can have several potential actions, but none of these will be in isolation from other reactive signaling molecules which have similar chemistry to NO. Therefore, NO metabolism will take place in an environment containing reactive oxygen species (ROS), hydrogen sulfide (H_2S), glutathione, other antioxidants and within a reducing redox state. Direct reactions with NO are likely to produce new signaling molecules such as peroxynitrite and nitrosothiols, and it is probable that chemical competitions will exist which will determine the ultimate end result of signaling responses. How NO is generated in plants cells and how NO fits into this complex cellular environment needs to be understood.

Keywords: antioxidants; hydrogen gas; hydrogen peroxide; hydrogen sulfide; nitric oxide; reactive oxygen species

1. Introduction

Since nitric oxide (NO) was mooted to be an important signaling molecule in animals in 1987 [1], and with the subsequent reporting of its role in plant signaling [2–4], there has been extensive work on investigating its function in plants.

Higher plants would have evolved through a lineage that would have been exposed to a range to toxic and reactive compounds and have therefore adapted to encompass them into their normal metabolism [5]. NO, along with reactive oxygen species (ROS) such as the superoxide anion ($O_2^{·-}$) and hydrogen peroxide (H_2O_2), along with hydrogen sulfide (H_2S), works as part of a suite of relatively reactive small molecules in cells which help to control the cell's activity and the function of proteins. NO has been implicated in seed germination [6], root development [7], stomatal closure [8], pathogen challenge [9], plant reproduction [10,11] and stress responses [12]. Therefore, how NO is produced, perceived and leads to a range of effects is important to unravel.

The generation of NO in plants remains controversial, as discussed below, while the measurement of NO [13] in plant materials is still contentious, and often it is not possible to give its sub-cellular location or quantification. This can itself lead to problems with interpretation, as it is not known if NO accumulates to significant, perhaps what could be referred to as threshold, levels, or whether the accumulation of NO is compartmentalized, as reported for other signaling molecules [14], such as cAMP [15,16] and Ca^{2+} [17] but also including ROS and redox signaling [18,19]. Therefore, the idea of compartmentalisation is important to consider here. It is often difficult, therefore, to interpret the data generated. On top of this, NO will react with other signaling molecules, and this makes it difficult to understand fully how NO integrates into a complex signaling pathway. This is also further discussed below.

2. Nitric Oxide Generation in Plant Cells

There seems to be little doubt that plant cells generate NO and are able to respond to it. Therefore, multiple routes to NO accumulation have been suggested, including some that are enzyme-dependent and others that are enzyme-independent [20,21].

One of the major sources of NO is the enzyme nitrate reductase (NR) [22,23]. It has been shown to be important, for example, in the control of stomatal closure [24]. *Arabidopsis thaliana* has two isoforms of NR, and it is thought that both are important in signaling [25,26]. Furthermore, other proteins may interact to create nitrite-dependent enzymes as well [22].

Another enzyme which can generate NO, albeit under hypoxic conditions, is xanthine oxidoreductase (XOR) [27], while other molybdenum-based enzymes may also be important [22]. However, the enzyme which has attracted most attention, perhaps not surprisingly, is nitric oxide synthase (NOS). Despite early reports of the isolation of a NOS from higher plants, it became apparent that the protein which directly produces NO was not identified [28]. To date, this remains controversial.

In lower plants, NOS homologues have been identified [29]. Two green algae genomes showed evidence of sequences for NOS, *Ostreococcus tauri* and *Ostreococcus lucimarinus* [30]. The *O. tauri* sequence was 45% similar to human NOS and the structure was most similar to eNOS. On characterising this enzyme, it was found that the k_m for L arginine, the likely substrate for this NOS enzyme, was found to be 12 ± 5 µM, suggesting that it might have physiological relevance [30]. Such data give hope for finding such an enzyme in higher plants. However, the literature on the nature of a plant NOS has been reviewed widely, and it has been argued previously that higher plants do not contain a NOS enzyme [31,32]. More convincing is the genomic search that was reported [33]. Here, the search involved data sets from the 1000 Plants (1KP) international consortium. No typical NOS sequences were found when 1087 sequenced transcriptomes from land plants were investigated. In contrast to this, 15 of the 265 algal species analyzed showed evidence of NOS sequences. The authors concluded that land plants must produce NO using a different mechanism to that found in animals [33]. This makes it hard to explain much of the data that has been published on NOS-like enzymes in plants, such as a recent study on barley root tips [34] where the NOS inhibitor N(ω)-nitro-L-arginine methyl ester (L-NAME) was shown to have effects. Such work leads researchers to refer to a NOS-like enzyme in plants, but as no homologue, at the gene or protein level, to a mammalian NOS has been reported in any higher plant, it is suggested here that the term NOS-like should not be used and such enzymes and proteins should be referred to as nitric oxide generating (NOGs).

If an enzyme were to generate NO in manner similar to that reported for mammalian NOS, there should be identifiable aspects. Butt et al. [35] used a proteomic approach to identify plant proteins which cross-reacted with mammalian NOS antibodies. Using 2-D gels of extracts from *Zea mays* L. they reported that 20 proteins were immunoreactive following Western blot analysis. Fifteen of the proteins were identified using matrix-assisted laser desorption/ionization time-of-flight mass spectrometry and found not to be related to NO metabolism. Although five proteins remained unidentified, the authors concluded that the immunological techniques so far used were not sufficient to infer the presence of a plant NOS protein [35]. For an up-to-date summary of the discussion of the presence of NOS in plants, see Santolini et al. [32] or Astier et al. [21].

However, it cannot be assumed that all elements of a mammalian NOS should be identifiable. OtNOS lacks the autoregulatory control element, suggesting that it is most closely related to the iNOS isoform in mammals [30]. This may also suggest that looking for a similar domain in a higher plant NOS is futile. Here, as an example of the sorts of aspects that could be looked for, a bioinformatics approach was used. This way of searching for a NOS-like enzyme is predicated on the fact that there should be domains or motifs which are important for NOS-like function, and therefore there should be some level of conservation in these sequences, albeit perhaps hard to find. The mammalian enzymes contain an oxidase domain, a reductase domain and regions which are able to interact with calmodulin [36]. Here, small stretches of sequence have been used to search for possible NOS-like candidates in Arabidopsis and Oryza (Table 1). These relatively short sequences have been derived from the work of others [37,38], such as those looking for NOS-interactions in rat NOS, as well as using the Prosite [39] NOS signature and sequences identified from alignments of the three human NOS proteins using ClustalOmega [40] (data not shown). The relative positions of such sequences within the rat nNOS peptide sequence is shown in Figure 1.

If a plant NOS-like enzyme were to function in a manner similar to a mammalian one, it should have a reductase domain capable of oxidizing NAPDH and having a flavin prosthetic group, such as flavin mononucleotide (FMN). The mammalian NOS reductase is homologous to that of the P450 family of enzymes [41]. Plants have reductases which are similar. Arabidopsis has two proteins which can identified as p450 reductases: NP_001190823 and NP_194750. Some of the short sequences used in Table 1, such as eNOS 952–980, found evidence of reductases in the plant genomes searched. It is therefore possible that any NOS-like enzyme in plants does not have a dedicated reductase, but can draw electrons from other reductases, which are possibly multifunctional.

If there is no need for a dedicated reductase domain, this is almost certainly not true for the oxidase domain. To generate NO, this is the active site that would need to exist. It is very possible that the plant NOS-like protein may only be an oxidase domain, lacking a reductase. It has been reported that bacterial NOS enzymes are indeed like this, lacking a reductase but using electron donation from a nonspecific reductase [42]. Therefore, a search for the oxidase domain is important.

Using the NOS signature from the rat nNOS sequence (NM_052799 XM_346438) in Blastp at NCBI had the highest score hit of hypothetical protein OsI_24933 [*Oryza sativa* Indica Group] ID: EAZ02807.1. This had the following match:

```
Query 1     RCVGRIQW 8
            RC G IQW
Sbjct 224   RCTGKIQW 231
```

The same match was found for an *Arabidopsis thaliana* hypothetical protein (amino acids 115–122: Table 1). Others on the Blastp output are annotated as F-box kelch-repeat proteins. Putting both the *Oryza* and *Arabidopsis* sequences through Prosite revealed nothing of significance; only phosphorylation sites and other Prorules for post-translational modifications. Therefore, these hypothetical proteins look unlikely to be able to act as part of a NOS protein.

Using the NOS signature from the Prosite ProRule data (PS60001) revealed nothing in plants of significance, but it did pull out NOS-like sequences from a range of other organisms, including Staphylococcus and insects. Therefore, there is little evidence of this short NOS signature sequence being in either Arabidopsis or Oryza databases, at least to date.

NOS is likely to interact with other peptides, and this would be a way to identify important functional regions. The calmodulin-interacting regions (CaM) from the rat sequence revealed nothing of note. When the three human NOS sequences were aligned, these CaM motifs were not represented in all NOS peptides, and therefore it could be argued that they are not essential, and not finding them does not rule out the presence of a plant NOS. Others have looked for other interacting regions as well [37,38]; for example, between the FMN and oxidase domains. Taking interesting sequences such as those thought to be involved in protein interactions from the literature also failed to reveal a likely NOS sequence in Arabidopsis or Oryza (Table 1).

It can be concluded so far, in that case, that there is no significant evidence from the sequence searching of a NOS-like protein in two plant sequences for which major genome sequencing projects have been undertaken [43,44].

Table 1. Sequences used to search for matches in *Arabidopsis* and *Oryza* using Blastp and tBlastn. Areas used from the rat nitric oxide synthase (nNOS) sequence are highlighted in Figure 1.

Source of Sequence	Sequence Name	Sequence	Reference/Source for Sequence	Significant Find/Comment
NCBI	NOS signature	[GR]-C-[IV]-G-R-[ILS]-x-W	prosite.expasy.org/PS60001	No significant sequence identified using ScanProsite/ previously found in a range of species including Staphylococcus, insects and mammals.
Rat nNOS	NOS signature	-RCVGRIQW-	[37]	• Hypothetical protein OsI_24933 [*Oryza sativa* Indica Group] ID: EAZ02807.1, Length: 515, Identity 75%, Query cover 100%. • Hypothetical protein AXX17_AT1G30490 [*Arabidopsis thaliana*] ID: OAP19388.1, Length: 398, Identity, 75%, Query cover 100%.
Rat nNOS	FMN subdomain 538–547	-PELVLEVPIR-	[37]	No significant sequence identified
Rat nNOS	FMN subdomain 582–605	-CPFSGWYMGTEIGVRDYCDNSRYN-	[37]	No significant sequence identified
Rat nNOS	Haem domain 80–102	-ALEVLRGIASETHVVLILRGPEG-	[37]	No significant sequence identified
Rat nNOS	Haem domain 187–203	-TKANLQDIGEHDELLKE-	[37]	No significant sequence identified
Rat nNOS	Haem domain 366–386	-YSSIKRFGSKAHMDRLEEVNK-	[37]	No significant sequence identified
Rat nNOS	Haem domain 396–465	-LKDTELIYGAKHAWRNASRCVGRIQW SKLQVFDARDCTTAHGMFNYICNHVKY ATNKGNLRSAITIFPQR-	[37]	No significant sequence identified /NOS consensus sequence underlined, but not found in plants.
Rat nNOS	Haem domain 471–485	-DFRVWNSQLIRYAGY-	[37]	No significant sequence identified
Rat nNOS	CaM domain 20–36	-LFKRKVGGLGFLVKERV-	[37]	No significant sequence identified
Rat nNOS	CaM domain 105–124	-THLETTFTGDGTPKTIRVTQ-	[37]	No significant sequence identified
Human iNOS	509–537	-KRREIPLKV1VKAVLFACMLMRKTMAS<u>RV</u>-	[38]	Poor homology in some *Oryza* sequences /R536 important in human (underlined)/ -SRV- present
Rat nNOS	725–753	-KRRAIGFKKLAEAVKFSAKLMCQAMAK<u>RV</u>-	[38]	Poor homology in some *Oryza* sequences /R752 important (underlined) in rat 1. Os08g0243500, partial [*Oryza sativa* Japonica Group] ID: BAF23260.1, Length: 651, Identity 32%, Query cover 82%. 2. NADPH-cytochrome P450 reductase [*Oryza sativa* Japonica Group] ID: XP_015650780.1, Length: 719, Identity 32%, Query cover 82%. 3. PREDICTED: NADPH-dependent diflavin oxidoreductase 1 [*Oryza brachyantha*] ID: XP_006659755.1, Length: 625, Identity 35%, Query cover 89%. 4. Hypothetical protein OsJ_30318 [*Oryza sativa* Japonica Group] ID: EEE70211.1, Length: 795, Identity 32%, Query cover 89%. Others similar can be identified.
Mouse iNOS/FMN domain	532–694	-VRATV … PKRFT-	Derived from [37]	

Table 1. Cont.

Source of Sequence	Sequence Name	Sequence	Reference/Source for Sequence	Significant Find/Comment
eNOS (human)	566–585	-LVLVVTSTFGNGPPENGES-	Derived from human Clustal Omega e/i/n NOS	No significant sequence identified
eNOS (human)	952–980	-EIHKTVAVLAYRTGDGLGPLHYGVCSTWL-	Derived from human Clustal Omega e/i/n NOS	Evidence of being part of a oxidoeductase or P450 reductase in plants: for example XP_015696451.1 & XP_006653834.1(both have 45% identical over 96% coverage) from Oryza: CAA46814.1 & NP_194183.1 (both 67% identical over 41% coverage) from Arabidopsis].
Nitric Oxide Synthase Related Proteins/Peptides				
Nostrin isoform 2 [*Homo sapiens*]: NP_001034813	Full sequence	MRDPLI ... NTATKA	https://www.ncbi.nlm.nih.gov/protein/NP_001034813.2/ & [45]	1. SH3 domain-containing protein 3 [*Arabidopsis lyrata* subsp. lyrata] ID: XP_002868013.1, Length: 351, Identity 37%, Query cover 15% 2. Os04g0539800 [*Oryza sativa* Japonica Group] ID: BAF15352.2, Length: 115, Identity 48%, Query cover 10% 3. Putative protein [*Arabidopsis thaliana*] ID: CAB53647.1, Length: 330, Identity 26%, query cover 30%.
Carboxyl-terminal PDZ ligand of neuronal nitric oxide synthase protein isoform 1 [*Homo sapiens*]. NP_055512	Full sequence	MPSKT ... DDEIAV	https://www.ncbi.nlm.nih.gov/protein/NP_055512 & [46]	No significant sequences identified/PH-like superfamily predicted.
Nitric oxide synthase-interacting protein isoform 1 [*Homo sapiens*]. NP_057037	Full sequence	MTRHG ... SRPVMGA	https://www.ncbi.nlm.nih.gov/protein/NP_057037 & [47]	1. PREDICTED: nitric oxide synthase-interacting protein [*Oryza brachyantha*] ID: XP_006649867.1, Length: 305, Identity 32%, Query cover 96% 2. E3 ubiquitin-protein ligase CSU1 [*Oryza sativa* Japonica Group] ID: XP_015630570.1, Length: 305, Identity 32%, Query cover 96%. 3. Phosphoinositide binding protein [*Arabidopsis thaliana*] ID: NP_564781.1, Length: 310, Identity 31%, Query cover 98%. 4. Nitric oxide synthase-interacting protein homolog [*Arabidopsis lyrata* subsp. lyrata] ID: XP_020890108.1, Length: 310, Identity 31%, Query cover 98%.

Rattus norvegicus nitric oxide synthase 1 (Nos1), mRNA. (nNOS)
ACCESSION   NM_052799 XM_346438

MEENTFGVQQ IQPNVISVRL FKRKVGGLGF LVKERVSKPP VIISDLIRGG
AAEQSGLIQA GDIILAVNDR PLVDLSYDSA LEVLRGIASE THVVLILRGP
EGFTTHLETI FTGDGTPKTI RVTQPLGPPT KAVDLSHQPS ASKDQSLAVD
RVTGLGNGPQ HAQGHGQGAG SVSQANGVAI DPTMKSTKAN LQDIGEHDEL
LKEIEPVLSI LNSGSKATNR GGPAKAEMKD TGIQVDRDLD GKSHKAPPLG
GDNDRVFNDL WGKDNVPVVL NNPYSEKEQS PTSGKQSPTK NGSPSRCPRF
LKVKNWETDV VLTDTLHLKS TLETGCTEHI CMGSIMLPSQ HTRKPEDVRT
KDQLFPLAKE FLDQYYSSIK RFGSKAHMDR LEEVNKEIES TSTYQLKDTE
LIYGAKHAWR NASRCVGRIQ WSKLQVFDAR DCTTAHGMFN YICNHVKYAT
NKGNLRSAIT IFPQRTDGKH DFRVWNSQLI RYAGYKQPDG STLGDPANVQ
FTEICIQQGW KAPRGRFDVL PLLLQANGND PELFQIPPEL VLEVPIRHPK
FDWFKDLGLK WYGLPAVSNM LLEIGGLEFS ACPFSGWYMG TEIGVRDYCD
NSRYNILEEV AKKMDLDMRK TSSLWKDQAL VEINIAVLYS FQSDKVTIVD
HHSATESFIK HMENEYRCRG GCPADWVWIV PPMSGSITPV FHQEMLNYRL
TPSFEYQPDP WNTHVWKGTN GTPTKRRAIG FKKLAEAVKF SAKLMGQAMA
KRVKATILYA TETGKSQAYA KTLCEIFKHA FDAKAMSMEE YDIVHLEHEA
*LVLVVTSTFG NGDPPEN*GEK FGCALMEMRH PNSVQEERKY PEPLRFFPRK
GPSLSHVDSE AHSLVAARDS QHRSYKVRFN SVSSYSDSRK SSGDGPDLRD
NFESTGPLAN VRFSVFGLGS RAYPHFCAFG HAVDTLLEEL GGERILKMRE
GDELCGQEEA FRTWAKKVFK AACDVFCVGD DVNIEKANNS LISNDRSWKR
NKFRLTYVAE APDLTQGLSN VHKKRVSAAR LLSRQNLQSP KSSRSTIFVR
LHTNGNQELQ YQPGDHLGVF PGNHEDLVNA LIERLEDAPP ANHVVKVEML
EERNTALGVI SNWKDESRLP PCTIFQAFKY YLDITTPPTP LQLQQFASLA
TNEKEKQRLL VLSKGLQEYE EWKWGKNPTM VEVLEEFPSI QMPATLLLTQ
LSLLQPRYYS ISSSPDMYPD *EVHLTVAIVS YHTRDGEGPV HHGVCSSWLN*
RIQADDVVPC FVRGAPSFHL PRNPQVPCIL VGPGTGIAPF RSFWQQRQFD
IQHKGMNPCP MVLVFGCRQS KIDHIYREET LQAKNKGVFR ELYTVYSREP
DRPKKYVQDV LQEQLAESVY RALKEQGGHI YVCGDVTMAA DVLKAIQRIM
TQQGKLSEED AGVFISRLRD DNRYHEDIFG VTLRTYEVTN RLRSESIAFI
EESKKDADEV FSS

Key:
 = Calmodulin binding regions
XXXXXX = Haem regions
XXXXXX = Flavin regions
XXXXXX = Region used from Xia *et al.* [38]
In red/BOLD = NOS consensus sequence region
Italics = areas equivalent to those identified from human NOS alignment

Figure 1. Areas of the rat NOS sequence used to search for higher plant NOS-like proteins. Findings shown in Table 1.

NOS in other species is not a stand-alone protein, but has interacting partners. NOS-interacting proteins can be found in the literature (Table 1), such as nostrin [45], carboxyl terminal PDZ ligand [46] and NOS-interacting protein [47]. Searching for evidence of such proteins in plant genomes may give circumstantial evidence of a NOS-like protein in plants. The nostrin sequence found plant proteins which can interact, perhaps through SH3 domains (Table 1). However, the most intriguing fact was that the *Homo sapiens* NOS interacting protein isoform (NP_057037) was revealed in both the Arabidopsis and Oryza data proteins, which have already been annotated as NOS-interacting proteins (XP_020890108.1 & XP_006649867.1, respectively). Such proteins may be used as lures to find interacting partners in plant extracts, some of which may have NOS-like activity.

Overall, the bioinformatic searching carried out here, although by no means exhaustive, showed no clear evidence of a NOS-like protein in plants, although elements such as a reductase do clearly exist. These data are not contrary to those found and reported by others [33].

3. Interactions of Nitric Oxide with Other Reactive Signals

When the chemistry of nitric oxide is discussed, it is often assumed that this involves the radical form: NO·. However, with the loss or gain of an electron, other forms are nitroxyl (NO^-) and nitrosonium (NO^+) ions [48]. It is important to appreciate that NO will not be generated in cells in isolation. It is often produced in response to a stress, and as such, other signals will be accumulating at the same time, including ROS and H_2S. If cadmium ion stress in plants is taken as an example, the cellular response includes the generation of NO and ROS [49], as well as H_2S [50], all presumably being accumulated in the same sub-cellular location, such as the cytoplasm. Therefore, it is important to consider how NO may interact with other compounds that are present.

One of the main downstream effects of NO is the post-translational modifications of thiols (Figure 2) and other amino acids such as tyrosine. S-nitrosation (often referred to as S-nitrosylation) is the modification of the –SH group to –SNO [51], which may cause a conformational change in the protein, with a concomitant change in activity or function. However, the thiol group may also be modified by oxidation, S-persulfidation by H_2S, glutathionylation by GSH, or reaction with another thiol to create a disulfide (reviewed previously [52]), and so a reaction with NO is not necessarily the outcome. With such a range of possible reactions, the actual resultant change seen will be dependent on the local concentrations of reactants and the kinetics of the possible reactions.

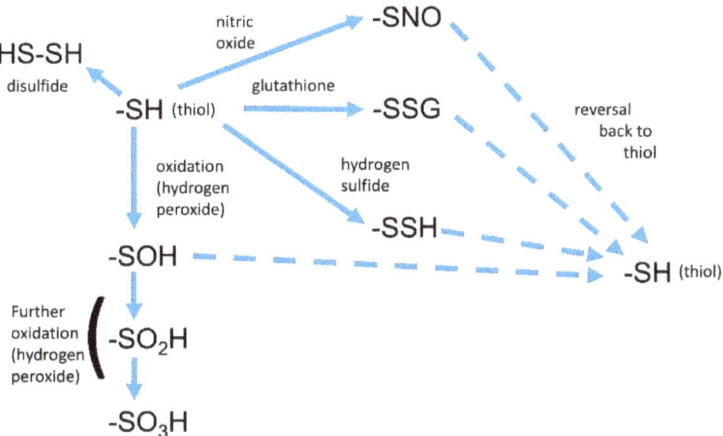

Figure 2. Some post-translational modifications of thiol groups. These include S-nitrosation and oxidation. Many modifications are reversible, and so are akin to phosphorylation.

The protein modification by NO is, however, an important signaling process. Many proteins in plants have been identified as being nitrosated [51], with a good example being glyceraldehyde 3-phosphate dehydrogenase (GAPDH). In mammalian cells, it has been shown that on S-nitrosation, the enzyme translocates to the nucleus, thus abandoning its role in glycolysis to take up a new role in the control of gene expression [53]. In plants, GAPDH has also been shown to be S-nitrosated, and cytosolic GAPDH can interact with nuclear DNA, specifically to a partial gene sequence of NADP-dependent malate dehydrogenase [54]. However, GAPDH can also be modified through oxidation by H_2O_2 [55] and in addition be S-persulfidated by H_2S [56], with the latter known to lead to its translocation to the nucleus. Clearly, there is competition between reactive signals in cells [52], and it cannot be assumed that NO signaling will dominate. However, methods to identify thiol modifications will help to unravel such signaling [57,58].

S-nitrosation also has a role in mediating the interplay between NO and other reactive signaling mechanisms, such as those involving ROS. For example, key enzymes which generate ROS, such as NAPDH oxidase, can be modified by NO. It has been reported that RBOHD is S-nitrosated at Cys890 which inactivates the enzyme and thus reduces its ROS-generating activity [59]. Therefore, NO has an important role in controlling ROS levels and hence the potential downstream signaling here.

The second important modification of proteins brought about by NO is tyrosine nitration [60], and again this may lead to alterations of function. As with S-nitrosation, NO interaction will lead to conformational changes in the protein and commensurate changes in activity, either increased or decreased. Some of these modifications may have the result of altering other signaling pathways mediated by other reactive signals; for example, tyrosine nitration can alter superoxide dismutase (SOD) activity and hence ROS signaling [61].

As NO and ROS are produced in the cell at the same time, it is important to consider their interaction and the ramifications of this chemistry. The most well-known reaction of NO and ROS it that between the superoxide anion and NO which produces peroxynitrite ($ONOO^-$) (Figure 3). This has two potentially important outcomes. Firstly, the reaction removes both $O_2^{\cdot-}$ and NO from the cell or the cell's environment, thereby reducing the bio-availability of both. Thus both ROS-dependent signaling, perhaps through H_2O_2, and NO signaling would be reduced. Secondly, there is a new compound produced which itself can act as a signaling molecule [62], perhaps giving a different response than would have resulted from ROS or NO signaling.

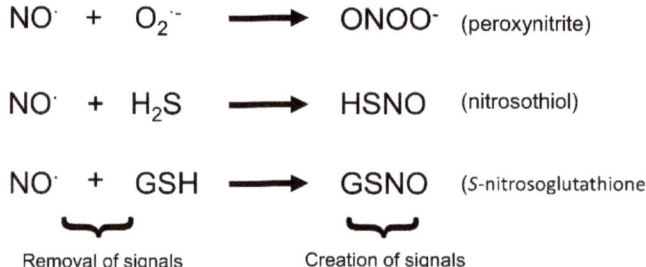

Figure 3. Some of the reactions in which NO can be involved, often leading to new signaling molecules.

The generation of NO will also be into an environment rich in antioxidants. NO may affect the activity of enzymatic antioxidants, as mentioned above, where NO, through a peroxynitrite-mediated mechanism, altered SOD activity [61], and hence reduced the cell's capacity to remove superoxide anions and produce H_2O_2, with the latter being important in signaling. In a similar manner, NO can alter catalase activity [63], thus lowering the cell's capacity to remove H_2O_2, perhaps prolonging ROS-mediated signaling.

A large part of the antioxidant capacity of the cell is due to the presence of low molecular-weight antioxidants. There are a range of small low molecular-weight thiols in cells [64], but one of the most important is glutathione [65]. This exists in the reduced state (GSH) and the oxidized state (GSSG), with the ratio of these compounds, along with the total GSH+GSSG concentration, being partly responsible for the maintenance of the intracellular environment in a very reduced state [65], probably below -200 mV. It is possible that the presence of NO—as it is a redox compound—will lead to the intracellular redox status being altered. It is known that the intracellular redox environment is not static and becomes more oxidizing if cells are in an apoptotic state [65], but it is also possible that the redox environment determines the state of any NO couple and hence the longevity of any NO species, as previously discussed [64,66]. In some cases, for example as the cell becomes more oxidizing, the presence of NO· will be prolonged, and so this will enhance NO·-mediated signaling.

Importantly, NO and glutathione can react together to produce GSNO. This potentially has the capacity to reduce GSH/GSSG levels in cells, and hence potentially alter the intracellular redox environment, especially if the reaction is compartmentalized. The reaction will also remove NO from directly partaking in further signaling. However, GSNO has important roles as well. GSNO can act as a donor and therefore a reservoir of NO, and it has been suggested that GSNO can mediate some NO effects [67], having distinct and overlapping molecular targets when compared to NO itself. GSNO has also been mooted as an important mechanism to transport NO around organisms [68], perhaps through the vasculature system of plants. To terminate GSNO-mediated signaling, it can be removed by the action of GSNO reductase [69,70], which would lower the bioavailability of NO.

Another reactive signal which may interfere with NO signaling is hydrogen sulfide (H_2S). H_2S has recently been found to be an important signaling molecule in both animals and plants [71–73]. It is produced in response to a range of stresses, such as cadmium ions, as mentioned above [50]. H_2S can react directly with NO to produce nitrosothiol (Figure 3). As discussed above, this will reduce the bioavailability of both H_2S and NO, but it will also create a new molecule with potential signaling effects [74]. H_2S will also increase GSH levels in cells [75] and therefore may have the potential to alter the accumulation of GSNO.

Lastly, it has recently been suggested that signaling in animals and plants may involve hydrogen gas (H_2) [76]. The presence of H_2 may alter antioxidant levels in cells [77] and so indirectly alter NO metabolism. However, there is also potential for a more direct interaction with H_2 and some nitrogen compounds [78,79]. Certainly, H_2 has been shown to have effects in plants [80,81] and has touted as a future treatment for plants [82]. NO has been reported to be needed for some of the H_2 gas effects [83,84] and no doubt more interactions between NO and H_2 signaling will be revealed in the future.

4. Conclusions and the Future

NO is a key signaling molecule in plants, being important in plant reproduction [85], development [86] and plant cell death [87]. However, the production of NO in plants remains controversial. Enzymes such as NR are known to be important [23], while others such as XOR may be involved. Much data points to the existence of a NOS-like enzyme being present in higher plants, and although there is such an enzyme in algae [29], the search for homologues in higher plants remains elusive [33], and it appears that such an enzyme really does not exist. It is possible that a novel peptide has oxidase-type activity which can produce NO, receiving electrons from a less specific reductase, as seen in bacteria [42], but if such a peptide does exist, it is very difficult to identify.

If there is an oxidase-type enzyme in higher plants it would need to obtain its electrons from somewhere; most likely a reductase, as seen with P450. As can be seen in Table 1, such reductases in plants do exist and might serve this function. Furthermore, putative NOS-interacting proteins have been identified in plants, as listed in Table 1. Therefore, by concentrating on proteins which are most likely to interact with an oxidase-like protein and using these as bait in purification experiments, it is possible that the future may see a novel NO-generating oxidase being discovered in higher plants. However, with divergent evolution of plants and animals, and the fact that plants appear to have other NO generating pathways such as nitrate reductase, it may be that such an NO-producing oxidase does not exist.

The role of NO is also complex and not fully understood. NO is made in plants cells in response to the same cues that initiate the generation of ROS and H_2S, and so NO will not work in isolation. The reaction of NO with ROS or H_2S will lower the bioavailability of NO, but also produce new signaling molecules, such as peroxynitrite [62] and nitrosothiols [74], which will have their own outcomes. The impact of NO on the cellular redox poise, especially if compartmentalized, needs to be considered, as does the impact of the redox environment on the NO metabolism that may ensue [64]. NO will interact with antioxidants, such as glutathione, which may even facilitate its organismic transport [68]. Furthermore, one of the main actions of NO is to chemically modify proteins, for example through *S*-nitrosation, but this may not be possible if other reactive compounds such as H_2S or ROS have already modified the relevant thiol. Therefore, the downstream actions of NO cannot always be assumed.

In conclusion, two major barriers exist to the progression of NO research in plants. Firstly, the controversy surrounding the presence of NO-generating enzymes needs to be resolved. Here, is it suggested that the term NOS-like is dropped to avoid continual confusion by drawing parallels with the mammalian system, as clearly the homology does not exist. The term nitric oxide-generating (NOG) would be more accurate. Secondly, the way NO is interwoven into the signaling of other important reactive chemicals needs to be understood. Is NO metabolism compartmentalized in such a way that ROS, GSH or H_2S do not interfere, or is there a competition between all these signals, keeping each other in check, as already been mooted [88]? Until such issues are resolved, the true nature of the role of NO in plants will remain elusive.

Author Contributions: Writing Original Draft Preparation, J.T.H.; Bioinformatic Analysis, J.T.H.; Conceptualization, J.T.H. and S.J.N.; Editing, J.T.H. and S.J.N.

Funding: This research received no external funding.

Acknowledgments: The authors would like to thank Eric Underbakke, Iowa State University, for email discussion and information sent.

Conflicts of Interest: The authors declare no conflict of interest.

References

1. Palmer, R.M.J.; Ferrige, A.G.; Moncada, S. Nitric oxide release accounts for the biological activity of endothelium-derived relaxing factor. *Nature* **1987**, *327*, 524–526. [CrossRef] [PubMed]
2. Laxalt, A.M.; Beligni, M.V.; Lamattina, L. Nitric oxide preserves the level of chlorophyll in potato leaves infected by *Phytophthora infestans*. *Eur. J. Plant Pathol.* **1997**, *103*, 643–651. [CrossRef]
3. Delledonne, M.; Xia, Y.; Dixon, R.A.; Lamb, C. Nitric oxide functions as a signal in plant diseas resistance. *Nature* **1998**, *394*, 585–588. [CrossRef]
4. Durner, J.; Wendehenne, D.; Klessig, D.F. Defense gene induction in tobacco by nitric oxide, cyclic GMP, and cyclic ADP-ribose. *Proc. Natl. Acad. Sci. USA* **1998**, *95*, 10328–10333. [CrossRef]
5. Hancock, J.T. Harnessing evolutionary toxins for signaling: Reactive oxygen species, nitric oxide and hydrogen sulfide in plant cell regulation. *Front. Plant Sci.* **2017**, *8*, 189. [CrossRef] [PubMed]
6. Arc, E.; Galland, M.; Godin, B.; Cueff, G.; Rajjou, L. Nitric oxide implication in the control of seed dormancy and germination. *Front. Plant Sci.* **2013**, *4*, 346. [CrossRef]
7. Sanz, L.; Albertos, P.; Mateos, I.; Sánchez-Vicente, I.; Lechón, T.; Fernández-Marcos, M.; Lorenzo, O. Nitric oxide (NO) and phytohormones crosstalk during early plant development. *J. Exp. Bot.* **2015**, *66*, 2857–2868. [CrossRef]
8. Gayatri, G.; Agurla, S.; Raghavendra, A.S. Nitric oxide in guard cells as an important secondary messenger during stomatal closure. *Front. Plant Sci.* **2013**, *4*, 425. [CrossRef]
9. Mur, L.A.; Carver, T.L.; Prats, E. NO way to live; the various roles of nitric oxide in plant-pathogen interactions. *J. Exp. Bot.* **2006**, *57*, 489–505. [CrossRef]
10. Hiscock, S.J.; Bright, J.; McInnis, S.M.; Desikan, R.; Hancock, J.T. Signaling on the stigma: Potential new roles for ROS and NO in plant cell signaling. *Plant Signal. Behav.* **2007**, *2*, 23–24. [CrossRef]
11. Kwon, E.; Feechan, A.; Yun, B.W.; Hwang, B.H.; Pallas, J.A.; Kang, J.G.; Loake, G.J. AtGSNOR1 function is required for multiple developmental programs in Arabidopsis. *Planta* **2012**, *236*, 887–900. [CrossRef]
12. Hu, J.; Yang, H.; Mu, J.; Lu, T.; Peng, J.; Deng, X.; Kong, Z.; Bao, S.; Cao, X.; Zuo, J. Nitric oxide regulates protein methylation during stress responses in plants. *Mol. Cell* **2017**, *67*, 702–710. [CrossRef] [PubMed]

13. Yamasaki, H.; Watanabe, N.S.; Sakihama, Y.; Cohen, M.F. An overview of methods in plant nitric oxide (NO) research: Why do we always need to use multiple methods? *Meth. Mol. Biol.* **2016**, *1424*, 1–14. [CrossRef]
14. McCormick, K.; Baillie, G.S. Compartmentalisation of second messenger signalling pathways. *Curr. Opin. Genet. Dev.* **2014**, *27*, 20–25. [CrossRef] [PubMed]
15. Zaccolo, M.; Magalhães, P.; Pozzan, T. Compartmentalisation of cAMP and Ca^{2+} signals. *Curr. Opin. Cell Biol.* **2002**, *14*, 160–166. [CrossRef]
16. Baillie, G.S.; Scott, J.D.; Houslay, M.D. Compartmentalisation of phosphodiesterases and protein kinase A: Opposites attract. *FEBS Lett.* **2005**, *579*, 3264–3270. [CrossRef] [PubMed]
17. Bononi, A.; Missiroli, S.; Poletti, F.; Suski, J.M.; Agnoletto, C.; Bonora, M.; De Marchi, E.; Giorgi, C.; Marchi, S.; Patergnani, S.; et al. Mitochondria-associated membranes (MAMs) as hotspot Ca^{2+} signaling units. *Adv. Exp. Med. Biol.* **2012**, *740*, 411–437. [CrossRef]
18. De Rezende, F.F.; Martins Lima, A.; Niland, S.; Wittig, I.; Heide, H.; Schröder, K.; Eble, J.A. Integrin α7β1 is a redox-regulated target of hydrogen peroxide in vascular smooth muscle cell adhesion. *Free Radic. Biol. Med.* **2012**, *53*, 521–531. [CrossRef]
19. Noctor, G.; Foyer, C.H. Intracellular redox compartmentation and ROS-related communication in regulation and signaling. *Plant Physiol.* **2016**, *171*, 1581–1592. [CrossRef]
20. Shapiro, A.D. Nitric oxide signaling in plants. *Vitam. Horm.* **2005**, *72*, 339–398.
21. Astier, J.; Gross, I.; Durner, J. Nitric oxide production in plants: An update. *J. Exp. Bot.* **2018**, *69*, 3401–3411. [CrossRef] [PubMed]
22. Rockel, P.; Strube, F.; Rockel, A.; Wildt, J.; Kaiser, W.M. Regulation of nitric oxide (NO) production by plant nitrate reductase in vivo and in vitro. *J. Exp. Bot.* **2002**, *53*, 103–110. [CrossRef]
23. Chamizo-Ampudia, A.; Sanz-Luque, E.; Llamas, A.; Galvan, A.; Fernandez, E. Nitrate reductase regulates plant nitric oxide homeostasis. *Trends Plant Sci.* **2017**, *22*, 163–174. [CrossRef] [PubMed]
24. Desikan, R.; Grifitths, R.; Hancock, J.; Neill, S. A new role for an old enzyme: Nitrate reductase-mediated nitric oxide generation is required for abscisic acid-induced stomatal closure in *Arabidopsis thaliana*. *Proc. Natl. Acad. Sci. USA* **2002**, *99*, 16314–16318. [CrossRef] [PubMed]
25. Hao, F.; Zhao, S.; Dong, H.; Zhang, H.; Sun, L.; Miao, C. *Nia1* and *Nia2* are involved in exogenous salicylic acid-induced nitric oxide generation and stomatal closure in Arabidopsis. *J. Integr. Plant Biol.* **2010**, *52*, 298–307. [CrossRef] [PubMed]
26. Zhao, C.; Cai, S.; Wang, Y.; Chen, Z.H. Loss of nitrate reductases NIA1 and NIA2 impairs stomatal closure by altering genes of core ABA signaling components in Arabidopsis. *Plant Signal. Behav.* **2016**, *11*, e1183088. [CrossRef] [PubMed]
27. Millar, T.M.; Stevens, C.R.; Benjamin, N.; Eisenthal, R.; Harrison, R.; Blake, D.R. Xanthine oxidoreductase catalyses the reduction of nitrates and nitrite to nitric oxide under hypoxic conditions. *FEBS Lett.* **1998**, *427*, 225–228. [CrossRef]
28. Zemojtel, T.; Fröhlich, A.; Palmieri, M.C.; Kolanczyk, M.; Mikula, I.; Wyrwicz, L.S.; Wanker, E.E.; Mundlos, S.; Vingron, M.; Martasek, P.; et al. Plant nitric oxide synthase: A never-ending story? *Trends Plant. Sci.* **2006**, *11*, 524–525. [CrossRef]
29. Astier, J.; Jeandroz, S.; Wendehenne, D. Nitric oxide synthase in plants: The surprise from algae. *Plant Sci.* **2018**, *268*, 64–66. [CrossRef]
30. Foresi, N.; Correa-Aragunde, N.; Parisi, G.; Caló, G.; Salerno, G.; Lamattina, L. Characterization of a nitric oxide synthase from the plant kingdom: NO generation from the green alga *Ostreococcus tauri* is light irradiance and growth phase dependent. *Plant Cell* **2010**, *22*, 3816–3830. [CrossRef]
31. Hancock, J.T.; Neill, S.J. NO Synthase in plants? *CAB Rev.* **2014**, *9*, 1–9. [CrossRef]
32. Santolini, J.; André, F.; Jeandroz, S.; Wendehenne, D. Nitric oxide synthase in plants: Where do we stand? *Nitric Oxide* **2017**, *63*, 30–38. [CrossRef] [PubMed]
33. Jeandroz, S.; Wipf, D.; Stuehr, D.J.; Lamattina, L.; Melkonian, M.; Tian, Z.; Zhu, Y.; Carpenter, E.J.; Wong, G.K.-S.; Wendehenne, D. Occurrence, structure, and evolution of nitric oxide synthase-like proteins in the plant kingdom. *Sci. Signal.* **2016**, *9*, re2. [CrossRef] [PubMed]
34. Tamás, L.; Demecsová, L.; Zelinová, V. L-NAME decreases the amount of nitric oxide and enhances the toxicity of cadmium via superoxide generation in barley root tip. *J. Plant Physiol.* **2018**, *224–225*, 68–74. [CrossRef] [PubMed]

35. Butt, Y.K.-C.; Lum, J.H.-K.; Lo, S.C.-L. Proteomic identification of plant proteins probed by mammalian nitric oxide synthase antibodies. *Planta* **2003**, *216*, 762–771. [PubMed]
36. Stuehr, D.; Vasquez-Vivar, J. Nitric oxide synthases- from genes to function. *Nitric Oxide* **2017**, *63*, 29. [CrossRef] [PubMed]
37. Smith, B.C.; Underbakke, E.S.; Kulp, D.W.; Schief, W.R.; Marletta, M.A. Nitric oxide synthase domain interfaces regulate electron transfer and calmodulin activation. *Proc. Natl. Acad. Sci. USA* **2013**, *110*, E3577–E3586. [CrossRef] [PubMed]
38. Xia, X.; Misra, I.; Iyanagi, T.; Kim, J.-J. Regulation of interdomain interactions by calmodulin in inducible nitric-oxide synthase. *J. Biol. Chem.* **2009**, *284*, 30708–30717. [CrossRef]
39. De Castro, E.; Sigrist, C.J.A.; Gattiker, A.; Bulliard, V.; Langendijk-Genevaux, P.S.; Gasteiger, E.; Bairoch, A.; Hulo, N. ScanProsite: Detection of PROSITE signature matches and ProRule-associated functional and structural residues in proteins. *Nucleic Acids Res.* **2006**, *34*, W362–W365. [CrossRef] [PubMed]
40. McWilliam, H.; Li, W.; Uludag, M.; Squizzato, S.; Park, Y.M.; Buso, N.; Cowley, A.P.; López, R. Analysis tool web services from the EMBL-EBI. *Nucleic Acids Res.* **2013**, *41*, W597–W600. [CrossRef]
41. Iyanagi, T. Structure and function of NADPH-cytochrome P450 reductase and nitric oxide synthase reductase domain. *Biochem. Biophys. Res. Commun.* **2005**, *338*, 520–528. [CrossRef] [PubMed]
42. Gusarov, I.; Starodubtseva, M.; Wang, Z.-Q.; McQuade, L.; Lippard, S.J.; Stuehr, D.J.; Nudler, E. Bacterial nitric-oxide synthases operate without a dedicated redox partner. *J. Biol. Chem.* **2008**, *283*, 13140–13147. [CrossRef] [PubMed]
43. The Arabidopsis Genome Initiative. Analysis of the genome sequence of the flowering plant *Arabidopsis thaliana*. *Nature* **2000**, *408*, 796–815. [CrossRef] [PubMed]
44. Eckardt, N.A. Sequencing the rice genome. *Plant Cell* **2000**, *12*, 2011–2018. [CrossRef]
45. Icking, A.; Matt, S.; Opitz, N.; Wiesenthal, A.; Müller-Esterl, W.; Schilling, K. NOSTRIN functions as a homotrimeric adaptor protein facilitating internalization of eNOS. *J. Cell Sci.* **2005**, *118*, 5059–5069. [CrossRef] [PubMed]
46. Jaffrey, S.R.; Snowman, A.M.; Eliasson, M.J.; Cohen, N.A.; Snyder, S.H. CAPON: A protein associated with neuronal nitric oxide synthase that regulates its interactions with PSD95. *Neuron* **1998**, *20*, 115–124. [CrossRef]
47. Dedio, J.; Konig, P.; Wohlfart, P.; Schroeder, C.; Kummer, W.; Muller-Esterl, W. NOSIP, a novel modulator of endothelial nitric oxide synthase activity. *FASEB J.* **2001**, *15*, 79–89. [CrossRef]
48. Lancaster, J., Jr. Nitric oxide: A brief overview of chemical and physical properties relevant to therapeutic applications. *Future Sci. OA* **2015**, *1*, FSO59. [CrossRef]
49. Rodríguez-Serrano, M.; Romero-Puertas, M.C.; Pazmiño, D.M.; Testillano, P.S.; Risueño, M.C.; Del Río, L.A.; Sandalio, L.M. Cellular response of pea plants to cadmium toxicity: Cross talk between reactive oxygen species, nitric oxide, and calcium. *Plant Physiol.* **2009**, *150*, 229–243. [CrossRef]
50. Mostofa, M.G.; Rahman, A.; Ansary, M.M.; Watanabe, A.; Fujita, M.; Tran, L.S. Hydrogen sulfide modulates cadmium-induced physiological and biochemical responses to alleviate cadmium toxicity in rice. *Sci. Rep.* **2015**, *5*, 14078. [CrossRef]
51. Lindermayr, C.; Saalbach, G.; Durner, J. Proteomic identification of S-nitrosylated proteins. *Plant Physiol.* **2005**, *137*, 921–930. [CrossRef] [PubMed]
52. Hancock, J.T.; Craig, T.; Whiteman, M. Competition of reactive signals and thiol modifications of proteins. *J. Cell Signal.* **2017**, *2*, 170. [CrossRef]
53. Sirover, M.A. Subcellular dynamics of multifunctional protein regulation: Mechanisms of GAPDH intracellular translocation. *J. Cell Biochem.* **2012**, *113*, 2193–2200. [CrossRef] [PubMed]
54. Holtgrefe, S.; Gohlke, J.; Starmann, J.; Druce, S.; Klocke, S.; Altmann, B.; Wojtera, J.; Lindermayr, C.; Scheibe, R. Regulation of plant cytosolic glyceraldehyde 3-phosphate dehydrogenase isoforms by thiol modifications. *Physiol. Plant* **2008**, *133*, 211–228. [CrossRef] [PubMed]
55. Hancock, J.T.; Henson, D.; Nyirenda, M.; Desikan, R.; Harrison, J.; Lewis, M.; Hughes, J.; Neill, S.J. Proteomic identification of glyceraldehyde 3-phosphate dehydrogenase as an inhibitory target of hydrogen peroxide in Arabidopsis. *Plant Physiol. Biochem.* **2005**, *43*, 828–835. [CrossRef] [PubMed]
56. Aroca, A.; Schneider, M.; Scheibe, R.; Gotor, C.; Romero, L.C. Hydrogen Sulfide Regulates the cytosolic/nuclear partitioning of glyceraldehyde-3-phosphate dehydrogenase by enhancing its nuclear localization. *Plant Cell Physiol.* **2017**, *58*, 983–992. [CrossRef] [PubMed]

57. Williams, E.; Pead, S.; Whiteman, M.; Wood, M.E.; Wilson, I.D.; Ladomery, M.R.; Teklic, T.; Lisjak, M.; Hancock, J.T. Detection of thiol modifications by hydrogen sulfide. *Methods Enzymol.* **2015**, *555*, 233–251. [CrossRef]
58. Williams, E.; Whiteman, M.; Wood, M.E.; Wilson, I.D.; Ladomery, M.R.; Allainguillaume, J.; Teklic, T.; Lisjak, M.; Hancock, J.T. Investigating ROS, RNS and H$_2$S sensitive signalling proteins. In *Redox Signal Transduction: Methods and Protocols*; Hancock, J.T., Conway, M., Eds.; Springer: Berlin, Germany, 2019; in press.
59. Yu, M.; Yun, B.W.; Spoel, S.H.; Loake, G.J. A sleigh ride through the SNO: Regulation of plant immune function by protein S-nitrosylation. *Curr. Opin. Plant Biol.* **2012**, *15*, 424–430. [CrossRef]
60. Kolbert, Z.; Feigl, G.; Bordé, Á.; Molnár, Á.; Erdei, L. Protein tyrosine nitration in plants: Present knowledge, computational prediction and future perspectives. *Plant Physiol. Bioch.* **2017**, *113*, 56–63. [CrossRef]
61. Holzmeister, C.; Gaupels, F.; Geerlof, A.; Sarioglu, H.; Sattler, M.; Durner, J.; Lindermayr, C. Differential inhibition of Arabidopsis superoxide dismutases by peroxynitrite-mediated tyrosine nitration. *J. Exp. Bot.* **2015**, *66*, 989–999. [CrossRef]
62. Speckmann, B.; Steinbrenner, H.; Grune, T.; Klotz, L.O. Peroxynitrite: From interception to signaling. *Arch Biochem. Biophys.* **2016**, *595*, 153–160. [CrossRef] [PubMed]
63. Bauer, G. Increasing the endogenous NO level causes catalase inactivation and reactivation of intercellular apoptosis signaling specifically in tumor cells. *Redox Biol.* **2015**, *6*, 353–371. [CrossRef] [PubMed]
64. Hancock, J.T.; Whiteman, M. Cellular redox environment and its influence on redox signalling molecules. *React. Oxyg. Species* **2018**, *5*, 78–85.
65. Schafer, F.Q.; Buettner, G.R. Redox environment of the cell as viewed through the redox state of the glutathione disulfide/glutathione couple. *Free Radic. Biol. Med.* **2001**, *30*, 1191–1212. [CrossRef]
66. Hancock, J.T. Considerations of the importance of redox state on reactive nitrogen species action. *J. Exp. Bot.* **2019**, in press.
67. Yun, B.W.; Skelly, M.J.; Yin, M.; Yu, M.; Mun, B.G.; Lee, S.U.; Hussain, A.; Spoel, S.H.; Loake, G.J. Nitric oxide and S-nitrosoglutathione function additively during plant immunity. *New Phytol.* **2016**, *211*, 516–526. [CrossRef]
68. Hogg, N.; Singh, R.J.; Kalyanaraman, B. The role of glutathione in the transport and catabolism of nitric oxide. *FEBS Lett.* **1996**, *382*, 223–228. [CrossRef]
69. Feechan, A.; Kwon, E.; Yun, B.W.; Wang, Y.; Pallas, J.A.; Loake, G.J. A central role for S-nitrosothiols in plant disease resistance. *Proc. Natl. Acad. Sci. USA* **2005**, *102*, 8054–8059. [CrossRef]
70. Lee, U.; Wie, C.; Fernandez, B.O.; Feelisch, M.; Vierling, E. Modulation of nitrosative stress by S-nitrosoglutathione reductase is critical for thermotolerance and plant growth in Arabidopsis. *Plant Cell* **2008**, *20*, 786–802. [CrossRef]
71. Lisjak, M.; Teklic, T.; Wilson, I.D.; Whiteman, M.; Hancock, J.T. Hydrogen sulfide: Environmental factor or signaling molecule? *Plant Cell Environ.* **2013**, *36*, 1607–1616. [CrossRef]
72. Olas, B. Hydrogen sulfide and signaling pathways. *Clin. Chim. Acta* **2015**, *439*, 212–218. [CrossRef]
73. Kimura, H. Hydrogen sulfide and polysulfide signaling. *Antioxid. Redox Signal.* **2017**, *27*, 619–621. [CrossRef] [PubMed]
74. Whiteman, M.; Li, L.; Kostetski, I.; Chu, S.H.; Siau, J.L.; Bhatia, M.; Moore, P.K. Evidence for the formation of a novel nitrosothiol from the gaseous mediators nitric oxide and hydrogen sulphide. *Biochem. Biophys. Res. Commun.* **2006**, *343*, 303–310. [CrossRef] [PubMed]
75. Kimura, Y.; Goto, Y.; Kimura, H. Hydrogen sulfide increases glutathione production and suppresses oxidative stress in mitochondria. *Antioxid. Redox Signal.* **2010**, *12*, 1–13. [CrossRef] [PubMed]
76. Wilson, H.R.; Veal, D.; Whiteman, M.; Hancock, J.T. Hydrogen gas and its role in cell signaling. *CAB Rev.* **2017**, *12*, 1–3. [CrossRef]
77. Ohta, S. Molecular hydrogen as a novel antioxidant: Overview of the advantages of hydrogen for medical applications. *Methods Enzymol.* **2015**, *555*, 289–317. [PubMed]
78. Buntkowsky, G.; Walaszek, B.; Adamczyk, A.; Xu, Y.; Limbach, H.-H.; Chaudret, B. Mechanisms of nuclear spin initiated *para*-H$_2$ to *ortho*-H$_2$ conversion. *Phys. Chem. Chem. Phys.* **2006**, *8*, 1929–1935. [CrossRef]
79. Hancock, J.T.; Hancock, T.H. Hydrogen gas, ROS metabolism and cell signaling: Are hydrogen spin states important? *React. Oxyg. Species* **2018**, *6*, 389–395. [CrossRef]

80. Lin, Y.; Zhang, W.; Qi, F.; Cui, W.; Xie, Y.; Shen, W. Hydrogen-rich water regulates cucumber adventitious root development in a heme oxygenase-1/carbon monoxide-dependent manner. *J. Plant Physiol.* **2014**, *171*, 1–8. [CrossRef]
81. Wu, Q.; Su, N.; Cai, J.; Shen, Z.; Cui, J. Hydrogen-rich water enhances cadmium tolerance in Chinese cabbage by reducing cadmium uptake and increasing antioxidant capacities. *J. Plant Physiol* **2015**, *175*, 174–182. [CrossRef]
82. Zeng, J.; Ye, Z.; Sun, X. Progress in the study of biological effects of hydrogen on higher plants and its promising application in agriculture. *Med. Gas Res.* **2014**, *4*, 1–7. [CrossRef] [PubMed]
83. Zhu, Y.; Liao, W.; Wang, M.; Niu, L.; Xu, Q.; Jin, X. Nitric oxide is required for hydrogen gas-induced adventitious root formation in cucumber. *J. Plant Physiol.* **2016**, *195*, 50–58. [CrossRef] [PubMed]
84. Zhu, Y.; Liao, W.; Niu, L.; Wang, M.; Ma, Z. Nitric oxide is involved in hydrogen gas-induced cell cycle activation during adventitious root formation in cucumber. *BMC Plant Biol.* **2016**, *16*, 146. [CrossRef] [PubMed]
85. Šírová, J.; Sedlářová, M.; Piterková, J.; Luhová, L.; Petřivalský, M. The role of nitric oxide in the germination of plant seeds and pollen. *Plant Sci.* **2011**, *181*, 560–572. [CrossRef] [PubMed]
86. Correa-Aragunde, N.; Graziano, M.; Lamattina, L. Nitric oxide plays a central role in determining lateral root development in tomato. *Planta* **2004**, *218*, 900–905. [CrossRef] [PubMed]
87. Clarke, A.; Desikan, R.; Hurst, R.D.; Hancock, J.T.; Neill, S.J. NO way back: Nitric oxide and programmed cell death in *Arabidopsis thaliana* suspension cultures. *Plant J.* **2000**, *24*, 667–677. [CrossRef] [PubMed]
88. Hancock, J.T.; Whiteman, M. Hydrogen sulfide and cell signaling: Team player or referee? *Plant Physiol. Biochem.* **2014**, *78*, 37–42. [CrossRef] [PubMed]

© 2019 by the authors. Licensee MDPI, Basel, Switzerland. This article is an open access article distributed under the terms and conditions of the Creative Commons Attribution (CC BY) license (http://creativecommons.org/licenses/by/4.0/).

Review

Impact of Nitric Oxide (NO) on the ROS Metabolism of Peroxisomes

Francisco J. Corpas *, Luis A. del Río and José M. Palma

Group of Antioxidants, Free Radicals and Nitric Oxide in Biotechnology, Food and Agriculture, Department of Biochemistry and Cell and Molecular Biology of Plants, Estación Experimental del Zaidín, Consejo Superior de Investigaciones Científicas (CSIC), Profesor Albareda 1, 18008 Granada, Spain; luisalfonso.delrio@eez.csic.es (L.A.d.R.); josemanuel.palma@eez.csic.es (J.M.P.)
* Correspondence: javier.corpas@eez.csic.es; Tel.: +34-958181600

Received: 24 January 2019; Accepted: 7 February 2019; Published: 10 February 2019

Abstract: Nitric oxide (NO) is a gaseous free radical endogenously generated in plant cells. Peroxisomes are cell organelles characterized by an active metabolism of reactive oxygen species (ROS) and are also one of the main cellular sites of NO production in higher plants. In this mini-review, an updated and comprehensive overview is presented of the evidence available demonstrating that plant peroxisomes have the capacity to generate NO, and how this molecule and its derived products, peroxynitrite ($ONOO^-$) and *S*-nitrosoglutathione (GSNO), can modulate the ROS metabolism of peroxisomes, mainly throughout protein posttranslational modifications (PTMs), including *S*-nitrosation and tyrosine nitration. Several peroxisomal antioxidant enzymes, such as catalase (CAT), copper-zinc superoxide dismutase (CuZnSOD), and monodehydroascorbate reductase (MDAR), have been demonstrated to be targets of NO-mediated PTMs. Accordingly, plant peroxisomes can be considered as a good example of the interconnection existing between ROS and reactive nitrogen species (RNS), where NO exerts a regulatory function of ROS metabolism acting upstream of H_2O_2.

Keywords: catalase; monodehydroascorbate reductase; tyrosine nitration; nitric oxide; peroxisome; reactive oxygen species; *S*-nitrosation; superoxide dismutase

1. Introduction

Peroxisomes are organelles with an essential oxidative metabolism present in almost all categories of eukaryotic cells. In higher plants, these organelles are recognized to have a versatile metabolism because their enzymatic composition can adapt to different cell and organ types, stages of development, and environmental conditions [1–6]. However, there is a common battery of enzymes that are present in all types of plant peroxisomes. This includes a set of antioxidant systems whose functions are to keep under control the internal active metabolism of reactive oxygen species (ROS), mainly superoxide radicals ($O_2^{·-}$) and hydrogen peroxide (H_2O_2). These ROS are generated under physiological conditions by different pathways, such as purine catabolism, fatty acid β-oxidation, and photorespiration [7–10]. These antioxidant systems acquire a special relevance in those situations where the ROS generation is intensified, like under plant stress conditions [11].

In recent years, different experimental data have demonstrated that plant peroxisomes also have the capacity to generate another free radical—nitric oxide (NO)—and a family of derived molecules designated as reactive nitrogen species (RNS), including peroxynitrite ($ONOO^-$) [12] and *S*-nitrosoglutathione (GSNO) [13]. The production of these two families of reactive species—ROS and RNS—raises new questions about their potential functions in peroxisomes, either as simple byproducts of the peroxisomal metabolism or perhaps having a regulatory function in the peroxisome and also outside these organelles, due to the characteristic signaling properties of ROS and RNS.

In this work, the interconnections existing between the metabolism of ROS and RNS in peroxisomes are presented. In this relationship, NO exerts a regulatory function by controlling the activity of some target enzymes through posttranslational modifications (PTMs), mainly S-nitrosation (or S-nitrosylation) and tyrosine nitration. It should be pointed out that the NO-generating capacity of peroxisomes may have significant implications in the cellular metabolism of plants under physiological conditions, including leaf senescence [14], pollen tube growth [15], and auxin-induced root organogenesis [16]. However, peroxisomal NO metabolism is particularly exacerbated under oxidative stress situations induced by abiotic conditions like salinity [17], and the heavy-metals cadmium [12,18], and lead [19].

2. Nitric Oxide Generation in Plant Peroxisomes

In higher plants, NO is a key signaling molecule [20,21] involved in numerous processes, including seed germination [22,23], primary and lateral root growth [24,25], plant development [26,27], stomatal closure [28], flowering [29], reproductive tissues [15,30,31], fruit ripening [32,33], senescence [14,34], abiotic stresses [35–39] and biotic stresses [40]. However, the enzymatic source(s) of NO in plant cells is still a controversial matter subject to intense discussions [41–43]. Different pieces of biochemical evidence have demonstrated the presence of L-arginine-dependent nitric oxide synthase (NOS)-like activity in plant peroxisomes. Data accumulated during the last twenty years indicate that the hypothetical protein responsible for the NO generation in peroxisomes has biochemical requirements similar to that of animal NOS, including substrate, cofactors and sensitivity to inhibitors [14,44], dependence on calcium and calmodulin [45], as well as dependence on the mechanism of the import system to peroxisomes through a peroxisomal targeting signal type 2 (PTS-2) [46]. The known biochemical properties of the protein responsible for NO generation in plant peroxisomes, in comparison with those described for animal NOS, are summarized in Table 1. Additionally, there are experimental data that have corroborated the presence of NO in plant peroxisomes and that were obtained by complementary approaches, including electron paramagnetic resonance (EPR) spectroscopy, ozone chemiluminescence, and NO-specific fluorescence probes [14,19]. It should be mentioned that in other cellular compartments a reductive NO generation involving nitrite/nitrate or nitrate reductase (NR) has been described, as well as a non-enzymatic production of NO at acidic pH in the presence of reductants like ascorbate [43,47]. However, peroxisomes have at oxidative metabolism and, to our knowledge, there is not any experimental evidence of the presence of nitrite/nitrate or NR in these plant organelles. Moreover, it has been reported that peroxisomes have an alkaline pH [48], what suggests that the mentioned non-enzymatic generation of NO in peroxisomes is not likely under normal physiological conditions.

Table 1. Biochemical requirements of the peroxisomal protein responsible for the L-arginine-dependent nitric oxide synthase (NOS)-like activity in higher plants.

Requirements	Peroxisomal NOS-Like Protein
Substrate	L-Arginine
Cofactor requirement	NADPH, Ca^{2+}, FAD, FMN, BH_4
Sensitivity to inhibitor	Aminiguanidine, L-NNA, L-NAME, L-NMMA
Peroxisomal targeting signal (PTS)	Type 2 (PTS2)
Dependence of peroxisomal protein import system	PEX5, PEX7, PEX12, PEX13, Ca^{2+}, CaM
Localization	Matrix

BH_4, tetrabiopterin; PEX, peroxin; L-NNA, L-NG-Nitroarginine; L-NAME, Nω-Nitro-L-arginine methyl ester hydrochloride; L-NMMA, N^G-Monomethyl-L-arginine, monoacetate salt; CaM, calmodulin.

Similarly, in animal peroxisomes, the presence of an inducible NOS isozyme [49,50], which is imported to the peroxisomal matrix using a PTS2 [51], has also been demonstrated. In conclusion, the above data indicated for the protein responsible for NO generation in peroxisomes from plant origin are in good agreement with the data reported for the animal peroxisomal NOS activity.

3. Peroxisomal Proteins: Targets of NO-mediated PTMs

At present, the number of potential targets that undergo NO-mediated PTMs is increasing. This is due to the identifications obtained by specific proteomic methodologies combined with biochemical analyses, such as the biotin switch method and labeling with isotope-coded affinity tags (ICAT). These approaches have also allowed confirming whether a specific protein is S-nitrosated and/or nitrated. In some cases, even the affected amino acid residues of the protein have been identified [52]. Furthermore, the existence of any NO-derived PTM is additional evidence of, at least, the presence of NO and its derived molecules in a specific subcellular compartment [53]. So far, the number of identified plant peroxisomal proteins susceptible to undergo a specific NO-derived PTM has also increased with the development of the mentioned methodologies. The characteristic peroxisomal proteins that have been identified as targets of NO in higher plants are summarized in Table 2. Among the different peroxisomal proteins undergoing NO-derived PTMs, in this article, we have focused on some of the key antioxidant enzymes of peroxisomes, including catalase (CAT), monodehydroascorbate reductase (MDAR), and copper-zinc superoxide dismutase (CuZnSOD).

Table 2. Some proteins from higher plant peroxisomes that undergo nitric oxide (NO)-derived posttranslational modifications (PTMs), either by S-nitrosation or tyrosine nitration.

Peroxisomal Enzyme	NO-Derived PTM	References
3-ketoacyl-CoA thiolase 1	S-nitrosation	[52]
Hydroxypyruvate reductase	S-nitrosation/nitration	[54–56]
Glycolate oxidase	S-nitrosation/nitration	[55,57,58]
Malate dehydrogenase	S-nitrosation/nitration	[55,59]
Catalase	S-nitrosation/nitration	[55,56,60,61]
CuZn superoxide dismutase (CSD3)	Nitration	[62]
Monodehydroascorbate reductase	S-nitrosation/nitration	[63]

3.1. Catalase (CAT, EC 1.11.1.6)

CAT is a heme-containing protein and one of the key H_2O_2-scavenging enzymes present in prokaryotic and eukaryotic cells [64–67]. Additionally, CAT is recognized as a constitutive enzyme of all kinds of peroxisomes from eukaryotic cells, being used as a biochemical marker of these organelles. The information available, at present, indicates that this enzyme is the main target of NO in animals and plants. In fact, initial in vitro assays showed that the bovine liver CAT was rapidly and reversibly inhibited by NO [68,69]. In plants, using purified tobacco CAT, similar studies demonstrated that both NO donors and $ONOO^-$ (a nitrating molecule) had the capacity to inhibit the enzyme activity [70]. More recently, studies carried out in different plant species have shown that CAT is a target of S-nitrosation in sunflower hypocotyls [60], pea leaves [55], and Arabidopsis [56], and of tyrosine nitration in pepper fruits [61]. Moreover, it was demonstrated that both S-nitrosation and tyrosine nitration inhibited CAT activity in pea leaves and pepper fruits [55,61]. It has been proposed that the potential target of S-nitrosation in Arabidopsis CAT is Cys86 [56], although this should be corroborated by specific mass spectrometry analyses. However, it must be taken into account that NO could also interact with the Fe atoms present in the heme groups of CAT, forming a metal nitrosyl complex, that perhaps could affect its activity, although, to our knowledge, there is no information on this mechanism in plant CAT. In any case, all the data available suggest that NO acts upstream of H_2O_2, thereby regulating CAT activity. This inhibition of CAT by NO could imply a lower capacity to remove

H_2O_2, and consequently it could be well correlated with those physiological or adverse processes that have associated an increase of their oxidative metabolism [18,61].

3.2. Monodehydroascorbate Reductase (MDAR, EC 1.6.5.4)

This enzyme is part of the ascorbate-glutathione (ASC-GSH) cycle, whose function is also to control the cellular content of H_2O_2 [71]. The ASC-GSH cycle is present in different subcellular compartments, including peroxisomes [72–74]. However, very little information is available on how RNS can regulate the specific isozymes of this cycle present in peroxisomes. MDAR catalyzes the NADH-dependent conversion of monodehydroascorbate to ascorbate, and peroxisomal MDAR has been characterized in pea leaves [75] and Arabidopsis [76]. Further in vitro analysis of recombinant MDAR from pea leaf peroxisomes in the presence of nitrating or S-nitrosylating agents ($ONOO^-$ or GSNO, respectively) demonstrated that both processes caused inhibition of the MDAR activity [63]. Mass spectrometric analysis and site-directed mutagenesis confirmed that Tyr345 was the primary site of nitration by $ONOO^-$ responsible for the inhibition of MDAR activity. On the other hand, in silico analysis of the MDAR indicated that Cys68 was the best candidate for S-nitrosylation [63]. This implies a possible modulation in peroxisomes of the ascorbate regeneration and the H_2O_2 scavenging by RNS.

3.3. Superoxide Dismutase (SOD; EC 1.15.1.1)

Superoxide dismutases (SODs) are a family of metalloenzymes that catalyze the disproportionation of $O_2^{·-}$ radicals into H_2O_2 and O_2. In higher plants, there are three main types of SODs, containing prosthetic metals Mn (Mn-SODs), Fe (Fe-SODs), or Cu plus Zn (Cu,Zn-SODs) [77,78]. The presence of SOD activity in peroxisomes was reported for the first time in plant tissues—in pea (*Pisum sativum* L.) leaves—in the early 1980s [79]. However, this report, in general, passed unnoticed and was even questioned until it was described in human cells years later [80]. Since then, the occurrence of different types of SODs in plant peroxisomes has been described in at least ten distinct plant species [11,78]. At present, SOD is considered a constitutive enzyme in all types of peroxisomes, although the family of isozyme present depends on the organ and plant species.

In relation to the susceptibility of SOD to different RNS-induced modifications, previous reports indicated that the recombinant human Mn-SOD and Cu,Zn-SOD were prone to be inactivated by $ONOO^-$ [81,82]. In the case of plant peroxisomes, recently the recombinant peroxisomal Cu,Zn-SOD (designated as CSD3) was obtained in Arabidopsis, and in vitro assays in the presence of nitrating or S-nitrosylating agents showed that 500 µM $ONOO^-$ provoked a 65% inhibition of the Cu,Zn-SOD activity, whereas GSNO did not cause any effect [62]. Regarding mass spectrometric analyses, Tyr115 was identified as the potential target of nitration [62]. Accordingly, SOD seems to be a relevant protein to be further investigated as a target of NO-mediated PTMs, since it appears to be sensitive to exert some discrimination between nitration and nitrosation processes.

4. Conclusions and Future Perspectives

Plant peroxisomes have relevant antioxidant systems comprised mainly of CAT, SOD, and the ASC-GSH cycle, which are present in all types of plant peroxisomes [7]. Likewise, results obtained in previous research works have demonstrated that besides an active ROS metabolism in peroxisomes, these organelles also have an active RNS metabolism. Although there are few specific studies on how distinct RNS can regulate the different peroxisomal antioxidant systems, the data available suggest the NO may act upstream of the H_2O_2 metabolism. A scheme based on previous reports [7,11,83,84], showing how NO can modulate the activity of peroxisomal antioxidant enzymes throughout either nitration or S-nitrosation, is presented in Figure 1. The peroxisomal xanthine oxidoreductase (XOR) activity catalyzes the oxidation of xanthine with the production of uric acid and $O_2^{·-}$ [85]. On the other hand, L-arginine-dependent NOS-like activity generates NO, which can react with $O_2^{·-}$ to produce $ONOO^-$, a powerful oxidant and strong nitrating molecule that can mediate PTMs through tyrosine nitration [86]. NO can also interact with reduced glutathione (GSH) to form GSNO, a NO donor that can mediate S-nitrosation of proteins [87]. Uric acid is a recognized inhibitor of ONOO⁻-mediated toxicity [88,89], and this brings out a new potential mechanism of peroxisomal auto-regulation through this powerful nitrating molecule. In this scenario, the identified targets of NO-derived PTMs in peroxisomes, CAT, CuZnSOD, and MDAR, which are either directly or indirectly linked to the H_2O_2 pool, are key points to be modulated by nitration or S-nitrosation.

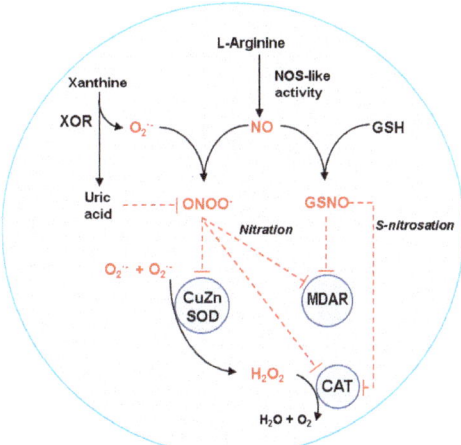

Figure 1. The interrelationship between nitric oxide (NO) metabolism and antioxidant enzymes in plant peroxisomes. Peroxisomal xanthine oxidoreductase (XOR) activity produces uric acid and superoxide radicals ($O_2^{·-}$). On the other hand, an L-arginine-dependent nitric oxide synthase (NOS)-like activity generates NO, which can react with $O_2^{·-}$ to give rise to peroxynitrite (ONOO⁻), which is a powerful oxidant and strong nitrating molecule that can mediate posttranslational modifications (PTMs), such as tyrosine nitration. NO can also interact with reduced glutathione (GSH) to form S-nitrosoglutathione (GSNO), a NO donor that can mediate S-nitrosation reactions. Uric acid is a recognized ONOO⁻ scavenger that could be part of a mechanism of peroxisomal auto-regulation. With all these components, the identified targets of NO-derived PTMs in peroxisomes, catalase (CAT), copper, zinc superoxide dismutase (CuZnSOD), and monodehydroascorbate reductase (MDAR) can undergo inhibition of their activity either by nitration or S-nitrosation.

In summary, the data presently available indicate that plant peroxisomes contain multiple elements of ROS and RNS metabolism, where NO seems to act upstream of H_2O_2 routes throughout the regulation of the peroxisomal antioxidant enzymes. Nevertheless, it should be taken into account that both NO and H_2O_2 could be released to the cytosol, acting as signal molecules among the different subcellular compartments. However, in plants under certain abiotic stress conditions an overproduction of H_2O_2 and NO could take place in peroxisomes, and a high accumulation of these signal molecules can mediate a nitro-oxidative stress in plant cells [11,90].

Author Contributions: F.J.C. conceived and wrote the manuscript. L.A.d.R. and J.M.P. critically revised the manuscript. All authors approved the final submitted version of the manuscript.

Funding: Research in our laboratory is supported by an ERDF-co-financed grant from the Ministry of Economy and Competitiveness (AGL2015-65104-P) and *Junta de Andalucía* (group BIO-192), Spain.

Conflicts of Interest: The authors declare no conflict of interest.

References

1. Nishimura, M.; Hayashi, M.; Kato, A.; Yamaguchi, K.; Mano, S. Functional transformation of microbodies in higher plant cells. *Cell Struct. Funct.* **1996**, *21*, 387–393. [CrossRef] [PubMed]
2. Hayashi, M.; Nishimura, M. Entering a new era of research on plant peroxisomes. *Curr. Opin. Plant Biol.* **2003**, *6*, 577–582. [CrossRef] [PubMed]
3. Hu, J.; Baker, A.; Bartel, B.; Linka, N.; Mullen, R.T.; Reumann, S.; Zolman, B.K. Plant peroxisomes: Biogenesis and function. *Plant Cell.* **2012**, *24*, 2279–2303. [CrossRef] [PubMed]
4. Sørhagen, K.; Laxa, M.; Peterhänsel, C.; Reumann, S. The emerging role of photorespiration and non-photorespiratory peroxisomal metabolism in pathogen defence. *Plant Biol. (Stuttg)* **2013**, *15*, 723–736. [CrossRef] [PubMed]
5. Goto-Yamada, S.; Mano, S.; Yamada, K.; Oikawa, K.; Hosokawa, Y.; Hara-Nishimura, I.; Nishimura, M. Dynamics of the light-dependent transition of plant peroxisomes. *Plant Cell Physiol.* **2015**, *56*, 1264–1271. [CrossRef] [PubMed]
6. Kao, Y.T.; González, K.L.; Bartel, B. Peroxisome function, biogenesis, and dynamics in plants. *Plant Physiol.* **2018**, *176*, 162–177. [CrossRef] [PubMed]
7. del Río, L.A.; Corpas, F.J.; Sandalio, L.M.; Palma, J.M.; Gómez, M.; Barroso, J.B. Reactive oxygen species, antioxidant systems and nitric oxide in peroxisomes. *J. Exp. Bot.* **2002**, *53*, 1255–1272. [CrossRef] [PubMed]
8. Weber, H. Fatty acid-derived signals in plants. *Trends Plant Sci.* **2002**, *7*, 217–224. [CrossRef]
9. del Río, L.A.; López-Huertas, E. ROS generation in peroxisomes and its role in cell signaling. *Plant Cell Physiol.* **2016**, *57*, 1364–1376. [CrossRef] [PubMed]
10. Corpas, F.J.; del Río, L.A.; Palma, J.M. Plant peroxisomes at the crossroad of NO and H_2O_2 metabolism. *J. Integr. Plant Biol.* **2019**. [CrossRef] [PubMed]
11. Corpas, F.J.; Barroso, J.B.; Palma, J.M.; Rodríguez-Ruiz, M. Plant peroxisomes: A nitro-oxidative cocktail. *Redox Biol.* **2017**, *11*, 535–542. [CrossRef] [PubMed]
12. Corpas, F.J.; Barroso, J.B. Peroxynitrite ($ONOO^-$) is endogenously produced in Arabidopsis peroxisomes and is overproduced under cadmium stress. *Ann. Bot.* **2014**, *113*, 87–96. [CrossRef] [PubMed]
13. Barroso, J.B.; Valderrama, R.; Corpas, F.J. Immunolocalization of S-nitrosoglutathione, S-nitrosoglutathione reductase and tyrosine nitration in pea leaf organelles. *Acta Physiol Plant* **2013**, *35*, 2635–2640. [CrossRef]
14. Corpas, F.J.; Barroso, J.B.; Carreras, A.; Quirós, M.; León, A.M.; Romero-Puertas, M.C.; Esteban, F.J.; Valderrama, R.; Palma, J.M.; Sandalio, L.M.; et al. Cellular and subcellular localization of endogenous nitric oxide in young and senescent pea plants. *Plant Physiol.* **2004**, *136*, 2722–2733. [CrossRef] [PubMed]
15. Prado, A.M.; Porterfield, D.M.; Feijó, J.A. Nitric oxide is involved in growth regulation and reorientation of pollen tubes. *Development* **2004**, *131*, 2707–2714. [CrossRef] [PubMed]
16. Schlicht, M.; Ludwig-Müller, J.; Burbach, C.; Volkmann, D.; Baluska, F. Indole-3-butyric acid induces lateral root formation via peroxisome-derived indole-3-acetic acid and nitric oxide. *New Phytol.* **2013**, *200*, 473–482. [CrossRef]
17. Corpas, F.J.; Hayashi, M.; Mano, S.; Nishimura, M.; Barroso, J.B. Peroxisomes are required for in vivo nitric oxide accumulation in the cytosol following salinity stress of *Arabidopsis* plants. *Plant Physiol.* **2009**, *151*, 2083–2094. [CrossRef]
18. Smiri, M.; Chaoui, A.; Rouhier, N.; Gelhaye, E.; Jacquot, J.P.; El Ferjani, E. Oxidative damage and redox change in pea seeds treated with cadmium. *C R Biol.* **2010**, *333*, 801–807. [CrossRef]
19. Corpas, F.J.; Barroso, J.B. Lead-induced stress, which triggers the production of nitric oxide (NO) and superoxide anion ($O_2^{·-}$) in Arabidopsis peroxisomes, affects catalase activity. *Nitric Oxide.* **2017**, *68*, 103–110. [CrossRef]
20. Wilson, I.D.; Neill, S.J.; Hancock, J.T. Nitric oxide synthesis and signalling in plants. *Plant Cell Environ.* **2008**, *31*, 622–631. [CrossRef]
21. Neill, S.; Bright, J.; Desikan, R.; Hancock, J.; Harrison, J.; Wilson, I. Nitric oxide evolution and perception. *J. Exp Bot.* **2008**, *59*, 25–35. [CrossRef] [PubMed]
22. Beligni, M.V.; Lamattina, L. Nitric oxide stimulates seed germination and de-etiolation, and inhibits hypocotyl elongation, three light-inducible responses in plants. *Planta* **2000**, *210*, 215–221. [CrossRef] [PubMed]
23. Bethke, P.C.; Gubler, F.; Jacobsen, J.V.; Jones, R.L. Dormancy of Arabidopsis seeds and barley grains can be broken by nitric oxide. *Planta* **2004**, *219*, 847–855. [CrossRef] [PubMed]

24. Pagnussat, G.C.; Lanteri, M.L.; Lombardo, M.C.; Lamattina, L. Nitric oxide mediates the indole acetic acid induction activation of a mitogen-activated protein kinase cascade involved in adventitious root development. *Plant Physiol.* **2004**, *135*, 279–286. [CrossRef] [PubMed]
25. Corpas, F.J.; Barroso, J.B. Functions of Nitric Oxide (NO) in roots during development and under adverse stress conditions. *Plants (Basel)* **2015**, *4*, 240–252. [CrossRef] [PubMed]
26. Sanz, L.; Albertos, P.; Mateos, I.; Sánchez-Vicente, I.; Lechón, T.; Fernández-Marcos, M.; Lorenzo, O. Nitric oxide (NO) and phytohormones crosstalk during early plant development. *J. Exp Bot.* **2015**, *66*, 2857–2868. [CrossRef] [PubMed]
27. Corpas, F.J.; Barroso, J.B.; Carreras, A.; Valderrama, R.; Palma, J.M.; León, A.M.; Sandalio, L.M.; del Río, L.A. Constitutive arginine-dependent nitric oxide synthase activity in different organs of pea seedlings during plant development. *Planta.* **2006**, *224*, 246–254. [CrossRef] [PubMed]
28. Bright, J.; Desikan, R.; Hancock, J.T.; Weir, I.S.; Neill, S.J. ABA-induced NO generation and stomatal closure in Arabidopsis are dependent on H_2O_2 synthesis. *Plant J.* **2006**, *45*, 113–122. [CrossRef]
29. Senthil Kumar, R.; Shen, C.H.; Wu, P.Y.; Suresh Kumar, S.; Hua, M.S.; Yeh, K.W. Nitric oxide participates in plant flowering repression by ascorbate. *Sci Rep.* **2016**, *6*, 35246. [CrossRef]
30. Prado, A.M.; Colaço, R.; Moreno, N.; Silva, A.C.; Feijó, J.A. Targeting of pollen tubes to ovules is dependent on nitric oxide (NO) signaling. *Mol Plant.* **2008**, *1*, 703–714. [CrossRef]
31. Zafra, A.; Rodríguez-García, M.I.; Alché Jde, D. Cellular localization of ROS and NO in olive reproductive tissues during flower development. *BMC Plant Biol.* **2010**, *10*, 36. [CrossRef] [PubMed]
32. Corpas, F.J.; Freschi, L.; Rodríguez-Ruiz, M.; Mioto, P.T.; González-Gordo, S.; Palma, J.M. Nitro-oxidative metabolism during fruit ripening. *J. Exp Bot.* **2018**, *69*, 3449–3463. [CrossRef] [PubMed]
33. Corpas, F.J.; Palma, J.M. Nitric oxide on/off in fruit ripening. *Plant Biol (Stuttg).* **2018**, *20*, 805–807. [CrossRef] [PubMed]
34. Du, J.; Li, M.; Kong, D.; Wang, L.; Lv, Q.; Wang, J.; Bao, F.; Gong, Q.; Xia, J.; He, Y. Nitric oxide induces cotyledon senescence involving co-operation of the NES1/MAD1 and EIN2-associated ORE1 signalling pathways in Arabidopsis. *J. Exp. Bot.* **2014**, *65*, 4051–4063. [CrossRef] [PubMed]
35. Signorelli, S.; Corpas, F.J.; Borsani, O.; Barroso, J.B.; Monza, J. Water stress induces a differential and spatially distributed nitro-oxidative stress response in roots and leaves of *Lotus japonicus*. *Plant Sci.* **2013**, *201–202*, 137–146. [CrossRef]
36. Manai, J.; Gouia, H.; Corpas, F.J. Redox and nitric oxide homeostasis are affected in tomato (Solanum lycopersicum) roots under salinity-induced oxidative stress. *J. Plant Physiol.* **2014**, *171*, 1028–1035. [CrossRef]
37. Feigl, G.; Lehotai, N.; Molnár, Á.; Ördög, A.; Rodríguez-Ruiz, M.; Palma, J.M.; Corpas, F.J.; Erdei, L.; Kolbert, Z. Zinc induces distinct changes in the metabolism of reactive oxygen and nitrogen species (ROS and RNS) in the roots of two Brassica species with different sensitivity to zinc stress. *Ann Bot.* **2015**, *116*, 613–625. [CrossRef]
38. Houmani, H.; Rodríguez-Ruiz, M.; Palma, J.M.; Corpas, F.J. Mechanical wounding promotes local and long distance response in the halophyte *Cakile maritima* through the involvement of the ROS and RNS metabolism. *Nitric Oxide.* **2018**, *74*, 93–101. [CrossRef]
39. Kharbech, O.; Houmani, H.; Chaoui, A.; Corpas, F.J. Alleviation of Cr(VI)-induced oxidative stress in maize (*Zea mays* L.) seedlings by NO and H_2S donors through differential organ-dependent regulation of ROS and NADPH-recycling metabolisms. *J. Plant Physiol.* **2017**, *219*, 71–80. [CrossRef]
40. Trapet, P.; Kulik, A.; Lamotte, O.; Jeandroz, S.; Bourque, S.; Nicolas-Francès, V.; Rosnoblet, C.; Besson-Bard, A.; Wendehenne, D. NO signaling in plant immunity: A tale of messengers. *Phytochemistry* **2015**, *112*, 72–79. [CrossRef]
41. Santolini, J.; André, F.; Jeandroz, S.; Wendehenne, D. Nitric oxide synthase in plants: where do we stand? *Nitric Oxide* **2017**, *63*, 30–38. [CrossRef] [PubMed]
42. Corpas, F.J.; Barroso, J.B. Nitric oxide synthase-like activity in higher plants. *Nitric Oxide* **2017**, *68*, 5–6. [CrossRef] [PubMed]
43. Astier, J.; Gross, I.; Durner, J. Nitric oxide production in plants: An update. *J. Exp. Bot.* **2018**, *69*, 3401–3411. [CrossRef] [PubMed]
44. Barroso, J.B.; Corpas, F.J.; Carreras, A.; Sandalio, L.M.; Valderrama, R.; Palma, J.M.; Lupiáñez, J.A.; del Río, L.A. Localization of nitric-oxide synthase in plant peroxisomes. *J. Biol. Chem.* **1999**, *274*, 36729–36733. [CrossRef]

45. Corpas, F.J.; Barroso, J.B. Calmodulin antagonist affects peroxisomal functionality by disrupting both peroxisomal Ca^{2+} and protein import. *J. Cell Sci.* **2018**, *131*. [CrossRef] [PubMed]
46. Corpas, F.J.; Barroso, J.B. Peroxisomal plant nitric oxide synthase (NOS) protein is imported by peroxisomal targeting signal type 2 (PTS2) in a process that depends on the cytosolic receptor PEX7 and calmodulin. *FEBS Lett.* **2014**, *588*, 2049–2054. [CrossRef]
47. Gupta, K.J.; Igamberdiev, A.U. The anoxic plant mitochondrion as a nitrite: NO reductase. *Mitochondrion* **2011**, *11*, 537–543. [CrossRef]
48. Shen, J.; Zeng, Y.; Zhuang, X.; Sun, L.; Yao, X.; Pimpl, P.; Jiang, L. Organelle pH in the Arabidopsis endomembrane system. *Mol. Plant.* **2013**, *6*, 1419–1437. [CrossRef]
49. Stolz, D.B.; Zamora, R.; Vodovotz, Y.; Loughran, P.A.; Billiar, T.R.; Kim, Y.M.; Simmons, R.L.; Watkins, S.C. Peroxisomal localization of inducible nitric oxide synthase in hepatocytes. *Hepatology* **2002**, *36*, 81–93. [CrossRef]
50. Loughran, P.A.; Stolz, D.B.; Vodovotz, Y.; Watkins, S.C.; Simmons, R.L.; Billiar, T.R. Monomeric inducible nitric oxide synthase localizes to peroxisomes in hepatocytes. *Proc. Natl. Acad. Sci. USA* **2005**, *102*, 13837–13842. [CrossRef]
51. Loughran, P.A.; Stolz, D.B.; Barrick, S.R.; Wheeler, D.S.; Friedman, P.A.; Rachubinski, R.A.; Watkins, S.C.; Billiar, T.R. PEX7 and EBP50 target iNOS to the peroxisome in hepatocytes. *Nitric Oxide* **2013**, *31*, 9–19. [CrossRef] [PubMed]
52. Fares, A.; Rossignol, M.; Peltier, J.B. Proteomics investigation of endogenous S-nitrosylation in Arabidopsis. *Biochem. Biophys. Res. Commun.* **2011**, *416*, 331–336. [CrossRef] [PubMed]
53. Heijnen, H.F.; van Donselaar, E.; Slot, J.W.; Fries, D.M.; Blachard-Fillion, B.; Hodara, R.; Lightfoot, R.; Polydoro, M.; Spielberg, D.; Thomson, L.; et al. Subcellular localization of tyrosine-nitrated proteins is dictated by reactive oxygen species generating enzymes and by proximity to nitric oxide synthase. *Free Radic. Biol. Med.* **2006**, *40*, 1903–1913. [CrossRef] [PubMed]
54. Corpas, F.J.; Leterrier, M.; Begara-Morales, J.C.; Valderrama, R.; Chaki, M.; López-Jaramillo, J.; Luque, F.; Palma, J.M.; Padilla, M.N.; Sánchez-Calvo, B.; et al. Inhibition of peroxisomal hydroxypyruvate reductase (HPR1) by tyrosine nitration. *Biochim. Biophys. Acta* **2013**, *1830*, 4981–4989. [CrossRef] [PubMed]
55. Ortega-Galisteo, A.P.; Rodríguez-Serrano, M.; Pazmiño, D.M.; Gupta, D.K.; Sandalio, L.M.; Romero-Puertas, M.C. S-Nitrosylated proteins in pea (*Pisum sativum* L.) leaf peroxisomes: Changes under abiotic stress. *J. Exp. Bot.* **2012**, *63*, 2089–2103. [CrossRef]
56. Puyaubert, J.; Fares, A.; Rézé, N.; Peltier, J.B.; Baudouin, E. Identification of endogenously S-nitrosylated proteins in Arabidopsis plantlets: effect of cold stress on cysteine nitrosylation level. *Plant Sci.* **2014**, *215–216*, 150–156. [CrossRef]
57. Abat, J.K.; Mattoo, A.K.; Deswal, R. S-nitrosylated proteins of a medicinal CAM plant *Kalanchoe pinnata*-ribulose-1,5-bisphosphate carboxylase/oxygenase activity targeted for inhibition. *FEBS J.* **2008**, *275*, 2862–2872. [CrossRef]
58. Tanou, G.; Job, C.; Rajjou, L.; Arc, E.; Belghazi, M.; Diamantidis, G.; Molassiotis, A.; Job, D. Proteomics reveals the overlapping roles of hydrogen peroxide and nitric oxide in the acclimation of citrus plants to salinity. *Plant J.* **2009**, *60*, 795–804. [CrossRef]
59. Lozano-Juste, J.; Colom-Moreno, R.; León, J. In vivo protein tyrosine nitration in *Arabidopsis thaliana*. *J. Exp. Bot.* **2011**, *62*, 3501–3517. [CrossRef]
60. Begara-Morales, J.C.; López-Jaramillo, F.J.; Sánchez-Calvo, B.; Carreras, A.; Ortega-Muñoz, M.; Santoyo-González, F.; Corpas, F.J.; Barroso, J.B. Vinyl sulfone silica: Application of an open preactivated support to the study of transnitrosylation of plant proteins by S-nitrosoglutathione. *BMC Plant Biol.* **2013**, *13*, 61. [CrossRef]
61. Chaki, M.; Álvarez de Morales, P.; Ruiz, C.; Begara-Morales, J.C.; Barroso, J.B.; Corpas, F.J.; Palma, J.M. Ripening of pepper (*Capsicum annuum*) fruit is characterized by an enhancement of protein tyrosine nitration. *Ann Bot.* **2015**, *116*, 637–647. [CrossRef] [PubMed]
62. Holzmeister, C.; Gaupels, F.; Geerlof, A.; Sarioglu, H.; Sattler, M.; Durner, J.; Lindermayr, C. Differential inhibition of Arabidopsis superoxide dismutases by peroxynitrite-mediated tyrosine nitration. *J. Exp. Bot.* **2015**, *66*, 989–999. [CrossRef] [PubMed]

63. Begara-Morales, J.C.; Sánchez-Calvo, B.; Chaki, M.; Mata-Pérez, C.; Valderrama, R.; Padilla, M.N.; López-Jaramillo, J.; Luque, F.; Corpas, F.J.; Barroso, J.B. Differential molecular response of monodehydroascorbate reductase and glutathione reductase by nitration and S-nitrosylation. *J. Exp. Bot.* **2015**, *66*, 5983–5996. [CrossRef] [PubMed]
64. Mhamdi, A.; Queval, G.; Chaouch, S.; Vanderauwera, S.; Van Breusegem, F.; Noctor, G. Catalase function in plants: A focus on Arabidopsis mutants as stress-mimic models. *J. Exp. Bot.* **2010**, *61*, 4197–4220. [CrossRef] [PubMed]
65. Mhamdi, A.; Noctor, G.; Baker, A. Plant catalases: Peroxisomal redox guardians. *Arch. Biochem. Biophys.* **2012**, *525*, 181–194. [CrossRef] [PubMed]
66. Nicholls, P. Classical catalase: ancient and modern. *Arch. Biochem. Biophys.* **2012**, *525*, 95–101. [CrossRef] [PubMed]
67. Glorieux, C.; Calderon, P.B. Catalase, a remarkable enzyme: Targeting the oldest antioxidant enzyme to find a new cancer treatment approach. *Biol. Chem.* **2017**, *398*, 1095–1108. [CrossRef] [PubMed]
68. Brown, G.C. Reversible binding and inhibition of catalase by nitric oxide. *Eur. J. Biochem.* **1995**, *232*, 188–191. [CrossRef] [PubMed]
69. Purwar, N.; McGarry, J.M.; Kostera, J.; Pacheco, A.A.; Schmidt, M. Interaction of nitric oxide with catalase: Structural and kinetic analysis. *Biochemistry* **2011**, *50*, 4491–4503. [CrossRef] [PubMed]
70. Clark, D.; Durner, J.; Navarre, D.A.; Klessig, D.F. Nitric oxide inhibition of tobacco catalase and ascorbate peroxidase. *Mol. Plant Microbe Interact.* **2000**, *13*, 1380–1384. [CrossRef] [PubMed]
71. Noctor, G.; Reichheld, J.P.; Foyer, C.H. ROS-related redox regulation and signaling in plants. *Semin. Cell Dev. Biol.* **2018**, *80*, 3–12. [CrossRef] [PubMed]
72. Jiménez, A.; Hernández, J.A.; del Río, L.A.; Sevilla, F. Evidence for the presence of the ascorbate-glutathione cycle in mitochondria and peroxisomes of pea leaves. *Plant Physiol.* **1997**, *114*, 275–284. [CrossRef]
73. Kuźniak, E.; Skłodowska, M. Compartment-specific role of the ascorbate-glutathione cycle in the response of tomato leaf cells to *Botrytis cinerea* infection. *J. Exp. Bot.* **2005**, *56*, 921–933. [CrossRef] [PubMed]
74. Palma, J.M.; Jiménez, A.; Sandalio, L.M.; Corpas, F.J.; Lundqvist, M.; Gómez, M.; Sevilla, F.; del Río, L.A. Antioxidative enzymes from chloroplasts, mitochondria, and peroxisomes during leaf senescence of nodulated pea plants. *J. Exp. Bot.* **2006**, *57*, 1747–1758. [CrossRef] [PubMed]
75. Leterrier, M.; Corpas, F.J.; Barroso, J.B.; Sandalio, L.M.; del Río, L.A. Peroxisomal monodehydroascorbate reductase. Genomic clone characterization and functional analysis under environmental stress conditions. *Plant Physiol.* **2005**, *138*, 2111–2123. [CrossRef] [PubMed]
76. Lisenbee, C.S.; Lingard, M.J.; Trelease, R.N. Arabidopsis peroxisomes possess functionally redundant membrane and matrix isoforms of monodehydroascorbate reductase. *Plant J.* **2005**, *43*, 900–914. [CrossRef] [PubMed]
77. Fridovich, I. Superoxide radical and superoxide dismutases. *Annu Rev Biochem.* **1995**, *64*, 97–112. [CrossRef] [PubMed]
78. del Río, L.A.; Corpas, F.J.; López-Huertas, E.; Palma, J.M. Plant superoxide dismutases: Function under abiotic stress conditions. In *Antioxidants and Antioxidant Enzymes in Higher Plants*; Gupta, D., Palma, J., Corpas, F., Eds.; Springer: Cham, Switzerland, 2018; pp. 1–26.
79. del Río, L.A.; Lyon, D.S.; Olah, I.; Glick, B.; Salin, M.L. Immunocytochemical evidence for a peroxisomal localization of manganese superoxide dismutase in leaf protoplasts from a higher plant. *Planta* **1983**, *158*, 216–224. [CrossRef]
80. Keller, G.A.; Warner, T.G.; Steimer, K.S.; Hallewell, R.A. Cu,Zn superoxide dismutase is a peroxisomal enzyme in human fibroblasts and hepatoma cells. *Proc. Natl. Acad. Sci. USA* **1991**, *88*, 7381–8735. [CrossRef]
81. Alvarez, B.; Demicheli, V.; Durán, R.; Trujillo, M.; Cerveñansky, C.; Freeman, B.A.; Radi, R. Inactivation of human Cu,Zn superoxide dismutase by peroxynitrite and formation of histidinyl radical. *Free Radic. Biol. Med.* **2004**, *37*, 813–822. [CrossRef]
82. Demicheli, V.; Quijano, C.; Alvarez, B.; Radi, R. Inactivation and nitration of human superoxide dismutase (SOD) by fluxes of nitric oxide and superoxide. *Free Radic. Biol. Med.* **2007**, *42*, 1359–1368. [CrossRef] [PubMed]
83. Corpas, F.J.; Barroso, J.B.; del Río, L.A. Peroxisomes as a source of reactive oxygen species and nitric oxide signal molecules in plant cells. *Trends Plant Sci.* **2001**, *6*, 145–150. [CrossRef]

84. del Río, L.A.; Sandalio, L.M.; Corpas, F.J.; Palma, J.M.; Barroso, J.B. Reactive oxygen species and reactive nitrogen species in peroxisomes. Production, scanvenging, and role in cell signaling. *Plant Physiol.* **2006**, *141*, 330–335. [CrossRef] [PubMed]
85. Corpas, F.J.; Palma, J.M.; Sandalio, L.M.; Valderrama, R.; Barroso, J.B.; del Río, L.A. Peroxisomal xanthine oxidoreductase: Characterization of the enzyme from pea (*Pisum sativum* L.) leaves. *J. Plant Physiol.* **2008**, *165*, 1319–1330. [CrossRef] [PubMed]
86. Ferrer-Sueta, G.; Campolo, N.; Trujillo, M.; Bartesaghi, S.; Carballal, S.; Romero, N.; Alvarez, B.; Radi, R. Biochemistry of peroxynitrite and protein tyrosine nitration. *Chem. Rev.* **2018**, *118*, 1338–1408. [CrossRef] [PubMed]
87. Broniowska, K.A.; Diers, A.R.; Hogg, N. S-nitrosoglutathione. *Biochim. Biophys. Acta.* **2013**, *1830*, 3173–3181. [CrossRef] [PubMed]
88. Alamillo, J.M.; García-Olmedo, F. Effects of urate, a natural inhibitor of peroxynitrite-mediated toxicity, in the response of *Arabidopsis thaliana* to the bacterial pathogen *Pseudomonas syringae*. *Plant J.* **2001**, *25*, 529–540. [CrossRef]
89. Signorelli, S.; Imparatta, C.; Rodríguez-Ruiz, M.; Borsani, O.; Corpas, F.J.; Monza, J. In vivo and in vitro approaches demonstrate proline is not directly involved in the protection against superoxide, nitric oxide, nitrogen dioxide and peroxynitrite. *Funct. Plant Biol.* **2016**, *43*, 870–879. [CrossRef]
90. del Río, L.A.; Palma, J.M.; Sandalio, L.M.; Corpas, F.J.; Pastori, G.M.; Bueno, P.; López-Huertas, E. Peroxisomes as a source of superoxide and hydrogen peroxide in stressed plants. *Biochem. Soc. Trans.* **1996**, *24*, 434–438. [CrossRef]

© 2019 by the authors. Licensee MDPI, Basel, Switzerland. This article is an open access article distributed under the terms and conditions of the Creative Commons Attribution (CC BY) license (http://creativecommons.org/licenses/by/4.0/).

Article
Isoform-Specific NO Synthesis by *Arabidopsis thaliana* Nitrate Reductase

Marie Agatha Mohn, Besarta Thaqi and Katrin Fischer-Schrader *

Institute of Biochemistry, Department of Chemistry, Zülpicher Str. 47, University of Cologne, 50674 Cologne, Germany; mmohn@smail.uni-koeln.de (M.A.M.); bthaqi@smail.uni-koeln.de (B.T.)
* Correspondence: k.schrader@uni-koeln.de; Tel.: +49-221-470-7474

Received: 6 February 2019; Accepted: 11 March 2019; Published: 16 March 2019

Abstract: Nitrate reductase (NR) is important for higher land plants, as it catalyzes the rate-limiting step in the nitrate assimilation pathway, the two-electron reduction of nitrate to nitrite. Furthermore, it is considered to be a major enzymatic source of the important signaling molecule nitric oxide (NO), that is produced in a one-electron reduction of nitrite. Like many other plants, the model plant *Arabidopsis thaliana* expresses two isoforms of NR (NIA1 and NIA2). Up to now, only NIA2 has been the focus of detailed biochemical studies, while NIA1 awaits biochemical characterization. In this study, we have expressed and purified functional fragments of NIA1 and subjected them to various biochemical assays for comparison with the corresponding NIA2-fragments. We analyzed the kinetic parameters in multiple steady-state assays using nitrate or nitrite as substrate and measured either substrate consumption (nitrate or nitrite) or product formation (NO). Our results show that NIA1 is the more efficient nitrite reductase while NIA2 exhibits higher nitrate reductase activity, which supports the hypothesis that the isoforms have special functions in the plant. Furthermore, we successfully restored the physiological electron transfer pathway of NR using reduced nicotinamide adenine dinucleotide (NADH) and nitrate or nitrite as substrates by mixing the N-and C-terminal fragments of NR, thus, opening up new possibilities to study NR activity, regulation and structure.

Keywords: nitrate reductase; NIA1; NIA2; nitric oxide; nitrite; nitrate; methyl viologen; benzyl viologen; NO analyzer; molybdenum cofactor; *Arabidopsis thaliana*

1. Introduction

In higher land plants, nitrate is the preferred nutrient for the nitrogen (N) assimilation pathway [1,2]. Nitrate reductase (NR, EC 1.7.1.1), which catalyzes the first intracellular and rate-limiting step in nitrate assimilation, is a homodimer of two approximately 100 kDa polypeptide chains, each of which binds three cofactors in individually folded domains. The enzyme functions as an internal electron transport chain [1]. The C-terminal domain carrying a flavine adenine dinucleotide (FAD) cofactor accepts two electrons from NADH or the phosphorylated form NADPH and passes them sequentially to the middle domain containing a b_5-type cytochrome heme. From here, the electrons are shuttled to the molybdenum cofactor (Moco)-containing catalytic site in the N-terminal domain, and it is here that substrate reduction takes place. Three non-conserved flexible regions are found in NR: an N-terminal peptide preceding the Moco-domain, and two linkers connecting the central heme-domain to the Moco-domain (hinge 1) and to the FAD-containing domain (hinge 2). Hinge 1 in NRs of higher plants has been demonstrated to be crucial for reversible inhibition of NR at the protein level [3,4]. A highly-conserved phosphoserine residue in hinge 1, as well as a motif rich in acidic residues in the N-terminal peptide, are the targets for binding of one of several 14-3-3 protein isoforms, which leads to inhibition by steric hindrance of the internal electron transfer between the heme and Moco [5,6].

In addition to its main anabolic function, NR has also been proposed to act as nitrite reductase in plants resulting in the formation of nitric oxide (NO) [7–9]. This universal signaling molecule is involved in various physiological processes in plants, such as development or stress responses (reviewed in [10–12]). Several NO sources are known in plants, among them the non-enzymatic, pH-dependent NO formation in the apoplast [13], an arginine-dependent oxidative reaction mechanism observed in peroxisomes and chloroplasts similar to the nitric oxide synthase activity of mammals [14,15], and NO synthesis based on heme-proteins or the mitochondrial respiratory electron transport chain [16]. All other members of the Moco-enzyme family, to which NR belongs, namely sulfite oxidase (SO), xanthine oxidoreductase/dehydrogenase (XOR), aldehyde oxidase (AO), and the amidoxime-reducing component (ARC) were shown to be capable of NO synthesis besides their respective name-giving functions (reviewed in [17]).

Several studies using plants, plant cells, or nitrate reductase purified from native tissue have in part quantified the NO formation by NR and suggested that it represents the major enzymatic source of NO [18–21]. Recent studies with the eukaryotic algae *Chlamydomonas reinhardtii* demonstrated that NR is also able to transfer electrons from its C-terminal FAD cofactor directly to other proteins, such as truncated hemoglobins (THB) or ARC [22,23]. While *Chlamydomonas* THB1 has an NO dioxygenase activity that consumes NO, ARC can act as an NO synthase. This finding, together with the observation that both NR and ARC are co-regulated on the transcriptional level, and that the NO synthesizing function of ARC is not inhibited by high nitrate concentrations (in contrast to plant NR, for which a $K_i^{nitrate}$ of 50 µM for the nitrite reductase activity was observed [24]), allowed the authors to propose that this physiologically relevant NO synthase in *Chlamydomonas* might be made up of two proteins, NR and ARC, forming a catalytic complex. Consequently, they suggested renaming ARC to NO-forming nitrite reductase (NOFNiR) [22]. Considering that NR is also involved in the removal of NO, these findings underline the complex role of NR in NO homeostasis (reviewed in [25]).

Interestingly, the function of NR in plants becomes even more complicated by the fact that many plants including *Nicotiana tabacum, Hordeum vulgare, Zea mays, Brassica napus, Glycine max, Oryza sativa* or *Arabidopsis thaliana*, possess two or more isoforms of NR, which might take over distinct functions. For some isoforms, it is known that they differ in their preference for the co-substrate NADH or NADPH [26,27], and in some plant species, the existence of both constitutively expressed and inducible NR isoforms was reported [28].

Focusing on the model plant *Arabidopsis thaliana*, the transcription of the two isoforms *NIA1* and *NIA2* is similar following the induction by nitrate, while several other factors including light or the cytokinin benzyladenine produce specific expression patterns for each isoform [29–31].

In plant extracts of *Arabidopsis thaliana*, it is impossible to differentiate between the proteins NIA1 and NIA2 that share 78% sequence identity, because only antibodies recognizing both isoforms are commercially available. Therefore, studies on the differences between NIA1 and NIA2 have mainly relied on mutant plants, in which one and/or the other *NIA* gene has been knocked out. Based on functional analyses of these mutant plants, some differences between NIA1 and NIA2 activity at the whole-plant level have been identified. For example, it was found that *nia2* knockout plants have only 10 to 20% residual nitrate reduction activity [32,33], or while ABA-induced NO synthesis to mediate guard cell closure was attributed to NIA1 [34], others report that both NR isoforms contributed to salicylic acid-induced NO production, mediating stomatal closure [35].

Information on the biochemical level about distinct functions of the NR isoforms is lacking to date. Therefore, we have established in vitro systems to analyze both the nitrate and nitrite reduction activities of plant NR. We produced functional proteins of the two NR isoforms from *A. thaliana* and subjected them to steady-state enzymatic studies to characterize their functional properties. We found that both isoforms are able to use either nitrate or nitrite as a substrate, with NIA2 having a clear preference for nitrate reductase activity, while NIA1 is the more efficient nitrite reductase, and the nitrite reducing activities of both were inhibited at low concentrations of nitrate.

2. Results

2.1. Nitrate Reduction Activity

NR is modularly folded and individual domains retain a partial activity of the full-length protein [36–38]. We have shown in the past that the N-terminal fragment of *Arabidopsis thaliana* NIA2 comprising the Moco- and heme-domains connected by hinge 1 (residues 1–625, NIA2-Mo-heme) exhibits similar nitrate reduction activity and 14-3-3 protein-mediated inhibition properties to the full-length NIA2 when the artificial electron donor reduced methyl viologen (MV) is supplied for nitrate reduction [5,6]. Therefore, we produced the corresponding N-terminal fragment of NIA1 (residues 1–627, NIA1-Mo-heme) to compare it to the kinetic properties of purified NIA2-Mo-heme.

Following successful purification of NIA1-Mo-heme and NIA2-Mo-heme, we first performed the nitrate reduction assay with reduced MV at different pH values and confirmed that NIA1 has the same pH-optimum at pH 7.0 as NIA2 and is also comparable to other NRs, e.g., from spinach [39,40] (Figure S1). Subsequently, we determined the steady-state kinetic parameters for a range of nitrate concentrations (Figure 1), yielding a $K_M^{nitrate} = 2120 \pm 160$ µM for NIA1-Mo-heme, which is approximately fivefold higher than the $K_M^{nitrate}$ for NIA2-Mo-heme (443 ± 26 µM), whereas the turnover number k_{cat} for NIA1-Mo-heme (51 ± 4 s^{-1}) is slightly but significantly lower than the one for NIA2-Mo-heme (69 ± 9 s^{-1}). These results reveal distinct catalytic efficiencies k_{cat}/K_M of 24 s^{-1} mM^{-1} for NIA1-Mo-heme and 155 s^{-1} mM^{-1} for NIA2-Mo-heme indicating that NIA2 is a far 'better' nitrate reductase, which can be mainly attributed to the lower K_M exhibited by NIA2.

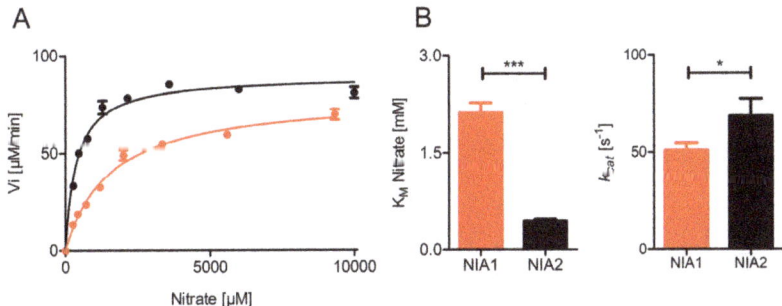

Figure 1. Nitrate reduction by NR-Mo-heme proteins. (**A**) Anaerobic Michaelis–Menten kinetics of NIA1-Mo-heme (red) and NIA2-Mo-heme (black) measured with the MV:nitrate assay. (**B**) Kinetic parameters of multiple batches of NIA1-Mo-heme (red) and NIA2-Mo-heme (black) determined in the MV:nitrate assay. The K_M and k_{cat} for NIA1-Mo-heme and NIA2-Mo-heme were compared via unpaired t-test (GraphPad Prism 5). The means ± SEM of n = 33 kinetic series for NIA1-Mo-heme (made with 23 protein batches) and n = 13 kinetic series for NIA2-Mo-heme (eight protein batches used) are shown. p-value: *** < 0.001 < ** < 0.01 < * < 0.05.

2.2. Re-Constitution of Full-Length NR activity In Vitro

While full-length NIA2 can be obtained in high purity by recombinant expression in *Pichia pastoris* cells [41,42], the expression of full-length NIA1 in *Pichia pastoris* has been of limited success, yielding only trace amounts of protein (unpublished results). The expression strategy used here was, therefore, adjusted, and recombinant expression of both NIA1 and NIA2 protein was performed as two separate fragments in *Escherichia coli*—the N-terminal NR-Mo-heme fragment described above and a C-terminal fragment containing hinge 2 and the FAD-domain (residues 628–917 for NIA1-FAD, residues 626–917 for NIA2-FAD) (Figure S1).

Different ratios of NR-Mo-heme and NR-FAD (1:1–200) were mixed to restore the original electron transfer path using NADH and nitrate as substrates (Figure 2A). As the NR-FAD fragment was found to exhibit substantial diaphorase (NADH:O$_2$ oxidoreductase) activity, the assay was performed under anaerobic conditions. With increasing NR-FAD concentrations, increasing enzyme-specific nitrate reductase activities were observed for both isoforms up to the maximal FAD concentrations of 10 or 20 µM (Figure 2B). Subsequently, we recorded nitrate-dependent NADH steady-state activity using an enzyme re-constituted with a ratio of 1:50 for both isoforms and found that the re-constitution of full-length activity was successfully achieved and that it was nitrate concentration dependent (Figure 2C). These kinetic series yielded an apparent $k_{cat} = 9.6\,\text{s}^{-1}$ for composite NIA1, and $k_{cat} = 13.4\,\text{s}^{-1}$ for composite NIA2. Both of these activities were somewhat lower compared to those obtained with MV as an electron donor and also compared to the reported parameters of full-length NIA2 ($k_{cat} = 33\,\text{s}^{-1}$) [41]. This can be explained by the lack of covalent contact between the heme- and the FAD-domain, which required the use of an excess of FAD fragment to increase the interaction between the separated protein fragments. Consequently, the respective apparent K_M values were found to be lower due to the reduction in k_{cat}, which is a result of the decreased electron transfer rate (apparent $K_M^{nitrate} = 17$ µM for NIA1, 35 µM for NIA2).

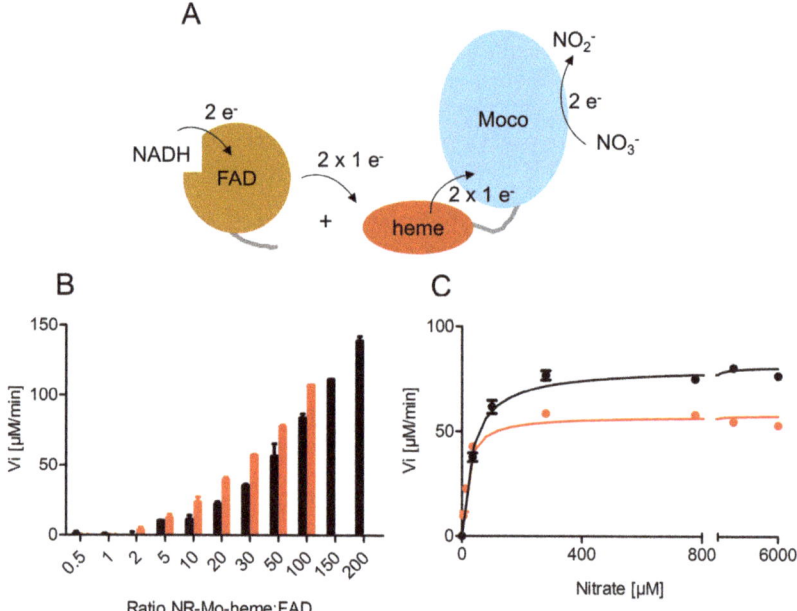

Figure 2. Re-constituted nitrate reductase activity. (**A**) Cartoon representation of the re-constitution of full-length NR activity by combination of the separate NR-Mo-heme and NR-FAD fragments in vitro. (**B**) Anaerobic NADH:nitrate assay of NR-Mo-heme (NIA1 red, NIA2 black) combined with increasing ratios of NR-FAD fragment. Increasing nitrate reductase activity was observed with increasing ratio of FAD-fragment. (**C**) Steady-state NADH:nitrate kinetics of re-constituted NIA1 (red) and NIA2 (black) activities.

2.3. Nitrite Reduction Activity

Reduced MV reacts non-enzymatically with nitrite at millimolar concentrations (Figure S2A). Therefore, an assay with an alternative electron donor had to be established for steady-state measurements of nitrite reductase activity. In contrast to MV, reduced benzyl viologen (BV) reacts non-enzymatically with nitrate, but is stable in the presence of nitrite at pH 7.5 (Figure S2), within the concentrations and time range required to conduct the experiments [43]. In initial tests, the optimal pH for nitrite reduction was determined for both NIA1- and NIA2-Mo-heme to be at pH 7.5 (Figure S3). Analogously to the MV:nitrate assay, the BV:nitrite assay had to be performed under anaerobic conditions to prevent non-enzymatic electron transfer from reduced BV to molecular oxygen. In contrast to MV which is known to donate electrons at the heme-domain [6], BV donates electrons directly to the Moco-domain. We could show this using a NIA2-Mo-heme mutant protein (H600A) with one of the heme-coordinating histidines mutated to an alanine, or a NIA1-fragment comprising only the Moco-domain, both of which showed nitrite reductase activity using BV as the artificial electron donor (Table 1).

Table 1. Kinetic parameters of nitrate reductase (NR)-fragments using the benzyl viologen (BV):nitrite assay.

Protein	$K_M^{nitrite}$ [µM]	$k_{cat}^{nitrite}$ [s^{-1}]
NIA1-Mo-heme	35.5 ± 2.7	19.8 ± 4.8
NIA1-Mo	35.3 ± 3.2	137.5 ± 2.5
NIA2-Mo-heme	13.7 ± 3.3	1.8 ± 0.3
NIA2-Mo-heme-H600A	10.3 ± 1.2	1.2 ± 0.02

For the calculation of the means ± SEM, two kinetic series (1 protein batch) were used for NIA1-Mo; one kinetic series of one protein batch were used for NIA2-Mo-heme-H600A; see Figure 3 for NIA1-Mo-heme and NIA2-Mo-heme.

The kinetic parameters of the nitrite reducing activities with BV as electron donors are $K_M^{nitrite}$ of 35.5 ± 2.7 µM and 13.7 ± 3.3 µM and k_{cat} values of 19.8 ± 4.8 s^{-1} and 1.8 ± 0.3 s^{-1} for NIA1-Mo-heme and NIA2-Mo-heme, respectively (Figure 3A,B, and Table 1) indicating that both proteins are able to act as nitrite reductases, with efficient substrate binding but slow turnover compared to the substrate nitrate. Nonetheless, their catalytic efficiencies differ significantly with 557 s^{-1}mM^{-1} (NIA1-Mo-heme) and 131 s^{-1}mM^{-1} (NIA2-Mo-heme). In this case, the difference in k_{cat}, which is approximately ten-fold higher for NIA1-Mo-heme than for NIA2-Mo-heme, mainly accounts for NIA1 being the more efficient nitrite reductase.

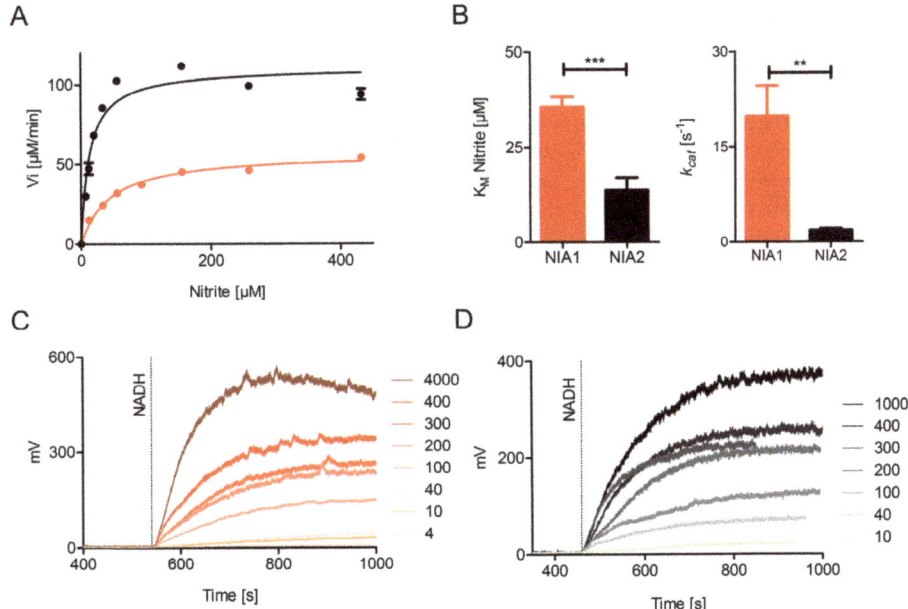

Figure 3. *Cont.*

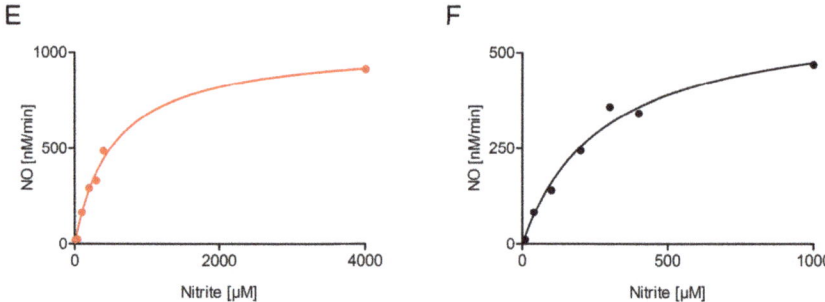

Figure 3. Nitrite reduction by NIA1 and NIA2. (**A**) Anaerobic Michaelis–Menten kinetics of 50 nM NIA1-Mo-heme (red) and 1 µM NIA2-Mo-heme (black) measured with the BV:nitrite assay. A higher concentration of NIA2-Mo-heme enzyme than NIA1-Mo-heme was needed to obtain reaction velocities in a similar order of magnitude. (**B**) Kinetic parameters of NIA1-Mo-heme (red) and NIA2-Mo-heme (black) determined in the BV:nitrite assay. The K_M and k_{cat} for the NIA1-Mo-heme (red) and NIA2-Mo-heme (black) are compared via unpaired t-test (GraphPad Prism 5). The means ± SEM of n = 21 kinetic series for NIA1-Mo-heme (made with 12 protein batches) and n = 10 kinetic series for NIA2-Mo-heme (made with eight protein batches) are shown. *p*-value: *** < 0.001 < **< 0.01 < * < 0.05. (**C,D**) Nitrite reductase activity by re-constituted NIA1 (**C**) and NIA2 (**D**) measured using an NO-analyzer at different nitrite concentrations (indicated by the numbers, µM). (**E,F**) Hyperbolic curve fit of the assays from (**C,D**).

To complement these results, an NO analyzer for direct quantification of the nitrite-dependent NO production by NR was used as it presents a very specific tool to record NO-release. However, despite its specificity for NO, this method can only give qualitative information about the kinetic parameters of enzyme-dependent NO production for two reasons: On the one hand, the NO is only quantified in the gas phase and not in the solution, where the reaction has taken place. This adds an unknown diffusion rate constant to the calculation. On the other hand, the weak interaction between the NR-Mo-heme and the NR-FAD fragments lowers the turnover number and consequently, also the K_M. At least four different batches of NIA1-Mo-heme and NIA2-Mo-heme were re-constituted with 50-fold excess of the respective NR-FAD, and nitrite-concentration dependent NO production was measured in the presence of saturating NADH concentrations. All batches could efficiently produce NO down to very low nitrite concentrations (Figure 3C–E), thus, clearly confirming the enzyme-specific nitrite reduction by both composite NIA1 and NIA2.

2.4. Nitrate Inhibition of Nitrite Reductase Activity

The NO analyzer and the re-constituted NRs allowed us to measure the impact of nitrate on nitrite reduction as a competing substrate for NR. This experiment cannot be performed using one of the viologen assays due to their non-enzymatic reaction with either substrate. Using a saturating nitrite concentration of 400 µM, increasing concentrations of nitrate (0–1 mM) were added to the reaction mix and the amount of NO produced over time was measured (Figure 4). The nitrate concentration resulting in half-maximal inhibition of NO generation rates was IC_{50} = 12 ± 1.7 µM for NIA1-Mo-heme and 36 ± 2.7 µM for NIA2-Mo-heme with a maximal inhibition of up to 97% for both isoforms. This confirms that nitrate is a potent inhibitor for both isoforms which is able to efficiently impair nitrite reduction already at nitrate concentrations that are far below the respective $K_M^{nitrate}$ values for NIA1 and NIA2.

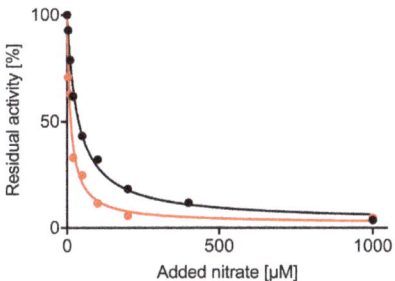

Figure 4. Inhibition of nitrite reductase activity by nitrate. Using re-constituted NR-activity, NO production by NIA1 (red) or NIA2 (black) was monitored on the NO analyzer with 400 μM nitrite. Increasing concentrations of nitrate were added simultaneously with the nitrite, and NO production decreased. The % residual activity was fitted with a hyperbolic curve (GraphPad Prism 7) and IC$_{50}$ and I$_{max}$ for nitrate determined.

3. Discussion

Nitrate reductase was originally recognized as the enzyme catalyzing the eponymous reaction of nitrate reduction, the first step in plant nitrogen anabolism from the inorganic nutrient nitrate [44]. Decades of research have been dedicated to examining this very important and tightly regulated process in the plant. The discoveries that NR is involved in NO synthesis and also in NO scavenging, are by comparison new but probably no less important [8,28].

The role of NR in plants is further complicated in many plant species by the existence of two or even more isoforms of NR, as well as the observation that NR may undergo various post-translational modifications, such as phosphorylation [3,4] or sumoylation [45], which may affect its activity. Only in soybean, have the different isoforms (some constitutively expressed, some inducible ones) been comparatively analyzed with respect to their catalytic properties in nitrate and nitrite reduction, revealing that there are significant differences between the isoforms that may result in distinct functions *in planta* [46–48]. Other studies on NR activity did not differentiate between the isoforms, when analyzing NR purified from a plant species comprising more than one isoform (e.g., from corn [24] or tobacco [20]). In particular, the individual isoforms of *Arabidopsis thaliana*, NIA1 and NIA2, have to our knowledge not yet been separately purified or recombinantly expressed and compared to date. It has been reported, however, that the NR isoforms have individual expression patterns that are distinctly affected by environmental conditions [29–31]. Furthermore, it has been reported that the NR isoforms may have distinct roles in *Arabidopsis*, e.g., stomatal closure that is mainly mediated by NO from NIA1 [34] or that the majority of nitrate-reducing activity is performed by NIA2 [32], but similar contributions to NO formation by both isoforms have also been described [35].

Therefore, the aim of our study was to analyze the functional properties of the recombinantly expressed NIA1 from *Arabidopsis thaliana* in comparison to the properties of NIA2 to reveal whether the distinct functions in the plant are due to distinct isoform-specific kinetic properties or due to specific expression and activation state of either isoform in different plant tissues. Using our well-defined in vitro activity assays, we were able to measure the substrate-dependent velocities without any inhibiting effects/modulators that might be present when using (partially) purified enzyme from plant tissue.

The nitrate-reducing activity measurements using reduced MV as electron donor revealed a large and significant difference between the $K_M^{nitrate}$ for NIA1-Mo-heme and NIA2-Mo-heme. These values are in a similar range to the different K_M values for the soybean NR isoforms [46] and result together with the turnover numbers in a six-fold higher catalytic efficiency (k_{cat}/K_M) of *Arabidopsis* NIA2 for nitrate as compared to NIA1. This lower catalytic efficiency of NIA1 is consistent with the observation that *nia2* knockout plants retained only about 10% of nitrate reduction activity [49].

However, considering the physiological cytosolic nitrate concentrations that lie in the low millimolar range [50–53], NIA1 also has the ability to act as an efficient nitrate reductase, which is manifested in *nia2* single-knockout plants that grow with a normal phenotype [49]. Furthermore, we found that both isoforms clearly prefer NADH over NADPH as a substrate (Figure S4), which indicates that the catalytic efficiencies are mainly due to differences at the catalytic site at the Moco-domain, and not at the FAD-domain where NAD(P)H binds.

To assess the nitrite-reducing capabilities of the *Arabidopsis* NR isoforms, we used two different methods. We first established an anaerobic assay using reduced BV as electron donor for the Moco domain of NR and nitrite as substrate. Measurement of the nitrite-reducing activity of NR using reduced BV is the first reported steady-state assay allowing direct and continuous measurement of the initial nitrite-dependent velocities of NO synthesis by NR and may also be useful for testing nitrite-reducing activity of NRs from other plant species in future. First, it has the advantage that reduced BV is stable in the presence of nitrite, in contrast to MV, which has been used previously to monitor nitrite reduction [28], but which reacts non-enzymatically with nitrite and, thus, causes significant background activity making it difficult to determine the nitrite-reduction velocities with varying nitrite concentrations. Second, the re-oxidation of BV due to electron transfer to the oxidized Mo center following nitrate reduction is directly monitored via a spectral change at 595 nm, which contrasts with the indirect NO quantification via the NO analyzer, in which the produced NO in the gas phase over the reaction mix is quantified and, thus, yields inexact NO synthesis rates. Nevertheless, as the NO analyzer specifically detects the released product NO, it serves as an important complementary method to confirm that the consumption of nitrite by NR indeed leads to NO formation.

The kinetic parameters of in vitro nitrite reducing activity determined with reduced BV clearly denote NIA1 as the more efficient nitrite reductase with a more than fourfold better catalytic efficiency compared to NIA2. The $K_M^{nitrite}$ values (35.5 µM for NIA1, 13.7 µM for NIA2) are considerably lower than those for nitrate, and are in a range similar to the physiological cytosolic nitrite concentrations. Nitrite concentrations in the plant cytosol may vary by two orders of magnitude depending on the environmental conditions but were determined not to exceed the low micromolar range [24,54]. In contrast to earlier reports, which described a significantly higher $K_M^{nitrite}$ for plant NR [24], the here determined values allow both NIA1 and NIA2 to bind nitrite as a substrate under physiological conditions.

Interestingly, very low nitrate concentrations are already sufficient to efficiently inhibit nitrite reductase activity of either NR isoform up to 97% ($IC_{50}^{nitrate}$ = 12 µM for NIA1, 36 µM for NIA2), which are roughly 200- and 10-fold lower than the $K_M^{nitrate}$ values for the respective isoforms. In light of the similarity of both substrates, a competitive inhibition mechanism by nitrite appears most likely.

Consequently, the question arises whether NR is at all able to directly produce NO under physiological conditions when nitrate is usually the much more highly centrated substrate compared to nitrite in the cytosol. The recent findings in *Chlamydomonas* [22] would support the hypothesis that NR may rather act in complex with NOFNiR as an indirect NO synthase, by donating electrons via its FAD-domain to NOFNiR which reduces nitrite to NO. However, this activity in higher land plants has not yet been confirmed. And if it were the case, it is as yet unclear what would trigger the switch in the electron transfer chain from the intra- to an inter-molecular pathway. In case of NIA2, it is possible that this trigger could be represented by a 14-3-3 protein binding to phosphorylated NIA2, which quickly inhibits the nitrate-reducing activity of NIA2 when the nutrient nitrate or reducing equivalents for the N assimilation become limiting. Then, binding of the 14-3-3 protein impairs the electron transfer from the heme cofactor to Moco by steric hindrance [5]. This would allow an immediate switch in the electron transfer pathway to NOFNiR, THB1 or other proteins yet to be identified, as the FAD-domain function is apparently not affected by the 14-3-3 protein binding. However, as no 14-3-3-mediated inhibition of NIA1 has been described until now, this trigger would be limited to NIA2, which would contrast with several reports that propose that NIA1 is the predominant isoform involved in NO synthesis [34,55,56]. More experiments are needed to support or refute this hypothesis: On the one hand, in vivo studies with mutant plants are needed that focus on the NR-NOFNiR interplay. On the other hand in vitro experiments with purified NIA1 are needed to analyze the impact of phosphorylation and 14-3-3 binding on NIA1 as well as with purified NIA1 and NIA2 to analyze the putative interaction with NOFNiR to produce NO.

Assuming that NR (NIA1 and/or NIA2) does not interact with NOFNiR, but is instead able to act as an NO synthase, leads again to the question how the two functions of NR are triggered *in planta*. Several arguments may help to answer this question: First, while nitrate reduction by NR is a crucial reaction for higher land plants, as growth and, thus, survival of the plant largely depends on the availability of nitrogen as a nutrient, the signaling molecule NO is only needed in trace amounts, so a rather slow NO synthesis rate by NR should be sufficient to meet the plant's demands. Second, NO release is usually associated with a spike in nitrite concentration in the tissue, such as upon transition from light into darkness [20,24]. This would point to the fact that a local increase in nitrite concentration enhances the nitrite-reducing activity of NR as nitrite competes with nitrate for binding in the catalytic site. Third, tiny local changes in pH might also play a role in rendering nitrite the substrate for NR: In our in vitro system we determined the pH optimum for nitrate reduction to be at pH 7.0, whilst the pH optimum for nitrite reduction was at pH 7.5 (Figures S1C and S3) suggesting that a slight pH increase might push the function of NR from nitrate reductase towards nitrite reductase activity. These factors in combination with isoform-specific differences in expression, protein activation and their distinct kinetic properties described here might be the determinants for NR to act as a nitrate or nitrite reductase.

Finally, with the successful re-constitution of nitrate reductase activity by mixing two NR fragments, we could demonstrate that the second linker of NR (hinge 2) is not essential for electron transfer from the FAD to the heme cofactor. This is consistent with our previous studies analyzing the electron transfer from FAD to heme in different viscous solutions that indicated no domain movement during intramolecular electron transfer [5] but is in contrast to a previous proposal that hinge 2 is essential for electron transfer activity within NR [57]. The interaction between the FAD and heme domains in the composite NR appears to be rather weak, resulting in activities below the maximum possible compared to full-length NR. Nonetheless, the successful restoration of both nitrate and nitrite reducing activities being able to use the physiological substrate NADH opens up new possibilities to study the structure, activity, and regulation of NR.

Taken together, this study presents the first comparison of the functional properties of the NR isoforms in *Arabidopsis thaliana* demonstrating that NIA2 functions mainly in nitrate reduction and NIA1 mainly in NO synthesis. However, more studies are needed to elucidate the complex interplay of nitrate reduction and nitrite reduction, in particular, whether an interaction with NOFNiR takes place, and the regulation of these processes in vitro as well as in vivo.

4. Materials and Methods

4.1. Recombinant Proteins

The N-terminal fragment of *Arabidopsis thaliana* NIA2 (AGI code: AT1G37130) (NIA2-Mo-heme) was expressed in *E. coli* TP1004 (kindly provided by Tracy Palmer, Newcastle University, UK) using the plasmid described before [5] in LB-medium supplemented with ampicillin (100 µg/mL), kanamycin (25 µg/mL), sodium molybdate (1 mM), magnesium chloride (2 mM), and iron (III) chloride (10 µM) incubated at 37 °C to an OD_{600} of 0.2 to 0.4 and then induced by the addition of 50 µM isopropyl-ß-D-thiogalactoside (IPTG). The culture temperature was reduced to 18 °C and culture was continued for 70 h, then harvested by centrifugation. All subsequent steps were performed at 4 °C and all buffers for immobilized metal affinity chromatography (IMAC) were supplemented with COmplete™ EDTA-free protease inhibitor cocktail (Roche, Mannheim, Germany). Cells were re-suspended in 10 mL lysis buffer (50 mM potassium phosphate pH 7.0, 200 mM sodium chloride) per gram wet cells and frozen at −80 °C. The suspension was then thawed and lysed using a Sonifier

250-D (BRANSON Ultrasonics Corporation, Danbury, CT, USA) and an EmulsiFlex-C5 (Avestin Europe GmbH, Mannheim, Germany). The raw lysate was supplemented with 10 µM hemin (from a 1 mM stock in 20 mM sodium hydroxide) [58]. Ni-NTA chromatography was performed in-batch for 30 min as the manufacturer describes (HisPur™, Thermo Scientific, Rockford, IL, USA). After pouring the resin into a column, a wash with lysis buffer including 5 mM imidazole was performed to remove unspecifically bound proteins. For elution, the imidazole concentration was increased to 200 mM. The deep-red NIA2-Mo-heme-containing fractions were pooled and subjected to size exclusion chromatography (SEC) using an Äkta Prime system (GE Healthcare Europe GmbH, Freiburg, Germany) using a 16/60 Superdex 200 prep grade column (GE Healthcare Europe GmbH) and SEC buffer (20 mM Tris/hydrochloric acid pH 7.5, 200 mM sodium chloride, 10 mM magnesium acetate, 0.05% Tween 20). The protein peak eluting at about 60 mL was pooled. The concentration of heme-containing protein was determined via absorption at 413 nm using the extinction coefficient $\epsilon_{413} = 120{,}000~M^{-1}~cm^{-1}$. Molybdenum co-factor (Moco) saturation was quantified after oxidation to Form A and subsequent HPLC analysis by comparison to a Form A standard as described [59]. Protein was shock frozen in droplets in liquid nitrogen and stored at $-80~°C$.

Using the RAFL plasmid pda08083 (RIKEN BRC, Ibaraki, Japan) as a template for the *NIA1* gene (AGI code: AT1G77760), the sequence corresponding to *Arabidopsis thaliana* NIA1-Mo-heme fragment (residues 1–627) was PCR-cloned into the SphI and SalI restriction sites of pQE80L plasmid. The expression was similar to NIA2-Mo-heme with the following differences: Growth phase and expression of the transformed cells was at 25 °C. Induction was at $OD_{600} = 0.4$ with 100 µM IPTG for a duration of 20 h. The pH of the lysis, wash and elution buffers was adjusted to 7.5. Wash of the immobilized metal affinity chromatography (IMAC) column was performed after addition of 20 mM imidazole to the lysis buffer, while for the elution step, 250 mM imidazole was added. After SEC, the fractions containing non-degraded NIA1-Mo-heme were pooled, and after cofactor quantifications, the protein was shock-frozen in aliquots and stored at $-80~°C$.

The gene sequence corresponding to the NIA1-Mo fragment (residues 1–488) was PCR-amplified from the pQE80L-NIA1-Mo-heme plasmid and cloned into the KpnI and SalI restriction sites of pQE80L. Growth and expression were in *E. coli* TP1004 as described above for NIA2-Mo-heme, but with 20 µM IPTG for induction at 30 °C and 30 h. The cells were suspended (1 g/10 mL) in lysis buffer (50 mM potassium phosphate pH 7.5, 200 mM sodium chloride, 10 mM dithiothreitol, 1 mM sodium molybdate, 10 mM imidazole, COmplete™ EDTA-free protease inhibitor cocktail). After one freeze-thaw cycle, the cells were lysed using an EmulsiFlex (Avestin). The His-tagged protein was first affinity-purified and then applied to an SEC as described for NIA2-Mo-heme. The NIA1-Mo peak was pooled, concentrated using an Amicon concentrator (Merck, Darmstadt, Germany) and Moco quantification via Form A, shock-frozen in aliquots and stored at 80 °C.

The NIA2-Mo-heme-H600A variant was expressed and purified as described elsewhere [5].

The DNA for the FAD-domains of NIA1 and NIA2 were PCR-amplified out of the respective full-length NR DNA sequences and had restriction sites introduced (BamHI and HindIII for NIA1-FAD, and PstI and HindIII for NIA2-FAD) for cloning into pQE80L plasmid. Expression was performed in *E. coli* BL21 Rosetta (Novagen, Darmstadt, Germany) using the same conditions for both FAD-fragments. Transformed cells were cultured at 37 °C to an OD_{600} of 0.4 and then induced by the addition of 400 µM IPTG. Induction was for 4 h at 37 °C. FAD-lysis buffer composition was 50 mM potassium phosphate pH 7.0, 200 mM sodium chloride, 5 mM imidazole, and COmplete™ EDTA-free protease inhibitor cocktail. Wash of the IMAC was performed using the lysis buffer supplemented with 20 mM imidazole, elution buffer was with 250 mM imidazole. After elution, buffer exchange was performed using PD-10 columns (GE Healthcare) and SEC buffer. Concentration determination was based on the FAD-cofactor specific absorption at 450 nm and using an extinction coefficient $\epsilon_{450} = 11{,}300~M^{-1}~cm^{-1}$.

4.2. SDS-PAGE and Western Blot

Protein samples were separated by SDS-PAGE [60] and visualized by Coomassie Brilliant Blue G250 staining [61]. Proteins for Western blotting were transferred after PAGE to a PVDF membrane [62] using a semi-dry blotter, blocked with fat-free milk powder solution in TBST buffer (20 mM Tris, 150 mM sodium chloride, 0.1% Twee 20) and probed using polyclonal NR-specific antibody diluted 1:10,000 (AS08310, Agrisera, Vännäs, Sweden) and as secondary anti-rabbit horse radish peroxidase-coupled antibody (1:5000 dilution, Thermo Scientific).

4.3. Enyzme Assays

All enzyme assays were performed in an anaerobic chamber at 22 °C to 25 °C (Coy Laboratory Products, Grass Lake, MI, USA), and enzyme-free negative controls were included in all experiments to confirm the enzyme-specific activities. For each single data point, three technical replicates were measured.

The MV:nitrate assay was performed with NIA1- or NIA2-Mo-heme as described [5] in a modified assay buffer (50 mM MOPS pH 7.0, 50 mM potassium chloride, 5 mM magnesium acetate, 1 mM calcium dichloride) in 96-well plates (Greiner-bio-one, Kremsmünster, Österreich) using a Sunrise plate reader (Tecan, Männedorf, Switzerland). Twenty-five nanomolar cofactor-saturated NR-Mo-heme protein was used in a final volume of 120 µL in the well. The slope of oxidizing MV was monitored at A_{595}, and the initial velocities v_i were calculated, with 2 mole MV consumed for 1 mole nitrate. Triplicate values were used to determine mean and standard error of the mean (SEM) and then plotted and fitted in GraphPad Prism 5 using the Michaelis–Menten curve fit to yield k_{cat} and K_M values. Activity assays were performed with multiple protein purification batches on multiple days (n = 33 for NIA1-Mo-heme and n = 13 for NIA2-Mo-heme).

The NADH:nitrate assay was performed using re-constituted NR. For re-constitution of NR activity, 100 nM NR-Mo-heme and 50 nM–10 or 20 µM NR-FAD (NIA1 and NIA2, respectively) were mixed in the pH 7.0 assay buffer (see above) in 96-well plates. Due to volume limitations in the experimental setup, we could not exceed a ratio of 1:100 for composite NIA1. The reaction was started by the addition of pre-mixed nitrate and NADH at a final saturating concentration of 220 µM (calculated based on its absorption at 340 nm and ϵ_{340} = 6220 M^{-1} cm^{-1}. For determination of the optimal Mo-heme:FAD ratio, a constant nitrate concentration of 2 mM was provided. For steady-state kinetic studies a range of nitrate concentrations from 0 to 6 mM were used and a constant FAD concentration of 5 µM (=50-fold excess). All measurements were performed in triplicate, and multiple NR-Mo-heme protein batches were used. The stoichiometric consumption of NADH was followed at A_{340}, and initial slopes were determined using the Magellan software (Tecan) to calculate the v_i and further evaluated using GraphPad Prism. The comparison of the co-substrate NADH and NADPH were performed using 100 nM NIA1- or NIA2-Mo-heme supplemented with 5 µM of the respective NR-FAD fragment, 2 mM nitrate and 220 µM NADH or NADPH in assay buffer (pH 7.0) as described for the titration experiments.

The BV:nitrite assay was performed in a similar fashion to the MV:nitrate assay using the Mo-heme fragments. The pH optimum for nitrite reduction was shown to be pH 7.5 (Figure S3). Therefore, the buffer composition for nitrite reduction was: 50 mM MOPS pH 7.5, 50 mM potassium chloride, 5 mM magnesium acetate, 1 mM calcium dichloride. A nitrite dilution curve (0–435 µM) was prepared from anaerobic sodium nitrite powder fresh daily. Typically, 50 nM NIA1-Mo-heme and 500 nM NIA2-Mo-heme protein (unless otherwise indicated in the figure legends) were added to measure the initial slopes of stoichiometric re-oxidation of BV at A_{595}. Activity assays were performed for multiple NR-Mo-heme purification batches on multiple days (n = 21 for NIA1-Mo-heme and n = 10 for NIA2-Mo-heme). The mean K_M and k_{cat} ± SEM was determined using GraphPad Prism.

4.4. NO Quantification Using the NO-Analyzer

For nitric oxide quantification, an NO analyzer (Sievers 280i, Analytix, Boldon, UK) and modified assay buffer at pH 7.5 (as for BV:nitrite assay) was used, supplemented with Antifoam Y30 (Sigma, Saint Louis, MO, USA) at a dilution of 1:2000. An oxygen-free argon gas stream was bubbled through the glass reaction vessel containing the reaction components in a volume of 3 mL. The mixture was pipetted in the following order: First, buffer was placed in the vessel and the argon pressure adjusted to be equivalent to the vacuum coming from the analyzer. The vessel was closed and allowed to bubble and become anaerobic. After 4 min, anaerobic sodium nitrite solution to yield final concentrations of 10 μM to 4 mM (or nitrite + nitrate for inhibition experiments) was added from a sealed vial using a Hamilton syringe, followed at 6 min by anaerobic protein mix (100 pmol NR-Mo-heme + 5000 pmol NR-FAD). At 8 min, NADH solution was added to a final concentration of 220 μM to start the reaction. Steady-state NO release was recorded up to 20 min (or longer).

For the evaluation, the areas under the steeply increasing start of the curve were determined (typically for 200 s) and converted to pmol NO by comparison with an NO standard curve that had been prepared as described elsewhere [63]. It was assumed that the amount of detected NO in the gas phase correlated with the concentration of NO in the solution. Therefore, by converting the amount of NO released from the 3 mL (at a given concentration of substrate nitrite) to NO concentration, resulted in an estimate of NO synthesis velocity (v_i). By plotting this against the substrate concentration, a Michaelis–Menten-like plot was generated. For the determination of inhibition of nitrite reduction by nitrate, the NO synthesis velocity with 400 μM nitrite was set to 100% activity, and the reduced activities in the reaction samples were compared with this.

Supplementary Materials: The following are available online at http://www.mdpi.com/2223-7747/8/3/67/s1, Figure S1: Recombinant proteins and pH optimum of nitrate-reducing activity with reduced MV as an electron donor. A. Purified NIA1-Mo-heme (0.8 μg) and NIA2-Mo-heme (2.2 μg) on 10% SDS-PAGE with Coomassie Brilliant Blue staining (C) and on a Western blot with anti-NR antibodies (WB). B. Purified FAD-fragment of NIA1 (3.6 μg) and NIA2 (0.9 μg) on 12% SDS-PAGE with Coomassie Brilliant Blue staining. C. The nitrate-reducing activity using reduced MV by NIA1-Mo-heme was monitored and had a pH optimum at pH 7.0. Figure S2: Unspecific re-oxidation of artificial electron donors (MV shown in blue, BV in brown) monitored at A595. A, B. Enzyme-free reaction mix of reduced MV, assay buffer and 0.157–200 mM nitrite (A) or nitrate (B). C, D. Enzyme-free reaction mix of reduced BV, assay buffer and 0.157–200 mM nitrite (C) or nitrate (D). Figure S3: Determination of pH optimum for the BV:nitrite steady-state kinetic assay. The pH optimum for both NIA1-Mo-heme (red, left panel) and NIA2-Mo-heme (black, right panel) for nitrite-reducing activity using reduced BV is at pH 7.5. Figure S4: Comparison of NADH or NADPH as substrates for composite NIA1 and NIA2. Initial velocities of 100 nM NIA1- (red) or NIA2- (black) Mo-heme supplemented with 5 μM of the respective NR-FAD at saturating nitrate and NADH or NADPH concentrations were recorded.

Author Contributions: M.M. designed the experiments, performed experiments, analyzed the data, prepared the figures and wrote the manuscript. B.T. performed experiments. K.F.S. designed the experiments, analyzed the data, wrote and revised the manuscript.

Funding: This research was funded by the 'Deutsche Forschungsgemeinschaft' (DFG), grant number SCHR1529/1-1 (K.F.S.).

Acknowledgments: We thank Monika Laurien for skilled technical support especially in protein expression and Günter Schwarz for critical reading of the manuscript.

Conflicts of Interest: The authors declare no conflict of interest.

References

1. Campbell, W.H. Nitrate reductase and its role in nitrate assimilation in plants. *Physiol. Plantarum.* **1988**, *74*, 214–219. [CrossRef]
2. Meyer, C.; Stitt, M. Nitrate reduction and signalling. In *Plant Nitrogen*; Lea, P.J., Morot-Gaudry, J.-F., Eds.; Springer Berlin Heidelberg: Berlin, Heidelberg, 2001; pp. 37–59.
3. Bachmann, M.; Shiraishi, N.; Campbell, W.H.; Yoo, B.C.; Harmon, A.C.; Huber, S.C. Identification of ser 543 as the major regulatory phosphorylation site in spinach leaf nitrate reductase. *Plant Cell* **1996**, *8*, 505–517. [CrossRef]

4. Su, W.; Huber, S.C.; Crawford, N.M. Identification in vitro of a post-translational regulatory site in the hinge 1 region of arabidopsis nitrate reductase. *Plant Cell* **1996**, *8*, 519–527. [CrossRef] [PubMed]
5. Lambeck, I.C.; Fischer-Schrader, K.; Niks, D.; Roeper, J.; Chi, J.C.; Hille, R.; Schwarz, G. Molecular mechanism of 14-3-3 protein-mediated inhibition of plant nitrate reductase. *J. Bio. Chem.* **2012**, *287*, 4562–4571. [CrossRef] [PubMed]
6. Chi, J.C.; Roeper, J.; Schwarz, G.; Fischer-Schrader, K. Dual binding of 14-3-3 protein regulates arabidopsis nitrate reductase activity. *J. Biol. Inorg. Chem.* **2015**, *20*, 277–286. [CrossRef] [PubMed]
7. Harper, J.E. Evolution of nitrogen oxide(s) during in vivo nitrate reductase assay of soybean leaves. *Plant Physiol.* **1981**, *68*, 1488–1493. [CrossRef] [PubMed]
8. Wildt, J.; Kley, D.; Rockel, A.; Rockel, P.; Segschneider, H.J. Emission of no from several higher plant species. *J. Geophys. Res.* **1997**, *102*, 5919–5927. [CrossRef]
9. Yamasaki, H.; Sakihama, Y. Simultaneous production of nitric oxide and peroxynitrite by plant nitrate reductase: In vitro evidence for the NR-dependent formation of active nitrogen species. *FEBS Lett.* **2000**, *468*, 89–92. [CrossRef]
10. Wendehenne, D.; Hancock, J.T. New frontiers in nitric oxide biology in plant. *Plant Sci. Int. J. Exp. Plant Biol.* **2011**, *181*, 507–508. [CrossRef] [PubMed]
11. Farnese, F.S.; Menezes-Silva, P.E.; Gusman, G.S.; Oliveira, J.A. When bad guys become good ones: The key role of reactive oxygen species and nitric oxide in the plant response to abiotic stress. *Front. Plant Sci.* **2016**, *7*, 471. [CrossRef]
12. Sanz-Luque, E.; Chamizo-Ampudia, A.; Llamas, A.; Galvan, A.; Fernandez, E. Understanding nitrate assimilation and its regulation in microalgae. *Front. Plant Sci.* **2015**, *6*, 899. [CrossRef] [PubMed]
13. Bethke, P.C.; Badger, M.R.; Jones, R.L. Apoplastic synthesis of nitric oxide by plant tissues. *Plant Cell* **2004**, *16*, 332–341. [CrossRef] [PubMed]
14. Barroso, J.B.; Corpas, F.J.; Carreras, A.; Sandalio, L.M.; Valderrama, R.; Palma, J.M.; Lupianez, J.A.; del Rio, L.A. Localization of nitric-oxide synthase in plant peroxisomes. *J. Biol. Chem.* **1999**, *274*, 36729–36733. [CrossRef]
15. Fröhlich, A.; Durner, J. The hunt for plant nitric oxide synthase (NOS): Is one really needed? *Plant Sci.* **2011**, *181*, 401–404. [CrossRef] [PubMed]
16. Alber, N.A.; Sivanesan, H.; Vanlerberghe, G.C. The occurrence and control of nitric oxide generation by the plant mitochondrial electron transport chain. *Plant Cell Environ.* **2017**, *40*, 1074–1085. [CrossRef] [PubMed]
17. Bender, D.; Schwarz, G. Nitrite-dependent nitric oxide synthesis by molybdenum enzymes. *FEBS Lett.* **2018**, *592*, 2126–2139. [CrossRef] [PubMed]
18. Gupta, K.J.; Fernie, A.R.; Kaiser, W.M.; van Dongen, J.T. On the origins of nitric oxide. *Trends Plant Sci.* **2011**, *16*, 160–168. [CrossRef] [PubMed]
19. Mur, L.A.; Mandon, J.; Persijn, S.; Cristescu, S.M.; Moshkov, I.E.; Novikova, G.V.; Hall, M.A.; Harren, F.J.; Hebelstrup, K.H.; Gupta, K.J. Nitric oxide in plants: An assessment of the current state of knowledge. *AoB Plants* **2013**, *5*, pls052. [CrossRef] [PubMed]
20. Planchet, E.; Jagadis Gupta, K.; Sonoda, M.; Kaiser, W.M. Nitric oxide emission from tobacco leaves and cell suspensions: Rate limiting factors and evidence for the involvement of mitochondrial electron transport. *Plant J. Cell Mol. Biol.* **2005**, *41*, 732–743. [CrossRef] [PubMed]
21. Modolo, L.V.; Augusto, O.; Almeida, I.M.; Magalhaes, J.R.; Salgado, I. Nitrite as the major source of nitric oxide production by arabidopsis thaliana in response to pseudomonas syringae. *FEBS Lett.* **2005**, *579*, 3814–3820. [CrossRef] [PubMed]
22. Chamizo-Ampudia, A.; Sanz-Luque, E.; Llamas, A.; Ocana-Calahorro, F.; Mariscal, V.; Carreras, A.; Barroso, J.B.; Galvan, A.; Fernandez, E. A dual system formed by the arc and nr molybdoenzymes mediates nitrite-dependent no production in chlamydomonas. *Plant Cell Environ.* **2016**, *39*, 2097–2107. [CrossRef] [PubMed]
23. Sanz-Luque, E.; Ocana-Calahorro, F.; de Montaigu, A.; Chamizo-Ampudia, A.; Llamas, A.; Galvan, A.; Fernandez, E. Thb1, a truncated hemoglobin, modulates nitric oxide levels and nitrate reductase activity. *Plant J. Cell Mol. Biol.* **2015**, *81*, 467–479. [CrossRef] [PubMed]
24. Rockel, P.; Strube, F.; Rockel, A.; Wildt, J.; Kaiser, W.M. Regulation of nitric oxide (NO) production by plant nitrate reductase in vivo and in vitro. *J. Exp. Bot.* **2002**, *53*, 103–110. [CrossRef]

25. Chamizo-Ampudia, A.; Sanz-Luque, E.; Llamas, A.; Galvan, A.; Fernandez, E. Nitrate reductase regulates plant nitric oxide homeostasis. *Trends Plant Sci.* **2017**, *22*, 163–174. [CrossRef] [PubMed]
26. Savidov, N.A.; Tokarev, B.I.; Lips, S.H. Regulation of mo-cofactor, NADH- and NAD(P)H-specific nitrate reductase activities in the wild type and two nar-mutant lines of barley (*Hordeum vulgare* L.). *J. Exp. Bot.* **1997**, *48*, 847–855. [CrossRef]
27. Wells, G.N.; Hageman, R.H. Specificity for nicotinamide adenine-dinucleotide by nitrate reductase from leaves. *Plant Physiol.* **1974**, *54*, 136–141. [CrossRef] [PubMed]
28. Dean, J.V.; Harper, J.E. The conversion of nitrite to nitrogen oxide(s) by the constitutive NAD(P)H-nitrate reductase enzyme from soybean. *Plant Physiol.* **1988**, *88*, 389–395. [CrossRef] [PubMed]
29. Yu, X.; Sukumaran, S.; Marton, L. Differential expression of the arabidopsis NIA1 and NIA2 genes. Cytokinin-induced nitrate reductase activity is correlated with increased NIA1 transcription and mrna levels. *Plant Physiol.* **1998**, *116*, 1091–1096. [CrossRef]
30. Cheng, C.L.; Acedo, G.N.; Dewdney, J.; Goodman, H.M.; Conkling, M.A. Differential expression of the two arabidopsis nitrate reductase genes. *Plant Physiol.* **1991**, *96*, 275–279. [CrossRef]
31. Lin, Y.; Cheng, C.L. A chlorate-resistant mutant defective in the regulation of nitrate reductase gene expression in arabidopsis defines a new hy locus. *Plant Cell* **1997**, *9*, 21–35. [CrossRef]
32. Wilkinson, J.Q.; Crawford, N.M. Identification of the arabidopsis chl3 gene as the nitrate reductase structural gene NIA2. *Plant Cell* **1991**, *3*, 461–471. [CrossRef] [PubMed]
33. Braaksma, F.J.; Feenstra, W.J. Isolation and characterization of nitrate reductase-deficient mutants of arabidopsis thaliana. *Theor. Appl. Genet.* **1982**, *64*, 83–90. [CrossRef] [PubMed]
34. Bright, J.; Desikan, R.; Hancock, J.T.; Weir, I.S.; Neill, S.J. Aba-induced no generation and stomatal closure in arabidopsis are dependent on H_2O_2 synthesis. *Plant J. Cell Mol. Biol.* **2006**, *45*, 113–122. [CrossRef] [PubMed]
35. Hao, F.S.; Zhao, S.L.; Dong, H.; Zhang, H.; Sun, L.R.; Miao, C. NIA1 and NIA2 are involved in exogenous salicylic acid-induced nitric oxide generation and stomatal closure in arabidopsis. *J. Integr. Plant Biol.* **2010**, *52*, 298–307. [CrossRef]
36. Kubo, Y.; Ogura, N.; Nakagawa, H. Limited proteolysis of the nitrate reductase from spinach leaves. *J. Biol. Chem.* **1988**, *263*, 19684–19689. [PubMed]
37. Campbell, W.H. Nitrate reductase structure, function and regulation: Bridging the gap between biochemistry and physiology. *Annu. Rev. Plant Physiol. Plant Mol. Biol.* **1999**, *50*, 277–303. [CrossRef] [PubMed]
38. Solomonson, L.P.; Barber, M.J. Assimilatory nitrate reductase: Functional properties and regulation. *Annu. Rev. Plant Phys.* **1990**, *41*, 225–253. [CrossRef]
39. Barber, M.J.; Notton, B.A. Spinach nitrate reductase: Effects of ionic strength and ph on the full and partial enzyme activities. *Plant Physiol.* **1990**, *93*, 537–540. [CrossRef] [PubMed]
40. Lambeck, I.C. Post-Translationale Regulation der Nitratreduktase Durch Phosphorylierung und 14-3-3-Protein-Bindung. Ph.D. Thesis, Universität zu Köln, Cologne, Germany, 2009.
41. Lambeck, I.; Chi, J.C.; Krizowski, S.; Mueller, S.; Mehlmer, N.; Teige, M.; Fischer, K.; Schwarz, G. Kinetic analysis of 14-3-3-inhibited arabidopsis thaliana nitrate reductase. *Biochemistry* **2010**, *49*, 8177–8186. [CrossRef] [PubMed]
42. Skipper, L.; Campbell, W.H.; Mertens, J.A.; Lowe, D.J. Pre-steady-state kinetic analysis of recombinant arabidopsis NADH:Nitrate reductase: Rate-limiting processes in catalysis. *J. Biol. Chem.* **2001**, *276*, 26995–27002. [CrossRef] [PubMed]
43. Hewitt, E.J.; James, D.M.; Eaglesham, A.R. The non-enzymic reduction of nitrite by benzyl viologen (free-radical) in the presence and absence of ammonium sulphate. *Mol. Cell Biochem.* **1975**, *6*, 101–105. [CrossRef] [PubMed]
44. Evans, H.J.; Nason, A. Pyridine nucleotide-nitrate reductase from extracts of higher plants. *Plant Physiol.* **1953**, *28*, 233–254. [CrossRef] [PubMed]
45. Park, B.S.; Song, J.T.; Seo, H.S. Arabidopsis nitrate reductase activity is stimulated by the E3 SUMO ligase ATSIZ1. *Nat. Commun.* **2011**, *2*, 400. [CrossRef] [PubMed]
46. Streit, L.; Nelson, R.S.; Harper, J.E. Nitrate reductases from wild-type and nr(1)-mutant soybean (*Glycine max* [L.] Merr.) Leaves: I. Purification, kinetics, and physical properties. *Plant Physiol.* **1985**, *78*, 80–84. [CrossRef] [PubMed]

47. Nelson, R.S.; Streit, L.; Harper, J.E. Nitrate reductases from wild-type and nr(1)-mutant soybean (*Glycine max* [L.] Merr.) Leaves: II. Partial activity, inhibitor, and complementation analyses. *Plant Physiol.* **1986**, *80*, 72–76. [CrossRef] [PubMed]
48. Jolly, S.O.; Campbell, W.; Tolbert, N.E. Nadph- and NADH-nitrate reductases from soybean leaves. *Arch. Biochem. Biophys.* **1976**, *174*, 431–439. [CrossRef]
49. Wilkinson, J.Q.; Crawford, N.M. Identification and characterization of a chlorate-resistant mutant of arabidopsis thaliana with mutations in both nitrate reductase structural genes NIA1 and NIA2. *Mol. Gen. Genet.* **1993**, *239*, 289–297. [PubMed]
50. Britto, D.T.; Kronzucker, H.J. Constancy of nitrogen turnover kinetics in the plant cell: Insights into the integration of subcellular n fluxes. *Planta* **2001**, *213*, 175–181. [CrossRef]
51. Britto, D.T.; Kronzucker, H.J. The case for cytosolic no3- heterostasis: A critique of a recently proposed model. *Plant Cell Environ.* **2003**, *26*, 183–188. [CrossRef]
52. Davenport, S.; Le Lay, P.; Sanchez-Tamburrrino, J.P. Nitrate metabolism in tobacco leaves overexpressing arabidopsis nitrite reductase. *Plant Physiol. Biochem.* **2015**, *97*, 96–107. [CrossRef]
53. Miller, A.J.; Smith, S.J. Nitrate transport and compartmentation in cereal root cells. *J. Exp. Bot.* **1996**, *47*, 843–854. [CrossRef]
54. Sugiura, M.; Georgescu, M.N.; Takahashi, M. A nitrite transporter associated with nitrite uptake by higher plant chloroplasts. *Plant Cell Physiol.* **2007**, *48*, 1022–1035. [CrossRef] [PubMed]
55. Ribeiro, D.M.; Desikan, R.; Bright, J.; Confraria, A.; Harrison, J.; Hancock, J.T.; Barros, R.S.; Neill, S.J.; Wilson, I.D. Differential requirement for no during aba-induced stomatal closure in turgid and wilted leaves. *Plant Cell Environ.* **2009**, *32*, 46–57. [CrossRef] [PubMed]
56. Wilson, I.D.; Ribeiro, D.M.; Bright, J.; Confraria, A.; Harrison, J.; Barros, R.S.; Desikan, R.; Neill, S.J.; Hancock, J.T. Role of nitric oxide in regulating stomatal apertures. *Plant Signal. Behav.* **2009**, *4*, 467–469. [CrossRef] [PubMed]
57. Barbier, G.G.; Campbell, W.H. Viscosity effects on eukaryotic nitrate reductase activity. *J. Biol. Chem.* **2005**, *280*, 26049–26054. [CrossRef] [PubMed]
58. Krainer, F.W.; Capone, S.; Jäger, M.; Vogl, T.; Gerstmann, M.; Glieder, A.; Herwig, C.; Spadiut, O. Optimizing cofactor availability for the production of recombinant heme peroxidase in pichia pastoris. *Microb. Cell Factor.* **2015**, *14*, 4. [CrossRef] [PubMed]
59. Schwarz, G.; Boxer, D.H.; Mendel, R.R. Molybdenum cofactor biosynthesis. The plant protein cnx1 binds molybdopterin with high affinity. *J. Biol. Chem.* **1997**, *272*, 26811–26814. [CrossRef] [PubMed]
60. Laemmli, U.K. Cleavage of structural proteins during the assembly of the head of bacteriophage t4. *Nature* **1970**, *227*, 680–685. [CrossRef]
61. Lawrence, A.M.; Besir, H.U. Staining of proteins in gels with coomassie g-250 without organic solvent and acetic acid. *J. Vis. Exp.* **2009**, *14*, 1350. [CrossRef]
62. Towbin, H.; Staehelin, T.; Gordon, J. Electrophoretic transfer of proteins from polyacrylamide gels to nitrocellulose sheets: Procedure and some applications. *Proc. Natl. Acad. Sci. USA* **1979**, *76*, 4350–4354. [CrossRef]
63. MacArthur, P.H.; Shiva, S.; Gladwin, M.T. Measurement of circulating nitrite and s-nitrosothiols by reductive chemiluminescence. *J. Chromatogr. B Anal. Technol. Biomed. Life Sci.* **2007**, *851*, 93–105. [CrossRef] [PubMed]

© 2019 by the authors. Licensee MDPI, Basel, Switzerland. This article is an open access article distributed under the terms and conditions of the Creative Commons Attribution (CC BY) license (http://creativecommons.org/licenses/by/4.0/).

Review

Role of Nitrate Reductase in NO Production in Photosynthetic Eukaryotes

Manuel Tejada-Jimenez, Angel Llamas, Aurora Galván and Emilio Fernández *

Departamento de Bioquímica y Biología Molecular, Campus de Rabanales y Campus Internacional de Excelencia Agroalimentario (CeiA3), Edif. Severo Ochoa, Universidad de Córdoba, 14071 Córdoba, Spain; manuel.tejada@uco.es (M.T.-J.); bb2llaza@uco.es (A.L.); bb1gacea@uco.es (A.G.)
* Correspondence: bb1feree@uco.es; Tel.: +34-957-218-591

Received: 16 January 2019; Accepted: 8 February 2019; Published: 6 March 2019

Abstract: Nitric oxide is a gaseous secondary messenger that is critical for proper cell signaling and plant survival when exposed to stress. Nitric oxide (NO) synthesis in plants, under standard phototrophic oxygenic conditions, has long been a very controversial issue. A few algal strains contain NO synthase (NOS), which appears to be absent in all other algae and land plants. The experimental data have led to the hypothesis that molybdoenzyme nitrate reductase (NR) is the main enzyme responsible for NO production in most plants. Recently, NR was found to be a necessary partner in a dual system that also includes another molybdoenzyme, which was renamed NO-forming nitrite reductase (NOFNiR). This enzyme produces NO independently of the molybdenum center of NR and depends on the NR electron transport chain from NAD(P)H to heme. Under the circumstances in which NR is not present or active, the existence of another NO-forming system that is similar to the NOS system would account for NO production and NO effects. PII protein, which senses and integrates the signals of the C–N balance in the cell, likely has an important role in organizing cell responses. Here, we critically analyze these topics.

Keywords: nitric oxide; nitrate reductase; NOFNiR; nitrogen metabolism

---

1. Introduction

Nitric oxide (NO) is a gaseous secondary messenger in humans, animals, plants, fungi, and bacteria. In plants, NO is involved in important physiological processes, such as growth, development, metabolism, leaf senescence, biotic and abiotic stress, defense processes, and plant–pathogen interactions, which have been extensively reviewed [1–5]. In particular, in algae, such as the green alga *Chlamydomonas reinhardtii*, NO also participates in fundamental cell functions, such as the regulation of N-metabolism, N- and S-starvation stress, chloroplast biogenesis, programmed cell death, and responses to darkness, hypoxia, or salt stress [6–11].

In the last two decades, it has been clarified that NO is a signaling molecule in plant defense during plant–pathogen interactions [12,13]. Since then, different strategies have been used to understand NO biosynthesis in plant cells, and this subject has not been short of controversies [5,14], with some aspects yet to be understood. Two main pathways, reductive and oxidative, appear to explain NO synthesis in plants. One is based on the reduction of nitrite, and the other involves the oxidation of aminated molecules, such as the amino acid arginine [15].

In spite of the seminal work of Foresi and collaborators, who identified the first NO synthase (NOS) from the plant kingdom in the green alga *Ostreococus taurii* [16], the existence of a plant NOS that has the characteristics of the animal NOS has been puzzling [14,17] since no plant genome contains such a conserved gene. In fact, Jeandrof and collaborators analyzed over 1000 species of land plants and algae and found no typical NOS sequences in the 1087 sequenced transcriptomes of land plants, but they did find said sequences in 15 of the 265 algal species. Thus, it was concluded that land plants had evolved a mechanism to synthesize NO in a manner that is different from that used in animals [18]. In this review, the different biosynthetic processes of nitric oxide formation are critically analyzed, together with their physiological relevance.

2. Nitrite: The Substrate for Reductive NO Production

Nitrite is a product of nitrate reductase (NR)-catalyzed nitrate reduction within the nitrate assimilation pathway. Nitrogen acquisition is a fundamental process for living beings, including plants in crops, in which N is usually a limiting factor that determines crop productivity [19]. Nitrate used to be the preferred form of inorganic N that was available in soils and, thus, was used in fertilizers [20].

The incorporation of nitrogen from nitrate first requires its acquisition from the medium by specific transporters, which are responsible for the sensing, uptake, storage, and distribution of nitrate among plant tissues. Plant nitrate transporters belong to several families: the nitrate transporter 1/peptide transporter/nitrate peptide transporter family (NRT1/PTR/NPF), NRT2/nitrate nitrite porter (NRT2/NNP), chloride channels (CLC), slow anion channel-associated 1 homolog 3 (SLAC1/SLAH), and aluminum-activated malate transporters (ALMT). These have all been reviewed in detail [21–25]. In the Chlamydomonas alga, this complexity is less but still significant, reflecting the importance of this step. Here, we highlight the three families of transporter proteins found in Chlamydomonas: NRT1/NPF, NRT2, and NAR1 (Figure 1). NRT1 has been described in Arabidopsis as a dual-affinity nitrate/nitrite transporter, and NRT2 (with the accessory protein NAR2) mediates the high-affinity transport (HAT) of nitrate and nitrite.

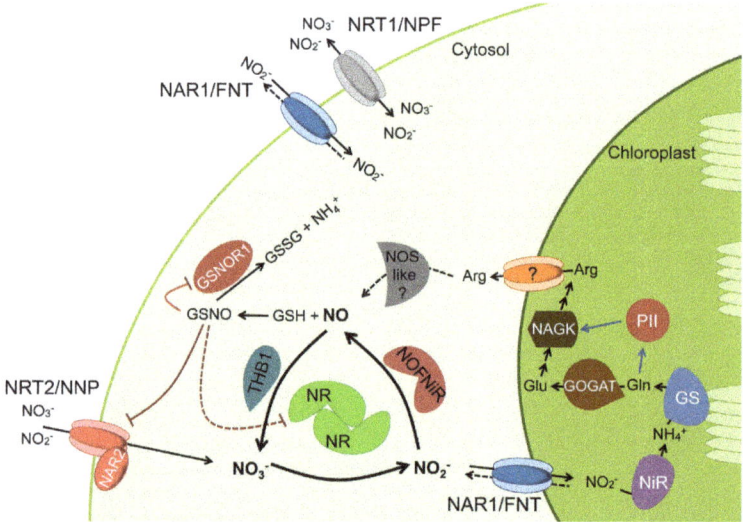

Figure 1. The schematic model for the coordinated regulation of Nitric Oxide (NO) synthesis and N metabolism. Blue arrows indicate activation and red lines indicate inhibition by trans-nitrosylation. Dashed lines represent hypothetical steps. The NOS-like component represents the L-Arg-dependent NOS activity reported in different plant species.

Nitrite in the cytosol, either produced from nitrate or absorbed from the medium, has to be transported to the chloroplast by a HAT system. In Chlamydomonas, HAT of nitrite is facilitated by NAR1, which belongs to the FNT family and is absent in land plants. In higher plants, nitrite transport to chloroplasts is typically mediated by members of the CLC family [26], and it is exported from the chloroplast by a transporter from the NRT1 family, as is found in cucumber, *Cucumis sativus* [27,28]. Regardless, nitrite concentrations in the cytosol are maintained at very low levels (micromolar range) [29] to prevent nitrite toxicity in the cell [30]. Once in the chloroplast, nitrite is reduced to ammonium in a reaction catalyzed by nitrite reductase (NiR). All of these genes (NRT2/NAR2/NR/NAR1/NiR) in Chlamydomonas are controlled by the master regulatory gene for nitrate assimilation: *NIT2* [31]. Orthologous regulatory genes in land plants—*NLP* genes—show a similar structural organization and signaling for nitrate [32–34]. Finally, ammonium is incorporated into C-skeletons in the form of glutamate by the glutamine synthetase/glutamate synthase (GS/GOGAT) cycle [35] (Figure 1).

It is important to point out here that in contrast to the low cytosolic concentrations of nitrite, those of nitrate are high (1–6 mM). Cytosolic nitrate levels are also more stably maintained than vacuolar concentrations (5–75 mM), considering that external nitrate concentrations may change by about 10,000-fold [36–38]. This is important for ensuring efficient nitrate assimilation, together with proper nitrate signaling in the tissues [38,39]. Nitrate homeostasis is the result of the membrane transporter-mediated supply of nitrate from vacuoles and the outer medium, as well as by nitrate efflux transporters such as NAXT1 [40]. NAXT1 belongs to the NRT1 family. In addition, the NR-catalyzed reaction has an effect, facilitating the conversion of nitrate into nitrite [37,38,41]. In the yeast *Hansenula polymorpha*, the sulfite transporter SSU2 and the nitrite transporter NAR1 have been characterized as essential components of the nitrate/nitrite efflux system [42].

3. Nitrate Reductase Is a Multidomain Protein

NR reduces nitrate to nitrite using electrons from NAD(P)H. The plant enzyme is about 200 KDa and contains two subunits, each bearing three prosthetic groups: FAD, heme b_{557}, and molybdenum. In an NR subunit, molybdenum is bound to a tricyclic pyranopterin and chelated by a dithiolene, which is named the molybdenum cofactor (Moco). These domains are joined by two protease-sensitive hinge regions. The domains are redox centers, and electrons flow from NAD(P)H→FAD→ heme →Moco, which is within the active site for nitrate reduction [43,44]. Interestingly, the enzyme has two partial activities, which can be assayed in vitro: diaphorase, which catalyzes the reduction of artificial acceptors (ferricyanide or cytochrome c with NAD(P)H), and terminal-NR, which catalyzes nitrate reduction using electrons supplied by FAD, FMN, viologens, or bromophenol blue, chemically reduced by dithionite [43,44]. The crystal structure of the dimerized form of NR and Moco domains was solved [45].

4. Does NR Catalyze Nitrite Reduction to NO?

The experimental data have led to the proposal that the molybdoenzyme NR is the main enzyme responsible for NO production in most plants. This proposal was based on the experimental findings described below.

The first evidence linking NR and NO production was their co-elution by NADH from Blue Sepharose columns loaded with Soybean extracts with both NR and NO(X) evolution activities. In their main conclusion, the authors inferred their linkage from the fact that inhibiting the partial activities of NR and NO(X) evolution activities led to the same pattern. However, they went further by indicating that the terminal molybdenum-containing portion of NR is involved in the reduction of nitrite to NO(X) [46]. Other studies showed that several plant species emitted in vivo NO when there was nitrate in the soil, and the function was abolished in all plants in the study when they were grown on ammonium-containing soil, indicating a role for NR [47]. In addition, isotopically labeled ^{15}N-nitrate resulted in the emission of ^{15}NO [48]. Moreover, using NR from corn led to the production of significant amounts of NO from both nitrite and nitrate [49]. Those findings reinforced the idea that NR reduces nitrate to nitrite and further converts de novo-generated nitrite into NO. The proposal would be feasible if it could account for several facts. First, the nitrite-reducing activity of NR is very low (only 1% of the nitrate-reducing activity), and the Km of nitrite is about 10 times higher than that of nitrate, in agreement with the competitive character of nitrite (Ki = 50 µM) in nitrate reduction [50,51].

Taking into consideration the intracellular concentrations of nitrite (in the micromolar range) and nitrate (in the millimolar range), this nitrite-reducing activity would be irrelevant.

Another piece of evidence indicating the involvement of NR in NO production originated from the use of tungstate—an NR inhibitor—and NR-deficient mutants. In addition to its uncontrolled effects, such as metal toxicity, tungstate is a very unspecific inhibitor of Mo-enzymes. When exchanging Moco for the inactive tungsten cofactor (Wco) [52], all molybdoenzymes are inactivated, including mitochondrial amidoxime-reducing component (mARC), which is presented below as the most important enzyme in NO production under phototrophic conditions. Thus, tungstate is a clear inhibitor of NO production because of the resulting mARC inhibition and not because of NR inhibition. In agreement with this, Moco-deficient mutants are deficient in molybdoenzymes, as described in barley with a thermo-sensitive, wilty phenotype [53]. The most used Arabidopsis NR mutant to support the involvement of NR in NO production is the double mutant deficient in *NIA1* and *NIA2* genes. This nia1/nia2 double mutant has only 0.5% of the activity of wild-type NR and grows very poorly on medium with nitrate as the only nitrogen source [54]. As described below, NR is in fact involved in NO production but not through its Moco-dependent activity.

All five molybdoenzymes in plants (nitrate reductase, xanthine oxidase reductase (XOR), aldehyde oxidase (AO), sulfite oxidase (SO), and mARC) are able to catalyze the one-electron reduction of nitrite to NO. The molybdoenzymes are classified depending on how Moco binds to the enzyme's active site: either covalently through an enzyme cysteine thiol group (NR, SO, and mARC) or with inorganic sulfur (XOR and AO) [55,56]. All of these enzymes show nitrite reductase activity to produce NO in vitro and in anaerobic conditions [44]. The four known human molybdenum-containing enzymes are the same as those in plants, except for NR, and they can also function as nitrite reductases under hypoxic conditions [56]. In mammals, two known pathways for NO formation are known: arginine oxidation under normoxic or aerobic conditions and nitrite reduction during hypoxia or anaerobiosis [56]. Plant SO seems to have a less potent nitrite reductase activity than human XOR and AO [57]. Whereas plant AO participates in the synthesis of phytohormones and contributes to reactive oxygen species (ROS) production, there is no information about its in vivo NO-producing activity [5].

Other processes that might produce NO from nitrite are associated with the plasma membrane-bound NR, which appears to be related to the mycorrhizal colonization of tobacco roots [58] and the mitochondrial electron transport chain (mETC), as demonstrated in several plants [59–62]. The mitochondrial complex III and IV are primarily implicated in the nitrite reaction (Km of 175 µM), which requires anaerobic conditions since oxygen is a strong inhibitor. Therefore, this reaction can occur in plant tissues exposed to hypoxia, such as roots, and its occurrence might be important to the plant by protecting the respiratory chain and mitochondrial metabolism when oxygen is lacking [63]. Recently, a role for Alternative Oxidase (AOX) in the production of large amounts of NO, observed under hypoxia, has been shown. In this pathway of NO production, AOX has a role in scavenging the NO and ROS linked to the hemoglobin–NO cycle, thus increasing energy efficiency without contributing to the formation of toxic peroxynitrite [64]. The implication of mitochondria in NO production from nitrite has also been shown in Chlamydomonas in the presence of high concentrations of nitrite [65].

In light of the biological importance of NO production, one would expect this process to be efficiently and finely regulated. Some of the Moco-dependent enzymes described above can mediate NO production under certain conditions, though it is difficult to currently envisage how to control this diversity of processes; even mitochondrial NO production requires anaerobiosis, suggesting a burst of NO synthesis from nitrite when this condition appears.

5. NR Does Not Catalyze In Vivo Nitrite Reduction to NO but Provides the Needed Electrons

Recently, NR was shown to be a necessary partner for NO production in a dual system, which, besides NR, includes another molybdoenzyme, mARC, since renamed NO-forming nitrite reductase (NOFNiR).

mARC has been extensively characterized in prokaryotic and eukaryotic organisms [66–69]. mARC proteins are about 35 KDa and require two electron transport proteins—NADH-cytochrome b5 reductase and cytochrome b5—to be functional. mARC is a very efficient reductase [70] for a number of N-oxygenated compounds, some of which are toxic or mutagenic [71–73]. This is why mARC has been related to cell detoxification processes. Both human mARC isoforms are associated with mitochondria, but mARC could also be located in peroxisomes because its two partners, NADH-cytochrome b5 reductase and cytochrome b5, were found in this organelle using a proteomic approach [74]. mARC was also connected to NO metabolism because of its activity on Nω-hydroxy-Nδ-methyl-L-arginine, an intermediate in the L-arginine-dependent biosynthesis of NO using NADH-cytochrome b5 reductase and cytochrome b5 [75]. In addition, human mARCs have nitric oxide synthase activity from nitrite with NADH and its two above-indicated partners [76].

In the eukaryotic alga Chlamydomonas, NO synthesis is carried out by a dual system comprising NR and NOFNiR (mARC). These two components are closely connected at both the transcriptional and activity levels, so mutants lacking one of them overexpress the other [77]. NR supplies NADH electrons to NOFNiR for the reduction of nitrite to NO much more efficiently than NADH-cytochrome b5 reductase and cytochrome b5. Both NR and NOFNiR are located in the cytosol [77].

NR mutants were found to be unable to provide electrons both in vitro and in vivo for NO synthesis. Some exceptions are NR mutants such as Chlamydomonas strain 301, whose NR lacks nitrate-reducing activity since its Moco domain is affected; however, it has functional diaphorase activity with intact FAD and heme domains [77]. So, the dual complex NR:NOFNiR produces NO independently of the molybdenum center of NR and depends on the NR electron transport chain from NAD(P)H to heme.

NR:NOFNiR has been proposed to be the main system producing NO during standard phototrophic, oxygenic plant growth, which corresponds to most of its aerial part [77]. In tissues exposed to hypoxia, such as roots, other molybdoenzymes or mETCs could be involved in the synthesis of NO, as discussed above.

6. NO Levels in the Cells Are Regulated

The first line of control of NO levels is the regulation of its synthesis and degradation within the NO cycle (Figure 1). Members of the hemoglobin (HB) superfamily can oxygenate NO to nitrate, as was shown for Chlamydomonas THB1 [78], a class 3 truncated hemoglobin (THB) [79]. In different plant species, nitrate, nitrite, and NO upregulate HB expression [80,81]. In maize roots, the coordinated expression of both NR and HB also occurs [82]. Similarly, in Chlamydomonas, the expression of two truncated HBs, THB1 and THB2, respond selectively to N signals (nitrate, nitrite, and NO) and, interestingly, also to NIT2, the major regulatory gene of the nitrate assimilation pathway [78,83]. This regulatory gene is also essential for NR upregulation by nitrate [19]. The activity of THB1 requires electrons to be supplied by the NADH-diaphorase of NR, and the electron flow is likely from NADH to FAD [78]. Similarly, the activity of NOFNiR also requires the NADH-diaphorase of NR, but the electron flow is now from NADH to heme b [77]. Thus, nitrate through NIT2 would stimulate NO production because of NR's increased expression (NOFNiR is not under NIT2 control), and in turn, NO degradation would occur as a result of stimulating both THB1 and NR. The homeostasis of NO is controlled by the activities of NR, NOFNiR, THB1, and THB2, which, in turn, depend on the relative concentrations of nitrate, nitrite, and NO, as well as NIT2.

So, just-synthesized NO, which is highly reactive, can react with different targets. Glutathione (γ-glutamylcysteinylglycine, GSH) is an essential metabolite in plants that participates in important functions, such as primary metabolism, redox signaling, and defense and detoxification processes [84].

GSH can react with NO to produce S-nitrosylated glutathione (GSNO). As a result, the half-life of NO in tissues available as a free radical gas changes from seconds to a few minutes. Thus, there are very sensitive mechanisms for regulating cellular processes. GSNO, which is considered the main reservoir for NO, provides NO signals to proteins. Protein S-nitrosation is considered the most important mode of action of NO. This is the covalent binding of NO to the thiol group of protein cysteine residues, and the creation of these bonds modifies the protein and can alter gene expression and/or lead to metabolic changes, all of which ultimately translate into physiological responses. Thousands of nitrosation sites have been identified in proteins [85]. GSNO is metabolized with GSNO reductase (GSNOR1) to convert it to glutathione disulfide (GSSG) and ammonia (NH3) [86,87]. GSNOR1 is a cytosolic enzyme that controls GSNO levels and, in this way, the nitrosation of proteins. GSNOR1 seems to be inhibited by NO, in which case, the scavenging of GSNO is prevented. Thus, NO controls its production and scavenging by regulating both nitrate assimilation efficiency and GSNOR1 activity [87,88] (Figure 1).

Other posttranslational modifications of proteins mediated by NO are the nitration of tyrosine and amine groups and the oxidation of thiols and tyrosine [89]. NO can react with superoxide (O_2^-) to form peroxinitrite ($ONOO^-$), which is a powerful oxidant contributing to the nitration of protein tyrosine residues to form 3-nitrotyrosine, which results in regulatory responses. Nitration seems to be a reversible process that might occur at specific tyrosine residues depending on the local environment and the secondary and tertiary structure of the protein. A putative specific denitrase removes the nitro group without degrading the protein, as has been shown in animal systems [89,90].

Reactive oxygen species (ROS), such as oxygen, singlet oxygen, hydroxyl radical, hydrogen peroxide, and superoxide anion, all of which are important signaling compounds produced under several environmental conditions, interact with NO and other reactive nitrogen species (RNS). They mediate the responses to different environmental situations, even promoting the systemic adaptation of plants to stress situations [2,91,92].

7. The NO Synthesis Systems Are Coordinated with Nitrogen Metabolism

Under circumstances in which NR is not present or active, another NO-forming system accounts for NO production and NO effects. Analysis of this topic and possible future directions are presented here.

With an ammonium medium as the sole N-source or in null NR-deficient mutants, NR is absent, and thus, the dual system NR–NOFNiR is not functional. Nevertheless, NO is being synthesized, probably due to the existence of a NOS-like activity in algae and plants. This activity can be inhibited by some compounds, which are primarily arginine analogs acting on the animal-type NOS [6,17].

Plant peroxisomes are single-membrane-bound organelles with an oxidative metabolism and a simple morphology, but they also have a complex composition of enzymes involved in the metabolism of oxygen free-radicals. Peroxisomes can generate ROS and nitric oxide and thus important signal molecules with implications for cellular metabolism in plants [91,93].

Many studies of different plant species have shown the presence of L-Arg-dependent NO synthase-like enzyme activity, which has biochemical requirements similar to animal NOS (L-Arg, NADPH, FMN, FAD, Calmodulin, and Ca^{2+}) [94,95]. Corpas and Barroso (2017) postulated that the L-Arg-dependent NO synthesis that occurs in plants could correspond to cooperation among discrete proteins, resulting in the formation of a protein complex with requirements for enzyme activity that are similar to animal NOS [96]. This would explain the lack of success in finding canonical NOS proteins at the molecular level.

On the other hand, coordination between nitrogen assimilation and the nutritional status within plant metabolism is a critical issue for plant viability. A very abundant family of N- and C-signaling proteins, widely distributed in nature, is the PII protein family from bacteria, archaea, and plants [97]. In plants, these nuclear-encoded PII proteins localize in the chloroplast and are not subject to the covalent modification reported for bacterial PII [98]. PII senses and integrates the signals of the C–N balance in the cells using 2-ketoglutarate as an indicator, together with the energy status by competitive ADP/ATP binding [99,100]. Plant PII proteins are conserved throughout the evolutionary history of the Chloroplastida—green algae and land plants [97]—and participate in a complex signal-transduction network that mediates nitrogen regulation [101]. In Arabidopsis, the PII protein controls arginine biosynthesis [102], and PII mutants show an enhanced nitrite uptake by the chloroplast [103]. By binding effector molecules, PII interacts with and signals to other chloroplast proteins, such as N-acetyl-L-glutamate kinase (NAGK), which catalyzes the committed step in the arginine biosynthesis pathway [102] (Figure 1). Among the effectors binding plant PII protein is glutamine, which was shown to bind to the termed Q-loop of the C-terminus crystalized plant PII protein [104]. The binding of glutamine to PII changes its conformation to one that interacts and activates NAGK for the synthesis of arginine and polyamines. PII–glutamine sensing is a widespread mechanism in the plant kingdom [104]. In Chlamydomonas, arginine is a component of the same repressive pathway as ammonium and CYG56, which is a guanylate cyclase activated by NO and needed to repress nitrogen assimilation genes [105].

Interestingly, plant PII seems to be connected to the regulation of nitrite metabolism, so PII knockout mutants show an increased sensitivity to nitrite and a decrease in total amino acids, especially glutamine. Lack of PII seems to increase the C/N balance, as previously shown in cyanobacteria [30]. Recently, Chlamydomonas PII levels have been shown to be tightly controlled by the nitrogen source and the physiological status of the cells [106]. In fact, PII expression is subject to positive (nitrate and nitrite) signaling and is downregulated by ammonium via an NO-mediated process that involves an NO-dependent guanylate cyclase, similar to the negative effect of ammonium on NR expression [6]. PII expression is very similar to that of nitrate assimilation genes [25], so an interaction/coordination between PII and nitrate assimilation pathways has been suggested [106].

Under standard phototropic conditions, nitrate assimilation takes place under the positive control of nitrate, mediated by the regulatory gene *NIT2* [25]. When both nitrate and ammonium are present, there is a balance between the positive and negative signals, and NR expression follows that balance; the NR transcript is detectable even in the presence of ammonium, provided that nitrate is also present [107]. The negative signal of ammonium on Chlamydomonas NIA1 gene expression depends on NO and its mediation by an NO-dependent guanylate cyclase (CYG56). So, NO concentrations in the cells increase with ammonium concentration, leading to complete NR repression [107] by means of a mechanism that is dependent on a possible NOS, which is inhibited by L-NAME.

In different plants, NO production is also sensitive to the mammalian NO synthase inhibitor L-NAME [2,5]. L-NAME was also shown to affect NO production by interfering with NR activity [108], which seems to question the mechanisms of action of this compound. However, it has to be considered that in the double nia1/nia2 Arabidopsis mutant commonly used to study NO effects [54], the content of nitrite, as expected from the NR deficiency, and of free amino acids, particularly L-arginine, are much lower than in wild-type plants [109]. Thus, both substrates for NO synthesis would be compromised.

In conclusion, PII proteins mediate the signaling of the N-source with respect to the carbon status (2-ketoglutarate). This N can be either oxidized (nitrate, nitrite) with a generally high C/N balance, or it can be reduced (ammonium) with a generally low C/N balance. With a high C/N balance, the nitrate assimilation pathway is operative, and NR expression would favor the increase in N capture and its incorporation into C-skeletons (2-ketoglutarate), which would be abundant. When the C/N balance is low, NR would be repressed, and the biosynthesis of arginine would be stimulated. So, under these two extreme conditions, the substrates to produce NO would change. In conditions of high C/N, nitrite would be efficiently produced; however, with low C/N, arginine biosynthesis would predominate with low nitrite production. So, it is proposed that the dual NR:NOFNiR system will preponderate at high C/N, whereas the NOS-like system will be mostly operative at low C/N. PII protein expression will follow a pattern similar to that of NR [110] to enhance N acquisition and balance the C/N ratio. These two extreme situations might be changed to intermediate ones, depending on the C/N balance of the cells.

8. Conclusions

Nitric oxide is such an important signaling molecule that its production and scavenging must be tightly regulated. Some of the biosynthetic mechanisms are starting to be disentangled, while others still require additional useful information for their elucidation. Part of the confusion regarding the primary source of NO might come from the fact that several pathways might function simultaneously to different extents, depending on nutritional and environmental conditions. Many points still have to be clarified for plants.

Author Contributions: Conceptualization, E.F.; writing-original draft preparation, E.F.; writing-review and editing M.T.-J., A.L., A.G. and E.F.; funding acquisition, E.F.

Funding: This work was funded by MINECO (Grant BFU2015-70649-P), the European FEDER program, Junta de Andalucía (BIO-502), the Plan Propio de la Universidad de Córdoba, and the U.E.INTERREG VA POCTEP-055_ALGARED_PLUS5_E.

Acknowledgments: We thank María Isabel Macías and Aitor Gómez for technical assistance.

Conflicts of Interest: The authors declare no conflict of interest.

References

1. Wendehenne, D.; Hancock, J.T. New frontiers in nitric oxide biology in plant. *Plant Sci.* **2011**, *181*, 507–508. [CrossRef] [PubMed]
2. Corpas, F.J.; Leterrier, M.; Valderrama, R.; Airaki, M.; Chaki, M.; Palma, J.M.; Barroso, J.B. Nitric oxide imbalance provokes a nitrosative response in plants under abiotic stress. *Plant Sci.* **2011**, *181*, 604–611. [CrossRef] [PubMed]
3. Mur, L.A.; Prats, E.; Pierre, S.; Hall, M.A.; Hebelstrup, K.H. Integrating nitric oxide into salicylic acid and jasmonic acid/ethylene plant defense pathways. *Front. Plant Sci.* **2013**, *4*, 215. [CrossRef] [PubMed]
4. Santolini, J.; Andre, F.; Jeandroz, S.; Wendehenne, D. Nitric oxide synthase in plants: Where do we stand? *Nitric Oxide* **2017**, *63*, 30–38. [CrossRef] [PubMed]
5. Astier, J.; Gross, I.; Durner, J. Nitric oxide production in plants: An update. *J. Exp. Bot.* **2018**, *69*, 3401–3411. [CrossRef] [PubMed]
6. De Montaigu, A.; Sanz-Luque, E.; Galvan, A.; Fernandez, E. A soluble guanylate cyclase mediates negative signaling by ammonium on expression of nitrate reductase in Chlamydomonas. *Plant Cell* **2010**, *22*, 1532–1548. [CrossRef] [PubMed]
7. Wei, L.; Derrien, B.; Gautier, A.; Houille-Vernes, L.; Boulouis, A.; Saint-Marcoux, D.; Malnoe, A.; Rappaport, F.; de Vitry, C.; Vallon, O.; et al. Nitric oxide-triggered remodeling of chloroplast bioenergetics and thylakoid proteins upon nitrogen starvation in *Chlamydomonas reinhardtii*. *Plant Cell* **2014**, *26*, 353–372. [CrossRef] [PubMed]
8. Yordanova, Z.P.; Iakimova, E.T.; Cristescu, S.M.; Harren, F.J.; Kapchina-Toteva, V.M.; Woltering, E.J. Involvement of ethylene and nitric oxide in cell death in mastoparan-treated unicellular alga *Chlamydomonas reinhardtii*. *Cell Biol. Int.* **2010**, *34*, 301–308. [PubMed]
9. Hemschemeier, A.; Duner, M.; Casero, D.; Merchant, S.S.; Winkler, M.; Happe, T. Hypoxic survival requires a 2-on-2 hemoglobin in a process involving nitric oxide. *Proc. Natl. Acad. Sci. USA* **2013**, *110*, 10854–10859. [CrossRef] [PubMed]
10. Chen, X.; Tian, D.; Kong, X.; Chen, Q.; Ef, A.A.; Hu, X.; Jia, A. The role of nitric oxide signalling in response to salt stress in *Chlamydomonas reinhardtii*. *Planta* **2016**, *244*, 651–669. [CrossRef] [PubMed]
11. De Mia, M.; Lemaire, S.D.; Choquet, Y.; Wollman, F.A. Nitric oxide remodels the photosynthetic apparatus upon S-starvation in *Chlamydomonas reinhardtii*. *Plant Physiol.* **2019**, *179*, 718–731. [CrossRef] [PubMed]
12. Delledonne, M.; Xia, Y.; Dixon, R.A.; Lamb, C. Nitric oxide functions as a signal in plant disease resistance. *Nature* **1998**, *394*, 585–588. [PubMed]
13. Durner, J.; Wendehenne, D.; Klessig, D.F. Defense gene induction in tobacco by nitric oxide, cyclic GMP, and cyclic ADP-ribose. *Proc. Natl. Acad. Sci. USA* **1998**, *95*, 10328–10333. [CrossRef] [PubMed]
14. Mur, L.A.; Mandon, J.; Persijn, S.; Cristescu, S.M.; Moshkov, I.E.; Novikova, G.V.; Hall, M.A.; Harren, F.J.; Hebelstrup, K.H.; Gupta, K.J. Nitric oxide in plants: An assessment of the current state of knowledge. *AoB Plants* **2013**, *5*, pls052. [CrossRef] [PubMed]
15. Maia, L.B.; Moura, J.J.G. Putting xanthine oxidoreductase and aldehyde oxidase on the NO metabolism map: Nitrite reduction by molybdoenzymes. *Redox Biol.* **2018**, *19*, 274–289. [CrossRef] [PubMed]

16. Foresi, N.; Correa-Aragunde, N.; Parisi, G.; Calo, G.; Salerno, G.; Lamattina, L. Characterization of a nitric oxide synthase from the plant kingdom: NO generation from the green alga *Ostreococcus tauri* is light irradiance and growth phase dependent. *Plant Cell* **2010**, *22*, 3816–3830. [CrossRef] [PubMed]
17. Astier, J.; Jeandroz, S.; Wendehenne, D. Nitric oxide synthase in plants: The surprise from algae. *Plant Sci.* **2018**, *268*, 64–66. [CrossRef] [PubMed]
18. Jeandroz, S.; Wipf, D.; Stuehr, D.J.; Lamattina, L.; Melkonian, M.; Tian, Z.; Zhu, Y.; Carpenter, E.J.; Wong, G.K.; Wendehenne, D. Occurrence, structure, and evolution of nitric oxide synthase-like proteins in the plant kingdom. *Sci. Signal* **2016**, *9*, re2. [CrossRef] [PubMed]
19. Fernandez, E.; Galvan, A. Nitrate assimilation in Chlamydomonas. *Eukaryot. Cell* **2008**, *7*, 555–559. [PubMed]
20. Crawford, N.M. Nitrate: Nutrient and signal for plant growth. *Plant Cell* **1995**, *7*, 859–868. [PubMed]
21. Dechorgnat, J.; Nguyen, C.T.; Armengaud, P.; Jossier, M.; Diatloff, E.; Filleur, S.; Daniel-Vedele, F. From the soil to the seeds: The long journey of nitrate in plants. *J. Exp. Bot.* **2011**, *62*, 1349–1359. [CrossRef] [PubMed]
22. Wang, Y.Y.; Tsay, Y.F. Arabidopsis nitrate transporter NRT1.9 is important in phloem nitrate transport. *Plant Cell* **2011**, *23*, 1945–1957. [CrossRef] [PubMed]
23. Krapp, A.; David, L.C.; Chardin, C.; Girin, T.; Marmagne, A.; Leprince, A.S.; Chaillou, S.; Ferrario-Mery, S.; Meyer, C.; Daniel-Vedele, F. Nitrate transport and signalling in Arabidopsis. *J. Exp. Bot.* **2014**, *65*, 789–798. [CrossRef] [PubMed]
24. O'brien, J.A.; Vega, A.; Onore Bouguyon, E.; Krouk, G.; Gojon, A.; Coruzzi, G.; Gutié Rrez, R.A. Nitrate Transport, Sensing, and Responses in Plants. *Mol. Plant* **2016**, *9*, 837–856. [CrossRef] [PubMed]
25. Calatrava, V.; Chamizo-Ampudia, A.; Sanz-Luque, E.; Ocana-Calahorro, F.; Llamas, A.; Fernandez, E.; Galvan, A. How Chlamydomonas handles nitrate and the nitric oxide cycle. *J. Exp. Bot.* **2017**, *68*, 2593–2602. [CrossRef] [PubMed]
26. Monachello, D.; Allot, M.; Oliva, S.; Krapp, A.; Daniel-Vedele, F.; Barbier-Brygoo, H.; Ephritikhine, G. Two anion transporters AtClCa and AtClCe fulfil interconnecting but not redundant roles in nitrate assimilation pathways. *New Phytol.* **2009**, *183*, 88–94. [CrossRef] [PubMed]
27. Sugiura, M.; Georgescu, M.N.; Takahashi, M. A nitrite transporter associated with nitrite uptake by higher plant chloroplasts. *Plant Cell Physiol.* **2007**, *48*, 1022–1035. [CrossRef] [PubMed]
28. Tsay, Y.-F.; Chiu, C.-C.; Tsai, C.-B.; Ho, C.-H.; Hsu, P.-K. Nitrate transporters and peptide transporters. *FEBS Lett.* **2007**, *581*, 2290–2300. [CrossRef] [PubMed]
29. Kawamura, Y.; Takahashi, M.; Arimura, G.; Isayama, T.; Irifune, K.; Goshima, N.; Morikawa, H. Determination of levels of nitrate, nitrite and ammonium ions in leaves of ions in leaves of various plants by capillary electrophoresis. *Plant Cell Physiol.* **1996**, *37*, 878–880. [CrossRef]
30. Ferrario-Mery, S.; Bouvet, M.; Leleu, O.; Savino, G.; Hodges, M.; Meyer, C. Physiological characterisation of Arabidopsis mutants affected in the expression of the putative regulatory protein PII. *Planta* **2005**, *223*, 28–39. [CrossRef] [PubMed]
31. Camargo, A.; Llamas, A.; Schnell, R.A.; Higuera, J.J.; Gonzalez-Ballester, D.; Lefebvre, P.A.; Fernandez, E.; Galvan, A. Nitrate signaling by the regulatory gene NIT2 in Chlamydomonas. *Plant Cell* **2007**, *19*, 3491–3503. [CrossRef] [PubMed]
32. Castaings, L.; Camargo, A.; Pocholle, D.; Gaudon, V.; Texier, Y.; Boutet-Mercey, S.; Taconnat, L.; Renou, J.P.; Daniel-Vedele, F.; Fernandez, E.; et al. The nodule inception-like protein 7 modulates nitrate sensing and metabolism in Arabidopsis. *Plant J.* **2009**, *57*, 426–435. [CrossRef] [PubMed]
33. Konishi, M.; Yanagisawa, S. Arabidopsis NIN-like transcription factors have a central role in nitrate signalling. *Nat. Commun.* **2013**, *4*, 1617. [PubMed]
34. Marchive, C.; Roudier, F.; Castaings, L.; Bréhaut, V.; Blondet, E.; Colot, V.; Meyer, C.; Krapp, A. Nuclear retention of the transcription factor NLP7 orchestrates the early response to nitrate in plants. *Nat. Commun.* **2013**, *4*, 1713. [CrossRef] [PubMed]
35. Forde, B.G.; Lea, P.J. Glutamate in plants: Metabolism, regulation, and signalling. *J. Exp. Bot.* **2007**, *58*, 2339–2358. [CrossRef] [PubMed]
36. Van der leij, M.; Smith, S.; Miller, J. Remobilisation of vacuolar stored nitrate in barley root cells. *Planta* **1998**, *205*, 64–72. [CrossRef]
37. Cookson, S.J.; Williams, L.E.; Miller, A.J. Light-dark changes in cytosolic nitrate pools depend on nitrate reductase activity in Arabidopsis leaf cells. *Plant Physiol.* **2005**, *138*, 1097–1105. [PubMed]

38. Miller, A.J.; Smith, S.J. Cytosolic nitrate ion homeostasis: Could it have a role in sensing nitrogen status? *Ann. Bot.* **2008**, *101*, 485–489. [CrossRef] [PubMed]
39. Miller, A.J.; Fan, X.; Orsel, M.; Smith, S.J.; Wells, D.M. Nitrate transport and signalling. *J. Exp. Bot.* **2007**, *58*, 2297–2306. [PubMed]
40. Segonzac, C.; Boyer, J.-C.; Ipotesi, E.; Szponarski, W.; Tillard, P.; Touraine, B.; Sommerer, N.; Rossignol, M.; Gibrat, R. Nitrate efflux at the root plasma membrane: Identification of an Arabidopsis excretion transporter. *Plant Cell* **2007**, *19*, 3760–3777. [CrossRef] [PubMed]
41. Fan, X.; Gordon-Weeks, R.; Shen, Q.; Miller, A.J. Glutamine transport and feedback regulation of nitrate reductase activity in barley roots leads to changes in cytosolic nitrate pools. *J. Exp. Bot.* **2006**, *57*, 1333–1340. [CrossRef] [PubMed]
42. Cabrera, E.; Gonzalez-Montelongo, R.; Giraldez, T.; Alvarez de la Rosa, D.; Siverio, J.M. Molecular components of nitrate and nitrite efflux in yeast. *Eukaryot. Cell* **2014**, *13*, 267–278. [CrossRef] [PubMed]
43. Campbell, W.H. Nitrate reductase structure, function and regulation: Bridging the Gap between Biochemistry and Physiology. *Annu. Rev. Plant Physiol. Plant Mol. Biol.* **1999**, *50*, 277–303. [CrossRef] [PubMed]
44. Chamizo-Ampudia, A.; Sanz-Luque, E.; Llamas, A.; Galvan, A.; Fernandez, E. Nitrate Reductase Regulates Plant Nitric Oxide Homeostasis. *Trends Plant Sci.* **2017**, *22*, 163–174. [CrossRef] [PubMed]
45. Fischer, K.; Barbier, G.G.; Hecht, H.J.; Mendel, R.R.; Campbell, W.H.; Schwarz, G. Structural basis of eukaryotic nitrate reduction: Crystal structures of the nitrate reductase active site. *Plant Cell* **2005**, *17*, 1167–1179. [CrossRef] [PubMed]
46. Dean, J.V.; Harper, J.E. The Conversion of Nitrite to Nitrogen Oxide(s) by the Constitutive NAD(P)H-Nitrate Reductase Enzyme from Soybean. *Plant Physiol.* **1988**, *88*, 389–395. [CrossRef] [PubMed]
47. Wildt, J.; Kley, D.; Rockel, A.; Rockel, P.; Segscheneider, H.J. Emission of NO from several higher plant species. *J. Geophys. Res.* **1997**, *102*, 5919–5927. [CrossRef]
48. Dean, J.V.; Harper, J.E. Nitric Oxide and Nitrous Oxide Production by Soybean and Winged Bean during the in Vivo Nitrate Reductase Assay. *Plant Physiol.* **1986**, *82*, 718–723. [CrossRef] [PubMed]
49. Yamasaki, H.; Sakihama, Y. Simultaneous production of nitric oxide and peroxynitrite by plant nitrate reductase: In vitro evidence for the NR-dependent formation of active nitrogen species. *FEBS Lett.* **2000**, *468*, 89–92. [CrossRef]
50. Rockel, P.; Strube, F.; Rockel, A.; Wildt, J.; Kaiser, W.M. Regulation of nitric oxide (NO) production by plant nitrate reductase in vivo and in vitro. *J. Exp. Bot.* **2002**, *53*, 103–110. [PubMed]
51. Meyer, C.; Lea, U.S.; Provan, F.; Kaiser, W.M.; Lillo, C. Is nitrate reductase a major player in the plant NO (nitric oxide) game? *Photosynth. Res.* **2005**, *83*, 181–189. [CrossRef] [PubMed]
52. Llamas, A.; Kalakoutskii, K.; Fernandez, E. Molybdenum cofactor amounts in *Chlamydomonas reinhardtii* depend on the Nit5 gene function related to molybdate transport. *Plant Cell Environ.* **2000**, *23*, 1247–1255. [CrossRef]
53. Walker-Simmons, M.; Kudrna, D.A.; Warner, R.L. Reduced Accumulation of ABA during Water Stress in a Molybdenum Cofactor Mutant of Barley. *Plant Physiol.* **1989**, *90*, 728–733. [CrossRef] [PubMed]
54. Wilkinson, J.Q.; Crawford, N.M. Identification and characterization of a chlorate-resistant mutant of Arabidopsis thaliana with mutations in both nitrate reductase structural genes NIA1 and NIA2. *Mol. Gen. Genet.* **1993**, *239*, 289–297. [PubMed]
55. Schwarz, G.; Mendel, R.R.; Ribbe, M.W. Molybdenum cofactors, enzymes and pathways. *Nature* **2009**, *460*, 839–847. [CrossRef] [PubMed]
56. Wang, J.; Keceli, G.; Cao, R.; Su, J.; Mi, Z. Molybdenum-containing nitrite reductases: Spectroscopic characterization and redox mechanism. *Redox Rep.* **2017**, *22*, 17–25. [CrossRef] [PubMed]
57. Wang, J.; Krizowski, S.; Fischer-Schrader, K.; Niks, D.; Tejero, J.; Sparacino-Watkins, C.; Wang, L.; Ragireddy, V.; Frizzell, S.; Kelley, E.E.; et al. Sulfite Oxidase Catalyzes Single-Electron Transfer at Molybdenum Domain to Reduce Nitrite to Nitric Oxide. *Antioxid. Redox Signal* **2015**, *23*, 283–294. [CrossRef] [PubMed]
58. Moche, M.; Stremlau, S.; Hecht, L.; Gobel, C.; Feussner, I.; Stohr, C. Effect of nitrate supply and mycorrhizal inoculation on characteristics of tobacco root plasma membrane vesicles. *Planta* **2010**, *231*, 425–436. [CrossRef] [PubMed]
59. Tischner, R.; Planchet, E.; Kaiser, W.M. Mitochondrial electron transport as a source for nitric oxide in the unicellular green alga Chlorella sorokiniana. *FEBS Lett.* **2004**, *576*, 151–155. [CrossRef] [PubMed]

60. Planchet, E.; Jagadis Gupta, K.; Sonoda, M.; Kaiser, W.M. Nitric oxide emission from tobacco leaves and cell suspensions: Rate limiting factors and evidence for the involvement of mitochondrial electron transport. *Plant J.* **2005**, *41*, 732–743. [CrossRef] [PubMed]
61. Gupta, K.J.; Stoimenova, M.; Kaiser, W.M. In higher plants, only root mitochondria, but not leaf mitochondria reduce nitrite to NO, in vitro and in situ. *J. Exp. Bot.* **2005**, *56*, 2601–2609. [CrossRef] [PubMed]
62. Gupta, K.J.; Kaiser, W.M. Production and scavenging of nitric oxide by barley root mitochondria. *Plant Cell Physiol.* **2010**, *51*, 576–584. [CrossRef] [PubMed]
63. Gupta, K.J.; Igamberdiev, A.U. The anoxic plant mitochondrion as a nitrite: NO reductase. *Mitochondrion* **2011**, *11*, 537–543. [CrossRef] [PubMed]
64. Vishwakarma, A.; Kumari, A.; Mur, L.A.J.; Gupta, K.J. A discrete role for alternative oxidase under hypoxia to increase nitric oxide and drive energy production. *Free Radic. Biol. Med.* **2018**, *122*, 40–51. [CrossRef] [PubMed]
65. Plouviez, M.; Wheeler, D.; Shilton, A.; Packer, M.A.; McLenachan, P.A.; Sanz-Luque, E.; Ocana-Calahorro, F.; Fernandez, E.; Guieysse, B. The biosynthesis of nitrous oxide in the green alga *Chlamydomonas reinhardtii*. *Plant J.* **2017**, *91*, 45–56. [CrossRef] [PubMed]
66. Havemeyer, A.; Bittner, F.; Wollers, S.; Mendel, R.; Kunze, T.; Clement, B. Identification of the missing component in the mitochondrial benzamidoxime prodrug-converting system as a novel molybdenum enzyme. *J. Biol. Chem.* **2006**, *281*, 34796–34802. [CrossRef] [PubMed]
67. Kozmin, S.G.; Leroy, P.; Pavlov, Y.I.; Schaaper, R.M. YcbX and yiiM, two novel determinants for resistance of Escherichia coli to N-hydroxylated base analogues. *Mol. Microbiol.* **2008**, *68*, 51–65. [CrossRef] [PubMed]
68. Chamizo-Ampudia, A.; Galvan, A.; Fernandez, E.; Llamas, A. The *Chlamydomonas reinhardtii* molybdenum cofactor enzyme crARC has a Zn-dependent activity and protein partners similar to those of its human homologue. *Eukaryot. Cell* **2011**, *10*, 1270–1282. [CrossRef] [PubMed]
69. Tejada-Jimenez, M.; Chamizo-Ampudia, A.; Calatrava, V.; Galvan, A.; Fernandez, E.; Llamas, A. From the Eukaryotic Molybdenum Cofactor Biosynthesis to the Moonlighting Enzyme mARC. *Molecules* **2018**, *23*, 3287. [CrossRef] [PubMed]
70. Yang, J.; Giles, L.J.; Ruppelt, C.; Mendel, R.R.; Bittner, F.; Kirk, M.L. Oxyl and hydroxyl radical transfer in mitochondrial amidoxime reducing component-catalyzed nitrite reduction. *J. Am. Chem. Soc.* **2015**, *137*, 5276–5279. [CrossRef] [PubMed]
71. Chamizo-Ampudia, A.; Galvan, A.; Fernandez, E.; Llamas, A. Study of Different Variants of Mo Enzyme crARC and the Interaction with Its Partners crCytb5-R and crCytb5-1. *Int. J. Mol. Sci.* **2017**, *18*, 670. [CrossRef] [PubMed]
72. Llamas, A.; Chamizo-Ampudia, A.; Tejada-Jimenez, M.; Galvan, A.; Fernandez, E. The molybdenum cofactor enzyme mARC: Moonlighting or promiscuous enzyme? *Biofactors* **2017**, *43*, 486–494. [CrossRef] [PubMed]
73. Kubitza, C.; Bittner, F.; Ginsel, C.; Havemeyer, A.; Clement, B.; Scheidig, A.J. Crystal structure of human mARC1 reveals its exceptional position among eukaryotic molybdenum enzymes. *Proc. Natl. Acad. Sci. USA* **2018**, *115*, 11958–11963. [CrossRef] [PubMed]
74. Islinger, M.; Luers, G.H.; Li, K.W.; Loos, M.; Volkl, A. Rat liver peroxisomes after fibrate treatment. A survey using quantitative mass spectrometry. *J. Biol. Chem.* **2007**, *282*, 23055–23069. [CrossRef] [PubMed]
75. Kotthaus, J.; Wahl, B.; Havemeyer, A.; Kotthaus, J.; Schade, D.; Garbe-Schonberg, D.; Mendel, R.; Bittner, F.; Clement, B. Reduction of N(omega)-hydroxy-L-arginine by the mitochondrial amidoxime reducing component (mARC). *Biochem. J.* **2011**, *433*, 383–391. [CrossRef] [PubMed]
76. Sparacino-Watkins, C.E.; Tejero, J.; Sun, B.; Gauthier, M.C.; Thomas, J.; Ragireddy, V.; Merchant, B.A.; Wang, J.; Azarov, I.; Basu, P.; et al. Nitrite reductase and nitric-oxide synthase activity of the mitochondrial molybdopterin enzymes mARC1 and mARC2. *J. Biol. Chem.* **2014**, *289*, 10345–10358. [CrossRef] [PubMed]
77. Chamizo-Ampudia, A.; Sanz-Luque, E.; Llamas, A.; Ocana-Calahorro, F.; Mariscal, V.; Carreras, A.; Barroso, J.B.; Galvan, A.; Fernandez, E. A dual system formed by the ARC and NR molybdoenzymes mediates nitrite-dependent NO production in Chlamydomonas. *Plant Cell Environ.* **2016**, *39*, 2097–2107. [CrossRef] [PubMed]
78. Sanz-Luque, E.; Ocana-Calahorro, F.; de Montaigu, A.; Chamizo-Ampudia, A.; Llamas, A.; Galvan, A.; Fernandez, E. THB1, a truncated hemoglobin, modulates nitric oxide levels and nitrate reductase activity. *Plant J.* **2015**, *81*, 467–479. [CrossRef] [PubMed]

79. Gupta, K.J.; Hebelstrup, K.H.; Mur, L.A.; Igamberdiev, A.U. Plant hemoglobins: Important players at the crossroads between oxygen and nitric oxide. *FEBS Lett.* **2011**, *585*, 3843–3849. [CrossRef] [PubMed]
80. Ohwaki, Y.; Kawagishi-Kobayashi, M.; Wakasa, K.; Fujihara, S.; Yoneyama, T. Induction of class-1 non-symbiotic hemoglobin genes by nitrate, nitrite and nitric oxide in cultured rice cells. *Plant Cell Physiol.* **2005**, *46*, 324–331. [CrossRef] [PubMed]
81. Hill, R.D. Non-symbiotic haemoglobins-What's happening beyond nitric oxide scavenging? *AoB Plants* **2012**, *2012*, pls004. [CrossRef] [PubMed]
82. Trevisan, S.; Manoli, A.; Begheldo, M.; Nonis, A.; Enna, M.; Vaccaro, S.; Caporale, G.; Ruperti, B.; Quaggiotti, S. Transcriptome analysis reveals coordinated spatiotemporal regulation of hemoglobin and nitrate reductase in response to nitrate in maize roots. *New Phytol.* **2011**, *192*, 338–352. [CrossRef] [PubMed]
83. Sanz-Luque, E.; Ocana-Calahorro, F.; Galvan, A.; Fernandez, E. THB1 regulates nitrate reductase activity and THB1 and THB2 transcription differentially respond to NO and the nitrate/ammonium balance in Chlamydomonas. *Plant Signal Behav.* **2015**, *10*, e1042638. [CrossRef] [PubMed]
84. Noctor, G.; Mhamdi, A.; Chaouch, S.; Han, Y.; Neukermans, J.; Marquez-Garcia, B.; Queval, G.; Foyer, C.H. Glutathione in plants: An integrated overview. *Plant Cell Environ.* **2012**, *35*, 454–484. [CrossRef] [PubMed]
85. Kovacs, I.; Lindermayr, C. Nitric oxide-based protein modification: Formation and site-specificity of protein S-nitrosylation. *Front. Plant Sci.* **2013**, *4*, 137. [CrossRef] [PubMed]
86. Ortega-Galisteo, A.P.; Rodriguez-Serrano, M.; Pazmino, D.M.; Gupta, D.K.; Sandalio, L.M.; Romero-Puertas, M.C. S-Nitrosylated proteins in pea (*Pisum sativum* L.) leaf peroxisomes: Changes under abiotic stress. *J. Exp. Bot.* **2012**, *63*, 2089–2103. [CrossRef] [PubMed]
87. Frungillo, L.; Skelly, M.J.; Loake, G.J.; Spoel, S.H.; Salgado, I. S-nitrosothiols regulate nitric oxide production and storage in plants through the nitrogen assimilation pathway. *Nat. Commun.* **2014**, *5*, 5401. [CrossRef] [PubMed]
88. Sanz-Luque, E.; Ocana-Calahorro, F.; Llamas, A.; Galvan, A.; Fernandez, E. Nitric oxide controls nitrate and ammonium assimilation in *Chlamydomonas reinhardtii*. *J. Exp. Bot.* **2013**, *64*, 3373–3383. [CrossRef] [PubMed]
89. Gow, A.J.; Farkouh, C.R.; Munson, D.A.; Posencheg, M.A.; Ischiropoulos, H. Biological significance of nitric oxide-mediated protein modifications. *Am. J. Physiol. Lung Cell. Mol. Physiol.* **2004**, *287*, L262–L268. [CrossRef] [PubMed]
90. Deeb, R.S.; Nuriel, T.; Cheung, C.; Summers, B.; Lamon, B.D.; Gross, S.S.; Hajjar, D.P. Characterization of a cellular denitrase activity that reverses nitration of cyclooxygenase. *Am. J. Physiol. Heart Circ. Physiol.* **2013**, *305*, H687–H698. [CrossRef] [PubMed]
91. Del Rio, L.A. ROS and RNS in plant physiology: An overview. *J. Exp. Bot.* **2015**, *66*, 2827–2837. [CrossRef] [PubMed]
92. Farnese, F.S.; Menezes-Silva, P.E.; Gusman, G.S.; Oliveira, J.A. When Bad Guys Become Good Ones: The Key Role of Reactive Oxygen Species and Nitric Oxide in the Plant Responses to Abiotic Stress. *Front. Plant Sci.* **2016**, *7*, 471. [CrossRef] [PubMed]
93. Corpas, F.J.; Barroso, J.B.; del Rio, L.A. Peroxisomes as a source of reactive oxygen species and nitric oxide signal molecules in plant cells. *Trends Plant Sci.* **2001**, *6*, 145–150. [CrossRef]
94. Barroso, J.B.; Corpas, F.J.; Carreras, A.; Sandalio, L.M.; Valderrama, R.; Palma, J.M.; Lupianez, J.A.; del Rio, L.A. Localization of nitric-oxide synthase in plant peroxisomes. *J. Biol. Chem.* **1999**, *274*, 36729–36733. [CrossRef] [PubMed]
95. Corpas, F.J.; Barroso, J.B.; Carreras, A.; Quiros, M.; Leon, A.M.; Romero-Puertas, M.C.; Esteban, F.J.; Valderrama, R.; Palma, J.M.; Sandalio, L.M.; et al. Cellular and subcellular localization of endogenous nitric oxide in young and senescent pea plants. *Plant Physiol.* **2004**, *136*, 2722–2733. [CrossRef] [PubMed]
96. Corpas, F.J.; Barroso, J.B. Nitric oxide synthase-like activity in higher plants. *Nitric Oxide* **2017**, *68*, 5–6. [CrossRef] [PubMed]
97. Chellamuthu, V.R.; Alva, V.; Forchhammer, K. From cyanobacteria to plants: Conservation of PII functions during plastid evolution. *Planta* **2013**, *237*, 451–462. [CrossRef] [PubMed]
98. Ninfa, A.J.; Jiang, P. PII signal transduction proteins: Sensors of alpha-ketoglutarate that regulate nitrogen metabolism. *Curr. Opin. Microbiol.* **2005**, *8*, 168–173. [CrossRef] [PubMed]
99. Huergo, L.F.; Chandra, G.; Merrick, M. P(II) signal transduction proteins: Nitrogen regulation and beyond. *FEMS Microbiol. Rev.* **2013**, *37*, 251–283. [CrossRef] [PubMed]

100. Zeth, K.; Fokina, O.; Forchhammer, K. Structural basis and target-specific modulation of ADP sensing by the Synechococcus elongatus PII signaling protein. *J. Biol. Chem.* **2014**, *289*, 8960–8972. [CrossRef] [PubMed]
101. Nunes-Nesi, A.; Fernie, A.R.; Stitt, M. Metabolic and signaling aspects underpinning the regulation of plant carbon nitrogen interactions. *Mol. Plant* **2010**, *3*, 973–996. [CrossRef] [PubMed]
102. Ferrario-Mery, S.; Besin, E.; Pichon, O.; Meyer, C.; Hodges, M. The regulatory PII protein controls arginine biosynthesis in Arabidopsis. *FEBS Lett.* **2006**, *580*, 2015–2020. [CrossRef] [PubMed]
103. Ferrario-Mery, S.; Meyer, C.; Hodges, M. Chloroplast nitrite uptake is enhanced in Arabidopsis PII mutants. *FEBS Lett.* **2008**, *582*, 1061–1066. [CrossRef] [PubMed]
104. Chellamuthu, V.R.; Ermilova, E.; Lapina, T.; Luddecke, J.; Minaeva, E.; Herrmann, C.; Hartmann, M.D.; Forchhammer, K. A widespread glutamine-sensing mechanism in the plant kingdom. *Cell* **2014**, *159*, 1188–1199. [CrossRef] [PubMed]
105. Gonzalez-Ballester, D.; Sanz-Luque, E.; Galvan, A.; Fernandez, E.; de Montaigu, A. Arginine is a component of the ammonium-CYG56 signalling cascade that represses genes of the nitrogen assimilation pathway in *Chlamydomonas reinhardtii*. *PLoS ONE* **2018**, *13*, e0196167. [CrossRef] [PubMed]
106. Zalutskaya, Z.; Kochemasova, L.; Ermilova, E. Dual positive and negative control of Chlamydomonas PII signal transduction protein expression by nitrate/nitrite and NO via the components of nitric oxide cycle. *BMC Plant Biol.* **2018**, *18*, 305. [CrossRef] [PubMed]
107. De Montaigu, A.; Sanz-Luque, E.; Macias, M.I.; Galvan, A.; Fernandez, E. Transcriptional regulation of CDP1 and CYG56 is required for proper NH4+ sensing in Chlamydomonas. *J. Exp. Bot.* **2011**, *62*, 1425–1437. [CrossRef] [PubMed]
108. Rasul, S.; Wendehenne, D.; Jeandroz, S. Study of oligogalacturonides-triggered nitric oxide (NO) production provokes new questioning about the origin of NO biosynthesis in plants. *Plant Signal Behav.* **2012**, *7*, 1031–1033. [CrossRef] [PubMed]
109. Modolo, L.V.; Augusto, O.; Almeida, I.M.; Pinto-Maglio, C.A.; Oliveira, H.C.; Seligman, K.; Salgado, I. Decreased arginine and nitrite levels in nitrate reductase-deficient Arabidopsis thaliana plants impair nitric oxide synthesis and the hypersensitive response to Pseudomonas syringae. *Plant Sci.* **2006**, *171*, 34–40. [CrossRef]
110. Zalutskaya, Z.; Ostroukhova, M.; Filina, V.; Ermilova, E. Nitric oxide upregulates expression of alternative oxidase 1 in *Chlamydomonas reinhardtii*. *J. Plant Physiol.* **2017**, *219*, 123–127. [CrossRef] [PubMed]

© 2019 by the authors. Licensee MDPI, Basel, Switzerland. This article is an open access article distributed under the terms and conditions of the Creative Commons Attribution (CC BY) license (http://creativecommons.org/licenses/by/4.0/).

Article

Nitric Oxide Overproduction by *cue1* Mutants Differs on Developmental Stages and Growth Conditions

Tamara Lechón, Luis Sanz, Inmaculada Sánchez-Vicente and Oscar Lorenzo *

Department of Botany and Plant Physiology, Instituto Hispano-Luso de Investigaciones Agrarias (CIALE), Facultad de Biología, Universidad de Salamanca, C/Río Duero 12, 37185 Salamanca, Spain; tlg@usal.es (T.L.); lusan@usal.es (L.S.); elfik@usal.es (I.S.-V.)
* Correspondence: oslo@usal.es; Tel.: +34-923294500-5117

Received: 29 September 2020; Accepted: 2 November 2020; Published: 4 November 2020

Abstract: The *cue1* nitric oxide (NO) overproducer mutants are impaired in a plastid phosphoenolpyruvate/phosphate translocator, mainly expressed in Arabidopsis thaliana roots. *cue1* mutants present an increased content of arginine, a precursor of NO in oxidative synthesis processes. However, the pathways of plant NO biosynthesis and signaling have not yet been fully characterized, and the role of CUE1 in these processes is not clear. Here, in an attempt to advance our knowledge regarding NO homeostasis, we performed a deep characterization of the NO production of four different *cue1* alleles (*cue1-1*, *cue1-5*, *cue1-6* and *nox1*) during seed germination, primary root elongation, and salt stress resistance. Furthermore, we analyzed the production of NO in different carbon sources to improve our understanding of the interplay between carbon metabolism and NO homeostasis. After in vivo NO imaging and spectrofluorometric quantification of the endogenous NO levels of *cue1* mutants, we demonstrate that CUE1 does not directly contribute to the rapid NO synthesis during seed imbibition. Although *cue1* mutants do not overproduce NO during germination and early plant development, they are able to accumulate NO after the seedling is completely established. Thus, CUE1 regulates NO homeostasis during post-germinative growth to modulate root development in response to carbon metabolism, as different sugars modify root elongation and meristem organization in *cue1* mutants. Therefore, *cue1* mutants are a useful tool to study the physiological effects of NO in post-germinative growth.

Keywords: nitric oxide homeostasis; *cue1/nox1*; reactive nitrogen species; germination; root development; stress responses; sugar metabolism

1. Introduction

Since the establishment of nitric oxide (NO) as an endogenous signaling molecule in plants over twenty years ago [1–4], a lot of progress has been made towards understanding NO synthesis and signaling in these organisms. In contrast to other eukaryotes, at least seven different sources of NO generation have been characterized in plants [5–8]. While in mammalian cells NO is synthesized from either nitrite or arginine oxidation in a reaction catalyzed by the enzyme NO synthase [9]; in plants, the existence of this last pathway is controversial [5,10]. Plant NO synthesis is controlled by a number of enzymatic synthesis reactions, catalyzed by the enzyme nitrate reductase, by the mitochondrial electron transport chain, or by the enzyme xanthine amine oxidoreductase during anaerobic conditions (reviewed in [5–11]). Moreover, NO can also be produced non-enzymatically, from acid solutions of nitrite in the presence of compounds that can act as antioxidants [12]. However, we only identified some of the components implicated in the different pathways and it is still unknown how they interact, how they fit in the larger context of carbon and nitrogen metabolism, and how much they contribute to the general NO homeostasis of the plant. The concentration of NO is also tightly regulated through the interactions of NO with other compounds, such as reactive oxygen species (ROS), proteins or lipids.

Because of its physicochemical characteristics, NO exerts its functions mainly through modification of these molecules, leading to changes in protein activity, gene expression, and modulation of the redox environment, both during physiological and stress responses (reviewed in [6–8,13]). In an attempt to answer these fundamental questions about when and where NO is produced, researchers have used both pharmacological and genetic approaches.

The use of pharmaceutical NO donors such as sodium nitroprusside (SNP), S-nitroso-N-acetyl-D,L-penicillamine (SNAP), or S-nitrosoglutathione (GSNO), although very extended, does not always replicate the endogenous effects of NO. SNP is in fact a nitrosonium cation donor that also generates cyanide [14]. On the other hand, SNAP and GSNO, in addition to releasing NO, can act mainly through trans-S-nitrosation reactions. Furthermore, the application of NO donors might result in nitrosative stress since there have been very few attempts to understand the kinetics of NO generation *in planta* by NO donors [15–17]. Thus, the use of mutants with altered endogenous NO content seems to be a more suitable way of assessing NO-modulated responses.

In *Arabidopsis thaliana*, only four groups of NO overproducer mutants have been described so far, *gsnor1/hot5*, *glb*, *argah*, and *cue1/nox1* mutants. *gsnor1/hot5* mutants accumulate GSNO, since they are defective in the enzyme S-nitrosoglutathione reductase 1 (GSNOR1) [18], responsible for the degradation of GSNO, a stable NO reservoir, to glutathione disulfide (GSSG) and ammonium [19]. Thus, *gsnor1* mutants accumulate both NO and GSNO [20]. The *glb* mutants are also impaired in a NO scavenging system, since they have reduced levels of non-symbiotic hemoglobins, which usually eliminate NO by binding it to their heme group [21]. The last two groups, *argah* and *cue1*, are both thought to be implicated in the oxidative synthesis of NO from arginine [22,23]. *argah* mutants have decreased arginase activity, which results in an increase in the available arginine pool, since arginases control the catabolism of this amino acid. This, in turn, leads to the accumulation of NO in these mutants [22].

The first *A. thaliana cue1* mutant was isolated in a screening to identify new mutants in light signaling components with an altered light-regulated expression of nuclear genes [24]. *CUE* stands for 'chlorophyll a/b binding protein (*CAB*) underexpressed' because mutations at this locus result in expression defects of photosynthesis genes in mesophyll cells, such as the light-harvesting chlorophyll a/b-binding protein 1 (*LHCB1*) of photosystem II, formerly known as *CAB*. Besides *cue1*, another eight *cue* mutants were isolated [25]. All of them are defective in greening and present an altered mesophyll structure. These defects result from a delayed differentiation of chloroplasts and a reduction in plastid size and granal stack size, along with defective etioplast development [25]. In *cue1*, *LHCB1* is expressed at low levels in the mesophyll cell layers but at wild-type levels in the bundle sheath cells, which causes a striking reticulate leaf phenotype with pale-green mesophyll cells and dark-green veins. At the same time, it entails a severe deficiency in the establishment of photoautotrophic growth because of the lack of sufficient carotenoids and chlorophylls, especially during early leaf development in response to light [24].

Of the nine *CUE* genes, only *CUE1* does not directly take part in the phytochrome-controlled expression of photosynthetic genes [25,26]. Instead, *CUE1* encodes plastid phosphoenolpyruvate (PEP)/phosphate translocator (PPT) expressed mainly in roots, but also in leaves and flowers [27]. PEP is the precursor for the shikimate pathway of aromatic amino acids and can be utilized as an alternative source for ATP in non-photosynthetic plastids. Most plastids either lack or have a very low expression of the complete set of glycolytic enzymes for the conversion of hexose and triose phosphates into PEP, so glycolysis cannot proceed further than 3-phosphoglycerate. PPT is thus the only source of PEP in the chloroplast stroma [28]. In line with this role, *cue1* mutants present an altered content of several amino acids and secondary metabolites. Among them, there is an increase in nitrate, arginine and two products of arginine catabolism, citrulline and urea [27].

Both nitrate and arginine are precursors of NO biosynthesis [29], so it was to be expected that *cue1* mutants also showed increased NO content. Indeed, in screening for NO overproducer (*nox*) mutants, the locus *NOX1* was identified as *CUE1* [23]. The isolation of putative *nox* mutants was based on their hypersensitivity to root growth inhibition by NO donor SNP. *nox1/cue1* was confirmed to have higher levels of NO in rosette leaves when analyzed with NO-sensitive dye 4,5-diaminofluorescein diacetate (DAF-2DA). Since then, *cue1* mutants have been described to have delayed flowering [23], smaller rosettes, leaves and cotyledons [30], increased stomatal development [31], reduced root length and meristem size [32–34], decreased mitosis and increased endoreduplication [32,35], reduced auxin response [32], decreased cytokinin content [36], higher iron uptake [37], better copper tolerance [38], and less pathogen resistance [39–41].

Because *cue1* mutants display a complex pleiotropic phenotype, understanding the role of CUE1/PPT1 in NO synthesis and signaling is not trivial. Here, we sought to establish the growth conditions in which *cue1* mutants overproduce NO and to identify which phenotypes can be unequivocally ascribed to an altered NO homeostasis in order to close the current gap between pharmacological and genetic studies. For this purpose, we analyzed the phenotype of four different *cue1* alleles during physiological processes and stress responses in which NO is known to have an important role: promotion of germination [42–45], inhibition of primary root elongation [23,32,33], and improvement of salt stress resistance [46–48]. This was done alongside the in vivo NO imaging and quantification of the NO content of each mutant by spectrofluorometry using the fluorescent NO-sensing dye 4-amino-5-methylamino-2′,7′-difluorofluorescein diacetate (DAF-FM DA) according to the method described in [49], to find out if there is a link between NO levels and the observed phenotypes. Furthermore, we analyzed the production of NO in different carbon sources to improve our understanding of the interplay between carbon metabolism and NO homeostasis, for which *cue1* mutants are an excellent model, as the primary role of *PPT1* is likely to be related to glycolysis and gluconeogenesis in the plastids.

2. Results

2.1. cue1 Mutants Accumulate Wild-Type NO Levels during Early Post-Germinative Plant Development

A rapid increase in NO levels appears in the endosperm of *A. thaliana* seeds after imbibition [50]. NO then promotes germination both by relieving dormancy [42,51] and by directly promoting embryo growth [45]. In order to study the response of *cue1* mutants during germination, we chose four different knockdown and knockout alleles that have been extensively used in the literature and are known to have increased NO in rosette leaves [23,31,36,39,40] (Figure 1A). Three of the alleles come from a mutagenized Col-0 line, *cue1-5*, *cue1-6*, and *nox1-1*. *cue1-5* is a weak allele with an Arg to Cys point mutation [27], *cue1-6* is a strong allele with a premature stop codon instead of a Trp caused by a point mutation [27], and *nox1-1* is a knockout mutant which was obtained by fast neutron mutagenesis and lacks most of the genomic CUE1/PPT1 sequence [23]. *cue1-1* is another deletion mutant, but it was obtained by gamma radiation mutagenesis of a transgenic line, pOCA108 (Be-0), that is also an alcohol dehydrogenase null mutant [24].

Although these mutants are routinely used as constitutive NO-overproducer mutants, there is no data on their NO levels during early plant development. While the link between *PPT1*, light signaling, and NO production is not understood, it is known that CUE1 strongly regulates the expression of *LHBC* and other photosynthetic genes when plants are initially exposed to light, but *cue1* mutants show a greater degree of plasticity at later stages of development [27]. Therefore, we quantified the NO content of the four alleles and their respective wild-type controls 4 days after completion of stratification (das), once the radicle was visible in most of the seeds in the population. Contrary to what was expected, all four *cue1* alleles had similar or less DAF fluorescence than their respective controls, suggesting that the mutants do not overproduce NO at this developmental stage. *cue1-6* and *nox1* had the same fluorescence intensity than Col-0 (0.77 ± 0.01 a.f.u./μg protein and 0.74 ± 0.01 a.f.u./μg protein compared to 0.72 ± 0.05 a.f.u./μg protein), whereas *cue1-5* (0.49 ± 0.05 a.f.u./μg protein) had 36% less than Col-0 ($p < 0.001$) and *cue1-1* (0.68 ± 0.06 a.f.u./μg protein), 25% less than pOCA108 (0.91 ± 0.05 a.f.u./μg protein; $p < 0.001$) (Figure 1B).

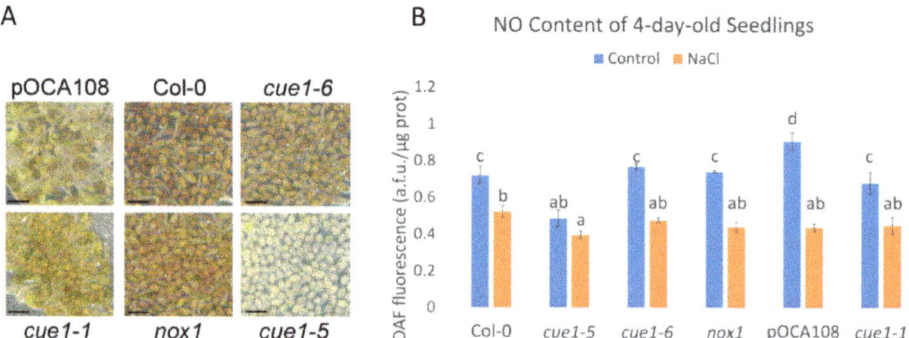

Figure 1. Phenotype (**A**) and nitric oxide (NO) content (**B**) of 4-day-old seedlings. Scale bar, 1 mm. DAF fluorescence intensity of extracts of 4-day-old seedlings grown on Murashige and Skoog (MS) with and without 100 mM NaCl. Values represent the mean ± CI (n = 3). Bars with common letters (a–d) do not show significant statistical differences. A two-way analysis of variance (ANOVA) with post-hoc Tukey's honestly significance difference (HSD) test showed a statistically significant interaction between genotype and salinity (($F_{(5,24)}$ = 23.13, p < 0.001, eta2[g] = 0.83). An analysis of simple main effects for each factor was performed with statistical significance after a Bonferroni correction (Table S2).

High salinity delays germination and impairs seedling establishment [52]. Endogenous NO is increased when plants are exposed to salt stress and might act as an antioxidant by quenching the ROS produced in response to salinity [53–56]. In order to see whether *cue1* mutants would accumulate NO during salt stress, we indirectly quantified the NO content of 4-day-old seedlings grown on a medium supplemented with 100 mM NaCl through quantification of DAF fluorescence. This salt concentration was chosen because it has been previously shown that it is enough to cause salt stress to *A. thaliana* seedlings [53]. Under our growth conditions, the DAF fluorescence significantly decreased in all the examined lines after salt stress, suggesting an inability to overproduce NO at this stage (Figure 1). In this case, *cue1-5* had 0.4 ± 0.02 a.f.u./μg protein, only 23% less than Col-0, which had 0.52 ± 0.03 a.f.u./μg protein. The rest of the lines had similar fluorescence levels to those of Col-0, ranging from 0.44 ± 0.02 a.f.u./μg protein to 0.47 ± 0.01 a.f.u./μg protein. The greatest decrease in the presence of high salinity was observed in pOCA108, which had 48% of the fluorescence intensity it accumulated in control conditions.

2.2. NO Is Necessary to Maintain Germination Vigor

Given the role of NO in the promotion of seed germination [45], we decided to analyze the germination rate of the different *cue1* alleles. While most of the Col-0, *cue1-6*, *nox1-1*, pOCA108, and *cue1-1* seeds had fully germinated after 4 das (between 96.5% and 99.5% of the population), the maximum germination (G_{max}) of *cue1-5* was only 67%, in accordance with the decrease in DAF fluorescence and supporting a reduced NO content at this stage (Figure 2A). In the presence of high salinity, the germination of all the lines decreased consistently with a reduction in endogenous NO. This germination delay was especially apparent for *cue1-5*, which had a G_{max} of 48% while the rest of the lines had similar G_{max} that varied between 66% and 88% (Figure 2A). Despite the stark decrease in DAF fluorescence after salt stress (Figure 1), the germination of pOCA108 was almost unaffected by the stress, with only an 11% decrease in its germination rate with respect to control conditions.

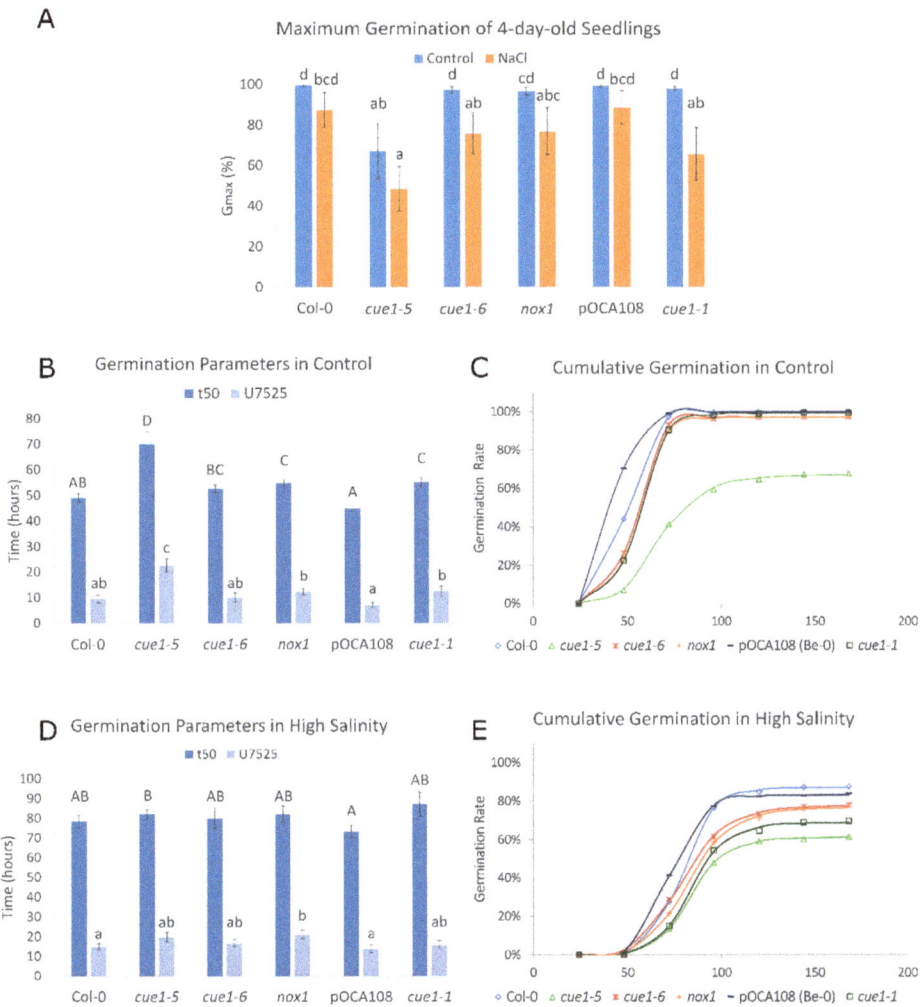

Figure 2. NO is required for germination vigor. Values represent the mean ± CI ($n = 12$). Bars with common letters do not show significant statistical differences as determined by the respective statistical tests: (**A**) Maximum germination of seedlings 4 das with and without 100 mM NaCl. Welch's one-way ANOVA with post-hoc Games–Howell test was $F_{(5,29)} = 7.98$. $p < 0.001$. (**B**) Germination parameters t_{50} and U_{7525} for seeds germinated in control conditions. (**C**) Representative cumulative germination curve for seeds germinated in control conditions as predicted by GERMINATOR. (**D**) Germination parameters t_{50} and U_{7525} for seeds exposed to high salinity. (**E**) Representative cumulative germination curve for seeds germinated in the presence of 100 mM NaCl as predicted by GERMINATOR.

The speed and uniformity of germination were analyzed using t_{50} and U_{7525}. t_{50} is a parameter that summarizes the time required for 50% of the viable seeds to germinate [57]. We found an inverse correlation between DAF fluorescence and speed germination parameter t_{50} ($r = -0.94$, $p < 0.001$). It took 69.99 ± 4.96 h for half of the *cue1-5* seeds to germinate, whereas the t_{50} of Col-0 was only 49.12 ± 1.69 h, 30% faster than the mutant. Equally, the t_{50} of pOCA108 was 44.92 ± 1.04 h, 19% faster than that of *cue1-1*, 55.19 ± 1.8 h (Figure 2B). U_{7525} measures the time interval between the germination of 25% and 75% of the viable seeds. A lower U_{7525} indicates greater uniformity [58]. The analysis of uniformity supported an inverse correlation between NO levels and germination ($r = -0.87, p < 0.001$). The U_{7525} of *cue1-5* was 22.65 ± 2.64 h, 13.16 h longer than its control, while on the other end of the spectrum pOCA108 had a U_{7525} of 6.9 ± 1.03 h (Figure 2B). These results suggest that a certain level of NO is necessary for seeds to germinate uniformly once dormancy is broken. Cumulative germination over 7 das showed a similar exponential curve for Col-0 and pOCA108, with a steeper slope for pOCA108, the line that showed the highest endogenous NO levels. On the other hand, all *cue1* mutants showed a longer lag phase that lasted up to 48 h after completion of stratification, which suggests that endogenous NO levels might be even lower in *cue1* mutants prior to radicle emergence (Figure 2C).

In the presence of high salinity, the t_{50} and U_{7525} of all the lines were similar (Figure 2D), in agreement with their similarity in NO production (Figure 1). While *cue1-1* seemed to have the highest t_{50} at 87.0 ± 6.1 h, this difference was not significant when compared to its control or other lines, which showed t_{50} ranging from 73.4 ± 3.3 h to 82.3 ± 2.0 h. Even though germination was generally slower under salt stress, uniformity of germination was more stable for all genotypes. Control lines had slightly more uniform germination with a U_{7525} of 15.0 ± 1.4 h for Col-0 and 13.8 ± 1.9 h for pOCA108, while the U_{7525} of the *cue1* mutants ranged between 15.6 ± 2.2 h and 21.0 ± 2.1 h. The analysis of cumulative germination showed that all the lines had a longer lag phase in the presence of salt stress than in control conditions (Figure 2E, compare to Figure 2C).

2.3. The Severe Germination Delay of cue1-5 Is Caused by Stabilization of the Germination Repressor ABI5

The plant hormone abscisic acid (ABA) maintains dormancy and post-germinative seedling arrest under unfavorable environmental conditions [59]. ABA exerts these functions mainly through the basic leucine zipper (bZIP) transcription factor ABSCISIC ACID INSENSITIVE5 (ABI5), a key repressor of seed germination [60,61]. The antagonistic effects of ABA and NO during germination occur through a crosstalk during the regulation of *ABI5* transcription [62] and protein stability [45]. NO-mediated S-nitrosation of ABI5 at Cys 153 facilitates the degradation of this germination repressor, and it has been proven that ABI5 protein levels are high in NO-deficient mutant backgrounds [45]. As well as during germination, ABA has a prominent role during the regulation of the response to most abiotic stresses [63]. The regulation of seed germination and seedling establishment in response to stress is also regulated by ABI5, as it has been shown that loss-of-function *abi5* mutants were able to germinate and green even in the presence of 200 mM NaCl [60].

In order to confirm that the germination deficiency observed in the *cue1-5* mutant is indeed due to its lack of sufficient NO, we analyzed the protein levels of ABI5 in 4-day-old seedlings. In agreement with the marked NO deficiency of *cue1-5* and its inability to reach a similar G_{max} to that of the other lines, we found that this mutant accumulated higher ABI5 protein levels (Figure 3). The quantity of ABI5 was comparable in the rest of lines, in accordance with their similar NO levels. Confirming the role of NO in the stability of the protein, we also found equally elevated ABI5 levels in all the lines in the presence of high salinity in the Col-0 background, consistent with the presence of abiotic stress and the decreased NO levels and germination rate.

2.4. Early Root Elongation Is Impaired in cue1 Mutants Independently of Their NO Levels, but Root Cell Patterning Is Not Altered

Studies carried out mainly with pharmacological NO donors and scavengers show that both excessive and insufficient NO result in the inhibition of root elongation [32,33,64,65]. Analysis of the root length of 5-day-old seedlings showed that all the *cue1* mutants had shorter roots than their respective controls (Figure 4A). The root lengths of Col-0 and pOCA108 were equivalent (0.85 ± 0.04 cm and 0.77 ± 0.05 cm, respectively), despite their differences in DAF measurements of NO at 4 das. At the same time, the root lengths of *cue1-6* and *nox1* were 40% and 28% shorter than Col-0, respectively, even though both lines showed as much DAF fluorescence as Col-0. On the other hand, *cue1-1* and *cue1-5*, the mutants that accumulated less NO according to the DAF measurements, also had the shortest primary roots.

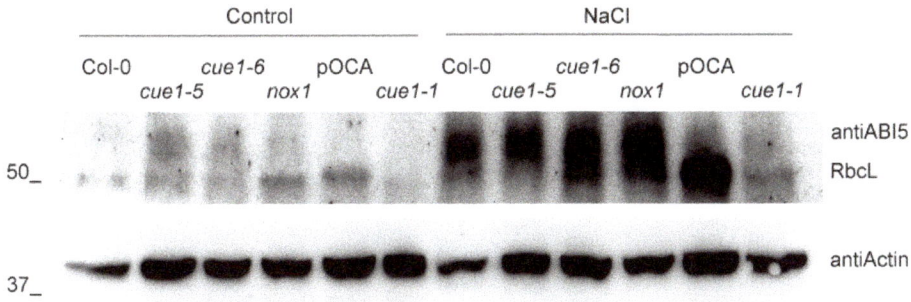

Figure 3. Germination inhibitor ABI5 is stabilized in *cue1-5* and in the presence of salt stress. Western blot analysis of ABI5 accumulation in the different *cue1* lines grown on MS with and without 100 mM NaCl for 4 das. Actin levels are shown as loading controls.

The root is organized into cell layers of different cell types that originate from a small set of cells at the core of the root, called the quiescent center (QC). QC cells divide infrequently and maintain the undifferentiated state of the adjacent cells, which are called stem cell initials. Stem cells continuously undergo asymmetric cell divisions that give rise to daughter cells that will divide symmetrically and start to differentiate [66]. This results in clear developmental zones along the longitudinal axis of the root. In the proximal meristem or meristematic zone (MZ), cells divide frequently until they are far enough from the stem cell niche. Once they reach the elongation/differentiation zone (EDZ) they start to expand quickly, beginning the process of differentiation. Maintaining these structures requires a balance between the generation of new cells in the meristem and the differentiation of cells in the EDZ. This balance determines the size of the root apical meristem (RAM) [67]. In accordance with their shorter roots, all the *cue1* mutants seemed to have a narrower vascular bundle than both wild-type lines, and a high accumulation of amyloplasts in the epidermis, cortex, and endodermis from the transition zone upwards (Figure 4B). A closer look at the meristematic zone showed that the altered organization of the stem cells around the QC that has been described to be caused by NO [32] was not apparent at this developmental stage (Figure 4B). This agrees with the fact that *cue1* mutants do not overproduce NO 4 days after completion of stratification.

2.5. Sugar Supplementation Modifies NO Production

In the absence of *CUE1*, plants are unable to establish photoautotrophic growth if they are not supplemented with exogenous metabolizable sugars, such as sucrose or glucose [24]. Chloroplasts are an important node of NO production [68], so we wanted to explore whether sugars have a role in NO homeostasis, since there are some reports that point to a possible role for NO in the regulation of energy production [69–71]. We tested the NO levels of *cue1-5*, *cue1-1*, and their respective controls after growing them for 7 das in a medium supplemented with either 2%(*w/v*) glucose or 0.75%(*w/v*) sucrose. Our results show that *cue1* mutants do indeed accumulate NO at this developmental stage (Figure 5A, Figure S1).

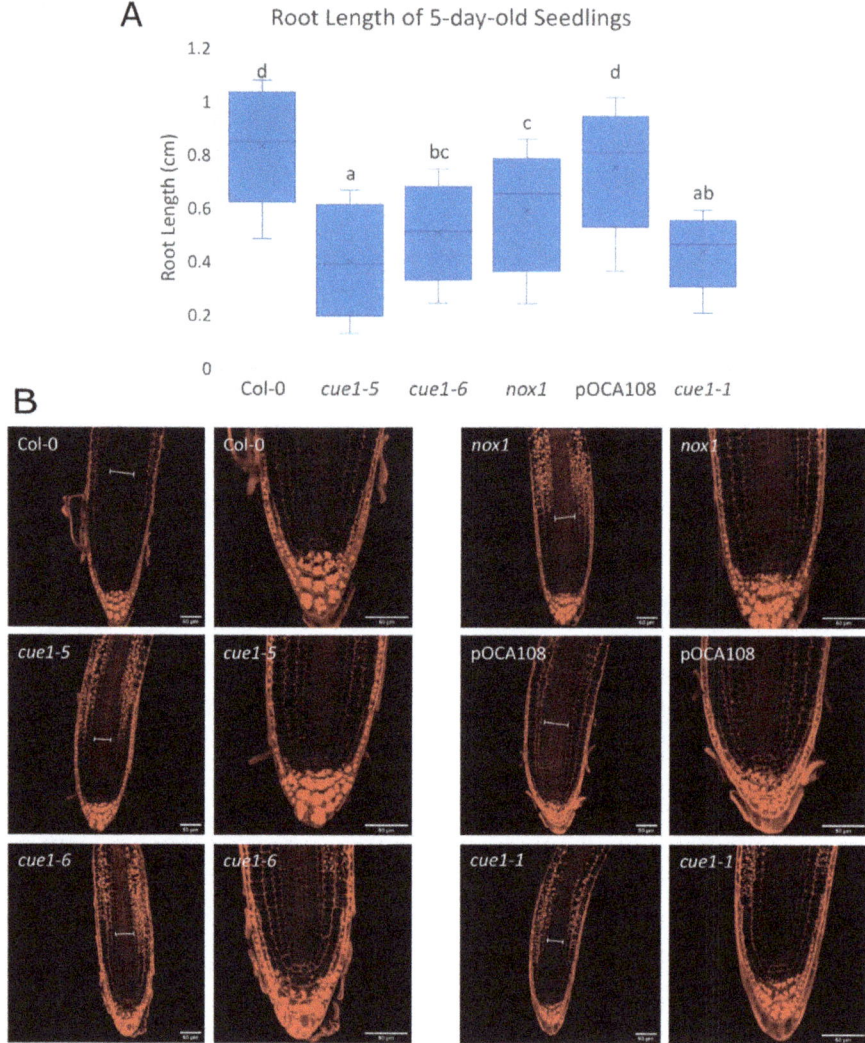

Figure 4. Root elongation is defective in all *cue1* mutants. (**A**) Primary root length of 5-day-old seedlings. The diagram shows data between the lower (Q1) and upper (Q3) quartiles, the median and the mean (x) for each genotype. Bars with common letters (a–d) do not show significant statistical differences as determined by one-way ANOVA with post-hoc Tukey's HSD test ($F_{(5,214)}$ = 57.8, $p < 0.001$). (**B**) Representative images of the root apex of 5-day-old seedlings stained with Schiff propidium iodide. Scale bars on the bottom right corner correspond to 50 μm. The width of the vascular bundle is indicated with a white line in each root.

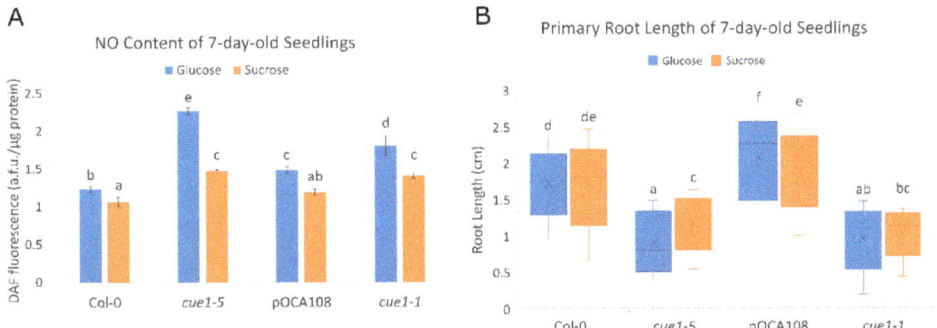

Figure 5. (**A**) The NO content of *cue1* alleles depends on their developmental stage. DAF fluorescence intensity of extracts of 7-day-old seedlings grown on MS-Root medium supplemented with either 2%(*w/v*) glucose or 0.75%(*w/v*) sucrose. Values represent the mean ± CI (*n* = 4). A two-way ANOVA with post-hoc Tukey's HSD test showed a statistically significant interaction between genotype and carbon source ((F$_{(3,24)}$ = 33.99, p < 0.001, eta2[g] = 0.81). An analysis of simple main effects for each factor was performed with statistical significance after a Bonferroni correction (Table S2). (**B**) Primary root length is influenced by NO and sugar. Primary root length of 7-day-old seedlings grown on MS-Root supplemented with either 2%(*w/v*) glucose or 0.75%(*w/v*) sucrose (*n* = 50). The diagram shows data between the lower (Q1) and upper (Q3) quartiles, the median and the mean (x) for each genotype. Common letters (a–f) indicate there are no significant statistical differences as determined by Welch's one-way ANOVA with post-hoc Games–Howell test (F$_{(7,571)}$ = 203.57, p < 0.001).

While the differences in NO levels are greater in the presence of glucose, the mutants also overproduce NO when grown with sucrose as a carbon source. *cue1-5* was the allele which exhibited greater DAF fluorescence, with 2.27 ± 0.05 a.f.u./µg protein in glucose and 1.48 ± 0.01 a.f.u./µg protein in sucrose. Compared to its control Col-0, this meant a 46% increase in NO content when grown with glucose, and a 23% increase in the presence of sucrose. *cue1-1* also showed a significant enhancement in NO levels compared to its control pOCA108, with 1.81 ± 0.14 a.f.u./µg protein in glucose compared to 1.49 ± 0.04 a.f.u./µg protein (18% increase), and 1.41 ± 0.04 a.f.u./µg protein in sucrose compared to 1.19 ± 0.03 a.f.u./µg protein (16% increase).

Since the mutants present an increased NO content in these growth conditions, we decided to measure primary root length to check whether there was a link between the carbon source, NO content, and root length in *cue1* mutants. We found that the root length of *cue1-5* and *cue1-1* was indeed significantly shorter (Figure 5B, Table S1). After 7 days of growth, *cue1-5* had a 1.2 ± 0.06 cm root in sucrose, which meant a 32% reduction compared to Col-0, with 1.78 ± 0.05 cm. The reduction was even more apparent in glucose, where the primary root of *cue1-5* only elongated to 0.86 ± 0.09 cm, a 50% reduction compared to Col-0, with a 1.73 ± 0.04 cm root. Primary root growth was also inhibited in *cue1-1* with respect to pOCA108, in sucrose (1.10 ± 0.06 cm compared to 1.85 ± 0.06 cm, 41% reduction) as well as in glucose (1.01 ± 0.07 cm compared to 2.20 ± 0.07 cm, 54% reduction). Previous studies have suggested that an optimal concentration of NO is needed for proper root development, and that both excessive and deficient NO levels are detrimental for the plant [32,33], which coincides with our results. Interestingly, the NO content of *cue1-5* and *cue1-1* in sucrose was similar to the NO content of pOCA108 in glucose, but the root length of the *cue1* mutants was vastly different to that of pOCA108, which suggests that both NO content and the sugar available as a carbon source have a role in the modulation of primary root growth.

We also explored the primary root apical meristems of these roots to analyze meristematic cortical cell length, meristem size, and meristem cell number of the mutants (Figure 6, Table S1). Morphological observation showed that all the lines accumulated amyloplasts in the cortical and epidermal cells upwards of the transition zone when they were grown in the presence of glucose (Figure 6A), but not in the presence of sucrose, where only *cue1-5* and *cue1-1* accumulated them (Figure 6B). Furthermore, we noticed that the *cue1* mutants presented amyloplasts in the columella stem cell layer, an indication of earlier differentiation. As described in [32], we observed disorganization of the cells surrounding the QC in some of the *cue1-1* and *cue1-5* plants, especially in the presence of glucose, where the endogenous NO levels were higher.

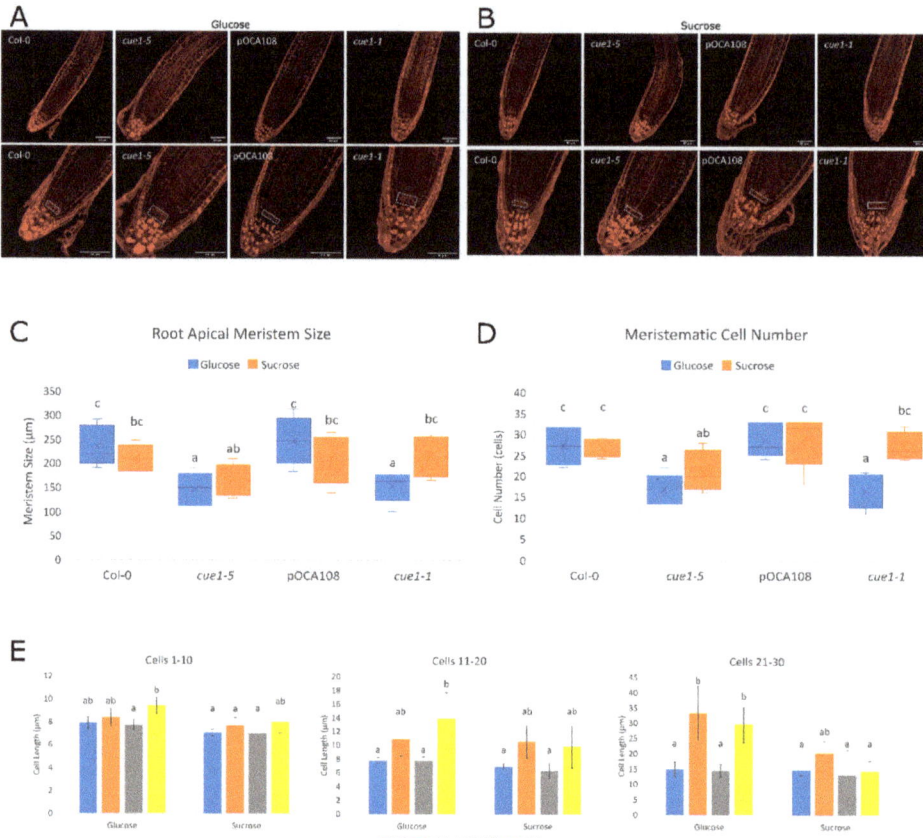

Figure 6. Increased NO causes disorganization of root meristematic stem cells. (A) Representative images of the root meristem of 7-day-old seedlings grown on MS-Root medium supplemented with 2%(*w/v*) glucose. (B) Representative images of the root meristem of 7-day-old seedlings grown on MS-Root medium supplemented with 0.75%(*w/v*) sucrose. Seedlings were stained with Schiff propidium iodide. Scale bars on bottom left corner correspond to 50 µm. A white rectangle was drawn around the stem cells. Differentiation parameters were quantified from these images. Statistical differences as determined by two-way ANOVA with post-hoc Tukey's HSD test are represented by different letters (a–c). (C) Meristematic size ($F_{(3,70)}$ = 6.65, $p < 0.001$, eta2[g] = 0.22). An analysis of simple main effects for each factor was performed with statistical significance after a Bonferroni correction (Table S2). (D) Number of cells in the root apical meristem ($F_{(3,70)}$ = 6.16, $p < 0.001$, eta2[g] = 0.21). An analysis of simple main effects for each factor was performed with statistical significance after a Bonferroni correction (Table S2). (E) Bars with common letters (a–c) are not significantly different according to Welch's ANOVA with post-hoc Tukey's HSD test. Average cortical cell sizes are shown for cortical cells 1–10 ($F_{(7,73)}$ = 3.99, $p < 0.001$), 11–20 ($F_{(7,73)}$ = 5.45, $p < 0.001$) and 21–30 ($F_{(7,73)}$ = 7.23, $p < 0.001$) counted from the QC. Values represent the mean ± CI (n = 10).

An analysis of root cortical cells showed that the differences among genotypes were clearer when the plants had been grown with glucose as a carbon source. In those conditions, *cue1-5* and *cue1-1* differentiated earlier than their controls, since the EDZ occurred closer to the initial cells than in their respective wild-type controls. While a statistical analysis showed that the carbon source only caused a significant difference in meristem size in *cue1-5* seedlings (Figure 6C), the variations observed in the rest of the lines supported the divergences observed in the root length of the mutants. Thus, the RAM size in *cue1-5* was 21% smaller than its control in glucose and 18% smaller in sucrose, while *cue1-1* presented only a 2% reduction in sucrose, but a 15% decrease in glucose compared to pOCA108. The same was true for the number of cortical cells in the apical meristem, as both *cue1-5* and *cue1-1* had significantly fewer cells than their respective controls in glucose, but not in sucrose (Figure 6D). This decrease in meristematic cell number was compensated by an increase in meristematic cell length (Figure 6E). These observations support the existence of an interplay between NO content and sugar metabolism in the modulation of primary root growth. At the same time, these results have allowed us to identify *cue1* mutants as suitable NO-overproducer plant lines to explore the role of NO during root growth.

3. Discussion

3.1. CUE1 Does Not Directly Contribute to the Rapid NO Synthesis during Seed Imbibition

NO is a stimulator molecule in plant photomorphogenesis, as it promotes seed germination and de-etiolation, and inhibits hypocotyl and internode elongation [42]. It has been demonstrated that NO-deficient mutants exhibit increased dormancy, hypersensitivity to ABA during seed germination and seedling establishment, as well as resistance to dehydration [72,73]. The decrease in NO during germination observed in *cue1-5* induces a general decline in germination parameters, since it presents a reduced maximum germination rate, decreased uniformity of germination among seeds of the same population, a delay in the initiation of germination and slower germination overall. Moreover, the rest of *cue1* mutants were unable to accumulate more NO than their controls at this developmental stage. It has been previously demonstrated that a NO burst is required for proper germination and post-germinative growth [12,50,74], possibly to counteract the inhibitory effect of the ABA-regulated transcription factor ABI5 [45,62,75], which acts as the main regulator of one of the earliest developmental checkpoints to spare the plant from pouring resources into growth when the environmental conditions are not optimal for the development of the seedling [60,61].

During germination, when the quiescent seed reactivates its metabolism, there is a surge in the production of NO that slows down three hours after imbibition [50,74], although NO synthesis can be detected in the aleurone layers as early as fifteen seconds after exogenous nitrate addition [12]. It is thought that the generation of NO by the seed is non-enzymatic because of its quick response, and that this non-enzymatic synthesis requires acidic pH and the presence of compounds that can act as antioxidants, so this synthesis pathway would be restricted to the apoplast of aleurone cells and maybe local areas of the root during transient acidification caused by alteration of nutrient supply. The aleurone layer fulfills both requisites, as its pH is usually between 3 and 4, and its plastids contain proanthocyanidins, phenolic compounds with antioxidant capacity [12]. The early stages of germination coincide with a depletion of oxygen, so it is assumed that seeds are in an anaerobic state until the radicle breaks through the testa, when oxygen gradually returns the seed to aerobic conditions [76]. Thus, it is highly unlikely that the seed can offer the oxidative environment necessary to obtain NO from arginine, and it would explain why *cue1* mutants, which are thought to be involved in the oxidative biosynthesis of NO, do not overproduce NO during germination.

Among all studied *cue1* mutants, only *cue1-5* mutants do not germinate as well as wild-type seeds in response to advantageous environmental conditions even after dormancy is broken by stratification. Our results show that this is due to its inability to de-stabilize ABI5, as shown by its increased ABI5 protein levels. The increase in ABI5 is a direct consequence of the reduced endogenous NO content of *cue1-5*. However, we do not think this is directly caused by the loss of function of *CUE1*, as the other alleles behave closer to their wild-type controls, both in endogenous NO levels and germination parameters. This mutant line contains an additional *transparent testa* mutation that affects seed pigmentation, testa solidity and germination [27].

Interestingly, data from the Arabidopsis Seed Coat eFP Browser [77] and the ePlant Browser [78] indicate that the expression of *CUE1* in the mature embryo and in dry seeds is almost negligible, but it increases steadily one day after imbibition, well after the rapid production of NO that starts germination. On the other hand, it is highly expressed in the ovaries of the plant, which will give rise to the seed coats. The expression of *CUE1* increases in the developing embryo until the walking stick stage, 7–8 days post-anthesis, at the end of the cellularization of the endosperm and before the accumulation of reserves. It is possible, then, that its role in NO generation during germination is carried out during embryo development and seed maturation, likely in an indirect fashion. Since the non-enzymatic synthesis of NO requires proanthocyanidins [12], which are generated as one of the end products of the shikimate pathway, it could be expected that *cue1* seeds have less of these polyphenolic compounds available for the quick apoplastic formation of NO and thus a longer lag phase. Indeed, *cue1* mutants have reduced flavonoids, hydroxycinnamic acids, and simple phenolics [27], in agreement with this hypothesis.

3.2. High Salinity Impairs Germination by Increasing ABI5 and Decreasing NO Levels

Plants respond to abiotic stress by increasing the production of ABA and reactive oxygen and nitrogen species, so we also tested the behavior of *cue1* mutants in response to salt stress during early plant development. Depending on its concentration, NO can protect plants against salt stress by lessening the secondary oxidative stress induced by high salinity [46,47,79,80], or it can enhance sensitivity to the stress if its accumulation is excessive, causing additional nitrosative stress [36,81,82]. It is thought that the protective role of NO is exerted mainly through redox modification or *S*-nitrosation of ROS scavenging enzymes, antioxidant systems, and respiratory pathways [83]. Reports on the accumulation of NO in *A. thaliana* in response to high salinity are sometimes contradictory, possibly because of differences in biological material, the extent of the stress, and the developmental stage at which the plant was subjected to salt stress [46,47,53–56,84,85].

In our growth conditions and developmental stage, the NO content of the seedlings decreased when seeds had been germinated on MS medium supplemented with 100 mM NaCl, regardless of their genotype and in agreement with [46]. Contrary to other salt stress treatments performed in *A. thaliana*, this may not be enough to cause secondary nitrosative stress. It has been demonstrated that the induction of NO by salt requires peroxisomal NO synthase activity [53], which needs oxygen to be able to oxidize arginine. Thus, it is possible that, in the anoxic state of the germinating seed, 100 mM NaCl is too mild a stress to generate enough ROS for creating the oxidative environment required for the function of NO synthase. Because of the reduced NO levels, germination was equally impaired in all the lines, as evidenced by the analysis of different germination parameters. This was explained by an increase in ABI5 protein levels, which was easily detected after germination during salt stress, since ABI5 is stabilized by ABA and degraded in a NO-mediated process [33], and it accumulates in seeds that undergo salt stress [60]. In fact, ABI5 expression is highly induced by abiotic stress at the transition from mature seeds to seedling growth. Salt delays the decline of ABI5 levels and promotes its expression throughout the seedling, while in unstressed conditions it would be undetectable [86].

3.3. Initial Seedling Establishment Is Impaired in cue1 Mutants Independently of Their NO Levels

In addition to the characterization of the role of *CUE1* during germination, we explored its role during early post-germinative growth by quantifying the primary root length and exploring the root apical meristem of *cue1* mutants 5 das. However, we were unable to find a relationship between their NO content and these parameters, since all *cue1* lines showed a significant reduction in root growth independently of the differences in their NO levels. This observation further stresses the need for a careful assessment of NO levels when working with NO mutants. Understanding the effect of the loss of a translocator involved in glycolysis and photosynthesis during root development can prove to be quite complex, as phenotypes are the result of an intricate interplay between hormones, other growth regulators, and environmental cues.

Early plant development is mostly supported by the storage reserves found in the endosperm [87], which in *A. thaliana* cannot last more than 4–5 days and are mostly used to elongate the hypocotyl until it reaches light, when cotyledons start greening and are converted into photosynthetic organs [88]. *cue1* mutants have been described to be defective in the maturation of eoplasts to chloroplasts and in the initial establishment of photoautotrophism in the absence of an exogenously supplied metabolizable sugar [24]. Sugars mobilized from the endosperm and synthesized in the green parts of the plant function as energy sources as well as signaling molecules. In particular, photosynthetic sugars delivered from the cotyledons to the root act as interorgan signals to initiate root growth and have a dominant role during the cotyledon stage of seedling development, even over the phytohormone auxin, which is thought to be essential for the regulation of root development [88]. Since all *cue1* mutants are photosynthetically defective, we cannot rule out that their initial inability to start root elongation stems from the lack of a photosynthetic signal.

3.4. Sugars Alter Root Development through Modulation of NO Homeostasis

After exposure to exogenous NO, the content of several glycolysis intermediates, metabolites of the TCA cycle and intermediates of the Calvin cycle were reduced [70]. On the other hand, the content of sucrose and different monosaccharides, disaccharides, amino and nucleotide sugars increased [70,71]. These metabolic changes are caused by the transcriptional upregulation of glycolytic enzymes and the downregulation of photosynthetic proteins [69], and by the regulation via *S*-nitrosation of enzymes involved in sugar metabolism, such as ATP synthase, enolase, or phosphoglycerate kinase [71].

CUE1/PPT1 has a central role in sugar metabolism because it is the only source of PEP into the chloroplast and its absence directly impacts carbon partitioning [27,89]. Interestingly, *cue1* mutants are unable to establish photoautotrophic growth right after germination and need to be exogenously supplemented with a fixed carbon photoassimilate [24]. These mutants have impaired light signaling and a reduced capacity of de-etiolation [24,26], possibly because they are defective in chloroplast maturation [24]. Essentially, *cue1* mutants behave as heterotrophs during early plant development.

cue1 mutants have not been the first NO mutants to be linked to sugar metabolism. *noa1*, which encodes a cGTPase necessary for assembling plastid ribosomes [90], was initially isolated as a mutant with less endogenous NO content and essential for its production [91]. Interestingly, the levels of NO in *noa1* mutants can be partially recovered by exogenous addition of sucrose [92], and the mutant presents chloroplast biogenesis defects [93] and reduced fumarate, even though its energy status and redox potential seem unaffected [92]. The NO deficiency in *noa1* was explained as an indirect effect of its reduced ability to generate photosynthates [93]. However, *cue1* mutants are also defective in chloroplast biogenesis and contain fewer photoassimilates than wild-type lines, whereas their NO levels are enhanced. Thus, a reduced photosynthetic capacity cannot solely be the reason for reduced NO content in *noa1*.

We analyzed the endogenous NO status of *cue1* mutants in the presence of two different sugars that can be used by the plant as a source of reduced carbon and energy, glucose and sucrose. If a general decrease in reduced carbon availability explained the alteration in NO content, the mutants would be expected to produce the same NO in any case. However, our results showed the differences between *cue1* mutants and their controls were greater when the seedlings had been grown in the presence of glucose. Sucrose did not revert the NO overproducing phenotype of *cue1* plants, but it did diminish the differences among the lines, as in the case of *noa1* [93]. The increased NO production in the presence of glucose in *cue1-5* and *cue1-1* caused a decrease in primary root elongation rate, meristem size, meristematic cell number, and longer root cell length, corroborating the findings of [29]. These differences were not as obvious when the plants were grown in the presence of sucrose, but the root growth of *cue1-5* and *cue1-1* was still affected. Our findings show that NO accumulation depends on sugar metabolism. Nitrate reductase (NR) is an enzyme that also participates in the production of NO [94] and it has been previously shown that the expression of *NIA2*, the gene encoding NR, is light-induced, while NR activity is linked to photosynthesis [95]. NR could be a good candidate linking NO homeostasis and carbon metabolism, but, surprisingly, *noa1* has higher NR levels [92], while *cue* mutants have reduced NR activity [26]. Understanding the role of carbon metabolism will help elucidate the molecular mechanisms underlying NO production in plant development, but further experiments are required.

4. Materials and Methods

4.1. Plant Lines

Arabidopsis thaliana cue1-5 [26], *cue1-6* [27] and *nox1-1* [23] plants are in the Columbia (Col-0) ecotype background, while *cue1-1* [24] is in the Bensheim (Be-0) ecotype background. All *cue1* mutants are defective in the *CUE1* locus (AT5G33320). The *cue1-5* (CS3156) and *cue1-6* (CS3168) alleles were generated by mutagenizing a Col-0 population with ethyl methanesulfonate (EMS) [27], while the *nox1-1* allele was generated by fast neutron mutagenesis [23]. Additionally, *cue1-5* seeds are yellow and lack brown pigments in the seed coat, as the mutant contains an additional *transparent testa/glabrous* mutation [96]. The *cue1-1* allele was generated by mutagenizing a population of line pOCA108-1 with gamma radiation [24]. pOCA108-1 is a single-insertion line that contains the reporter construct pOCA108 on chromosome 2. This construct contains the alcohol dehydrogenase (*ADH*) gene under the control of chlorophyll *a/b* binding protein (*CAB3*) promoter, and was transformed into Bensheim line R002, which contains a null mutation in the endogenous *ADH* gene [24].

4.2. Plant Growth Conditions

Seeds grown in vitro were surface sterilized using a bleach solution (25% bleach, 0.1% Tween). Sterilized seeds were stratified in water for 48–72 h at 4 °C to help synchronize germination. Stratified seeds used for root elongation assays and NO quantification in different carbon sources were grown on plates containing a modified Murashige and Skoog (MS) medium [97] optimized for root growth, MS-Root [2.3 g/L MS (Duchefa Biochemie, Haarlem, The Netherlands), 15 g/L agar], supplemented with either sucrose or glucose as indicated. The greater agar content allows for vertical growth of seedlings on the surface of the medium. For the rest of the experiments, plants were grown on plates containing 4.9 g/L MS, 2% glucose, 6 g/L agar. Stratified seeds used for NO quantification, germination and high salinity assays were grown on plates containing MS medium supplemented with 2% glucose. To analyze sensitivity to salt stress, MS was supplemented with 100 mM NaCl (PanReac AppliChem, Darmstadt, Germany). Plants were grown under a 16 h light/8 h dark photoperiod at a constant temperature of 21 °C and 50–60% humidity.

4.3. Germination Assays

Germination was determined as radicle emergence at indicated times. The analysis of germination parameters was carried out using the GERMINATOR software [58]. The parameters used are t_{50}, the time it takes for 50% of the viable seeds to germinate, for speed of germination, and U_{7525}, the elapsed time from 25% to 75% of the viable seeds to germinate, for uniformity of germination. Germination was also represented with a time course graph as cumulative germination.

4.4. Western Blotting

Total proteins were extracted from 4-day-old stratified and imbibed seeds (500–600 per genotype) for western blot analysis. Tissue was homogenized using a Silamat S6 homogenizer (Ivoclar Vivadent, Madrid, Spain) until all tissue was completely powdered. Samples were incubated with an extraction buffer containing 100 mM Tris-HCl, 150 mM NaCl, 0.25% NP-40 and 1× cOmplete EDTA-free Protease

Inhibitor Cocktail (Sigma, Saint Louis, MO, USA). Protein concentration was determined by the Bio-Rad Protein Assay (Bio-Rad, Hercules, CA, USA) based on the Bradford method [98]. An amount of 90 µg of total protein was loaded per well in 10% SDS-acrylamide/bisacrylamide gel electrophoresis using Tris–glycine–SDS buffer. Proteins were electrophoretically transferred to an Immobilon-P polyvinylidene difluoride membrane (Merck Millipore, Burlington, VT, USA) using the semi-dry Trans-Blot Turbo Transfer system (Bio-Rad). Membranes were blocked in Tris-buffered saline-0.1% Tween 20 containing 5% Blocking Agent and probed with antibodies diluted in blocking buffer. Anti-ABI5 Purified Rabbit Immunoglobulin (Biomedal, Sevilla, Spain, 1:10,000) and anti-Actin clone 10-B3 Purified Mouse Immunoglobulin (Sigma-Aldrich, A0480, 1:10,000) antibodies were used in the Western blot analyses. Detection was performed using ECL Advance Western Blotting Detection Kit (Amersham, Chicago, IL, USA) and the chemiluminescence was detected using an Intelligent Dark-Box II, LAS-1000 scanning system (Fujifilm, Tokyo, Japan).

4.5. Detection of NO Production

Freshly prepared protein extracts prepared as for western blotting were used to assay NO content. An amount of 20 µL of each protein extract was incubated with 180 µL of a solution containing 10 µM of DAF-FM DA (Sigma-Aldrich) in 50 mM HEPES buffer pH 7.5 in microtiter plates, following the method described in [49]. Samples were incubated at 37 °C for 2 h in the dark. After incubation, the emitted fluorescence of each well was measured in a Varioskan LUX Multimode Microplate Reader (ThermoFisher Scientific, Waltham, MA, USA). Samples were normalized by their total protein content (Table S3) and against a control condition in each experiment. Blanks included in all experiments behaved similarly and emitted a negligible signal, that was subtracted from all experimental samples.

Detection of NO through confocal microscopy was performed using the same experimental conditions and DAF-FM DA staining protocol for the former spectrofluorometry measurements. λ scan 500–666 nm was used to set the emission window and FIRE LUT to represent a fluorescence heatmap of intensity.

4.6. Root Growth Analysis

After full germination, root growth was captured by scanning plates with an Epson flatbed scanner and a resolution set to 600 ppi. Primary root length of individual seedlings was then measured using Fiji [99]. Average root elongation rate (mm/d) was calculated as an average of daily root elongation rates following the protocol described in [100].

Primary root apical meristems were analyzed to measure meristematic cortical cell length (µm), meristem size (µm), and meristem cell number by performing a propidium iodide (PI) stain following the protocol described in [101]. Root tips were examined using a Leica SP2 confocal microscope with a 40× oil immersion objective. The resulting image data was processed with semi-automatic image analysis software, Cell-O-Tape [102].

4.7. Statistical Analysis

For each dataset, the distribution was initially assessed by plotting all the values of the dependent variable as a histogram. Normality and homoscedasticity of the populations were determined using the Shapiro–Wilk and Levene's tests, respectively. An appropriate statistical model was then selected depending on the number of independent variables, the distribution of the dependent variable, and whether it was categorical or continuous. To account for type I error, data was presented with a 95% confidence interval (CI). Generalized linear models (ANOVAs) and Pearson's product-moment correlation tests were performed using R Statistical Software (R version 4.0.2, R Foundation for Statistical Computing) [103] in the RStudio environment (RStudio version 1.3.959, PBC) [104]. Excel (Microsoft Office 365 ProPlus, v.1902) was used for other statistical tests and graph plotting. The statistical power of the chosen tests was performed using G*Power v.3.1.9.2 (Franz Faul, Universität Kiel).

5. Conclusions

Despite the ample research performed to elucidate the role of NO during development, no systematic study of the production of NO during different developmental stages has been performed to date. The reports discussed in this article point to specific roles that would require tightly controlled spatio-temporal NO accumulation. To our knowledge, the only analysis comparing the production of NO at two different plant developmental stages was published using *Medicago truncatula* [105]. In this study it was shown that senescing plants had an increased sensitivity to nitrosative stress, as well as repression of nitrate uptake and NR activity, suggesting that accumulation of NO and regulation of its homeostasis depends on the developmental stage. Our results support this statement, as a careful characterization of the NO production of different *cue1* mutants, routinely used as NO overproducer mutants, proved that *cue1* mutants do not accumulate NO during early plant development, but they do at later stages. Since most NO mutants are defective in proteins involved in primary metabolism [68,106], we recommend that NO quantification be performed for NO mutants at early developmental stages, given that their alteration of NO homeostasis might stem from unexpected effects of their mutations. In conclusion, our results demonstrate that *cue1* is a useful tool to study the physiological functions of NO, since this mutant accumulates NO under controlled experimental conditions that require awareness of the developmental stage and growth conditions of the plants, especially in terms of stress trade-off.

Supplementary Materials: The following are available online at http://www.mdpi.com/2223-7747/9/11/1484/s1, Figure S1: (A) Detection of NO by confocal microscopy using root tips of 7-day-old Col-0 seedlings stained with DAF-FM DA incubation after treatment with NO scavenger (cPTIO) and donor (GSNO). cPTIO is able to scavenge DAF and GSNO increases local maxima. (B) DAF fluorescence by confocal microscopy using root tips of 7-day-old seedlings from Col-0, pOCA108 and *cue1* alleles grown on MS-Root medium supplemented with either 2%(*w/v*) glucose or 0.75%(*w/v*) sucrose. FIRE LUT was used as a fluorescence heatmap. Table S1: Sugar and NO affect root meristem size. Meristem size parameters of 7-day-old seedlings grown on MS-Root medium supplemented with either 2% glucose or 0.75% sucrose. Values represent the mean ± CI (N = 104). All units were in µm and 10 roots were analyzed per genotype and carbon source. Table S2: Two-way ANOVA statistical parameters and simple main effects results. Table S3: Protein content of 4-day-old and 7-day-old seedlings (µg/mL).

Author Contributions: Conceptualization, T.L., L.S. and O.L.; methodology, T.L., L.S., I.S.-V. and O.L.; validation, T.L., L.S., I.S.-V. and O.L.; formal analysis, T.L.; investigation, T.L. and I.S.-V.; resources, O.L.; writing—original draft preparation, T.L.; writing—review and editing, T.L., I.S.-V., L.S., O.L.; visualization, T.L.; supervision, L.S. and O.L.; project administration, O.L.; funding acquisition, O.L. All authors have read and agreed to the published version of the manuscript.

Funding: This research was funded by grants BIO2017-85758-R from the Ministerio de Ciencia, Innovación y Universidades (Spain), and SA313P18 from Junta de Castilla y León and Escalera de Excelencia CLU-2018-04 co-funded by the P.O. FEDER of Castilla y León 2014–2020 Spain (to O.L.). Fundación Solórzano FS/16 2019 (to I.S.-V.). T.L. was supported by a FPU predoctoral fellowship awarded by Ministerio de Educación, Cultura y Deporte (Spain), FPU13/05569.

Acknowledgments: We thank Dolores Rodriguez, Walter Dewitte, Isabel Mateos and Pablo Albertos for critical comments of the manuscript and technical assistance, and the Spanish network BIO2015-68957-REDT and RED2018-102397-T for stimulating discussions. We also thank Lucas Frungillo for providing the *cue1-6* seeds. *nox1* mutant was a kind gift from Zhen-Ming Pei and *cue1-1* and *cue1-5* were obtained from the Arabidopsis Biological Resource Center (ABRC).

Conflicts of Interest: The authors declare no conflict of interest. The funders had no role in the design of the study; in the collection, analyses, or interpretation of data; in the writing of the manuscript, or in the decision to publish the results.

References

1. Cueto, M.; Hernández-Perera, O.; Martín, R.; Bentura, M.L.; Rodrigo, J.; Lamas, S.; Golvano, M.P. Presence of nitric oxide synthase activity in roots and nodules of *Lupinus albus*. *FEBS Lett.* **1996**, *398*, 159–164. [CrossRef]
2. Lazalt, A.M.; Beligni, M.V.; Lamattina, L. Nitric oxide preserves the level of chlorophyll in potato leaves infected by *Phytophthora infestans*. *Eur. J. Plant Pathol.* **1997**, *103*, 643. [CrossRef]
3. Delledonne, M.; Xia, Y.; Dixon, R.A.; Lamb, C. Nitric oxide functions as a signal in plant disease resistance. *Nature* **1998**, *394*, 585–588. [CrossRef] [PubMed]

4. Durner, J.; Wendehenne, D.; Klessig, D.F. Defense gene induction in tobacco by nitric oxide, cyclic GMP, and cyclic ADP-ribose. *Proc. Natl. Acad. Sci. USA* **1998**, *95*, 10328–10333. [CrossRef]
5. Astier, J.; Gross, I.; Durner, J. Nitric oxide production in plants: An update. *J. Exp. Bot.* **2018**, *69*, 3401–3411. [CrossRef]
6. Bruand, C.; Meilhoc, E. Nitric oxide in plants: Pro-or anti-senescence. *J. Exp. Bot.* **2019**, *70*, 4419–4427. [CrossRef]
7. Sánchez-Vicente, I.; Fernández-Espinosa, M.G.; Lorenzo, O. Nitric oxide molecular targets: Reprogramming plant development upon stress. *J. Exp. Bot.* **2019**, *70*, 4441–4460. [CrossRef]
8. Rai, K.K.; Pandey, N.; Rai, S.P. Salicylic acid and nitric oxide signaling in plant heat stress. *Physiol. Plant.* **2020**, *168*, 241–255. [CrossRef]
9. Kapil, V.; Khambata, R.S.; Jones, D.A.; Rathod, K.; Primus, C.; Massimo, G.; Fukuto, J.M.; Ahluwalia, A. The noncanonical pathway for in vitro nitric oxide generation; the nitrate-nitrite-nitric oxide pathway. *Pharmacol. Rev.* **2020**, *72*, 692–766. [CrossRef]
10. Jeandroz, S.; Wipf, D.; Stuehr, D.J.; Lamattina, L.; Melkonian, M.; Tian, Z.; Zhu, Y.; Carpenter, E.J.; Wong, G.K.-S.; Wendehenne, D. Occurrence, structure, and evolution of nitric oxide synthase-like proteins in the plant kingdom. *Sci. Signal.* **2016**, *9*, re2. [CrossRef]
11. Hasanuzzaman, M.; Oku, H.; Nahar, K.; Bhuyan, M.H.M.B.; Mahmud, J.A.; Baluska, F.; Fujita, M. Nitric oxide-induced salt stress tolerance in plants: ROS metabolism, signaling, and molecular interactions. *Plant Biotechnol. Rep.* **2018**, *12*, 77–92. [CrossRef]
12. Bethke, P.C.; Badger, M.R.; Jones, R.L. Apoplastic Synthesis of Nitric Oxide by Plant Tissues. *Plant Cell* **2004**, *16*, 332–341. [CrossRef]
13. Begara-Morales, J.C.; Chaki, M.; Valderrama, R.; Sánchez-Calvo, B.; Mata-Pérez, C.; Padilla, M.N.; Corpas, F.J.; Barroso, J.B. Nitric oxide buffering and conditional nitric oxide release in stress response. *J. Exp. Bot.* **2018**, *69*, 3425–3438. [CrossRef] [PubMed]
14. Mur, L.A.J.; Mandon, J.; Persijn, S.; Cristescu, S.M.; Moshkov, I.E.; Novikova, G.V.; Hall, M.A.; Harren, F.J.M.; Hebelstrup, K.H.; Gupta, K.J. Nitric oxide in plants: An assessment of the current state of knowledge. *AoB Plants* **2013**, *5*, pls052. [CrossRef] [PubMed]
15. Broniowska, K.A.; Diers, A.R.; Hogg, N. *S*-Nitrosoglutathione. *Biochim. Biophys. Acta-Gen. Subj.* **2013**, *1830*, 3173–3181. [CrossRef]
16. Ederli, L.; Reale, L.; Madeo, L.; Ferranti, F.; Gehring, C.; Fornaciari, M.; Romano, B.; Pasqualini, S. NO release by nitric oxide donors in vitro and *in planta*. *Plant Physiol. Biochem.* **2009**, *47*, 42–48. [CrossRef] [PubMed]
17. Melvin, A.C.; Jones, W.M.; Lutzke, A.; Allison, C.L.; Reynolds, M.M. *S*-Nitrosoglutathione exhibits greater stability than *S*-nitroso-*N*-acetylpenicillamine under common laboratory conditions: A comparative stability study. *Nitric Oxide* **2019**, *92*, 18–25. [CrossRef]
18. Feechan, A.; Kwon, E.; Yun, B.-W.; Wang, Y.; Pallas, J.A.; Loake, G.J. A central role for *S*-nitrosothiols in plant disease resistance. *Proc. Natl. Acad. Sci. USA* **2005**, *102*, 8054–8059. [CrossRef] [PubMed]
19. Sakamoto, A.; Ueda, M.; Morikawa, H. *Arabidopsis* glutathione-dependent formaldehyde dehydrogenase is an *S*-nitrosoglutathione reductase. *FEBS Lett.* **2002**, *515*, 20–24. [CrossRef]
20. Lee, U.; Wie, C.; Fernandez, B.O.; Feelisch, M.; Vierling, E. Modulation of nitrosative stress by *S*-nitrosoglutathione reductase is critical for thermotolerance and plant growth in *Arabidopsis*. *Plant Cell* **2008**, *20*, 786–802. [CrossRef]
21. Hebelstrup, K.H.; Jensen, E.Ø. Expression of NO scavenging hemoglobin is involved in the timing of bolting in *Arabidopsis thaliana*. *Planta* **2008**, *227*, 917–927. [CrossRef]
22. Flores, T.; Todd, C.D.; Tovar-Mendez, A.; Dhanoa, P.K.; Correa-Aragunde, N.; Hoyos, M.E.; Brownfield, D.M.; Mullen, R.T.; Lamattina, L.; Polacco, J.C. Arginase-negative mutants of *Arabidopsis* exhibit increased nitric oxide signaling in root development. *Plant Physiol.* **2008**, *147*, 1936–1946. [CrossRef]
23. He, Y.; Tang, R.-H.; Hao, Y.; Stevens, R.D.; Cook, C.W.; Ahn, S.M.; Jing, L.; Yang, Z.; Chen, L.; Guo, F.; et al. Nitric oxide represses the *Arabidopsis* floral transition. *Science* **2004**, *305*, 1968–1971. [CrossRef] [PubMed]
24. Li, H.; Culligan, K.; Dixon, R.A.; Chory, J. CUE1: A Mesophyll Cell-Specific Positive Regulator of Light-Controlled Gene Expression in *Arabidopsis*. *Plant Cell* **1995**, *7*, 1599–1610. [CrossRef]
25. López-Juez, E.; Jarvis, R.P.; Takeuchi, A.; Page, A.M.; Chory, J. New *Arabidopsis cue* mutants suggest a close connection between plastid and phytochrome regulation of nuclear gene expression. *Plant Physiol.* **1998**, *118*, 803–815. [CrossRef]

26. Vinti, G.; Fourrier, N.; Bowyer, J.R.; López-Juez, E. *Arabidopsis cue* mutants with defective plastids are impaired primarily in the photocontrol of expression of photosynthesis-associated nuclear genes. *Plant Mol. Biol.* **2005**, *57*, 343–357. [CrossRef] [PubMed]
27. Streatfield, S.J.; Weber, A.; Kinsman, E.A.; Häusler, R.E.; Li, J.; Post-Beittenmiller, D.; Kaiser, W.M.; Pyke, K.A.; Chory, J. The Phosphoenolpyruvate/Phosphate Translocator Is Required for Phenolic Metabolism, Palisade Cell Development, and Plastid-Dependent Nuclear Gene Expression. *Plant Cell* **1999**, *11*, 1609–1621. [CrossRef]
28. Fischer, K.; Kammerer, B.; Gutensohn, M.; Arbinger, B.; Weber, A.; Häusler, R.E.; Flügge, U.I. A new class of plastidic phosphate translocators: A putative link between primary and secondary metabolism by the phosphoenolpyruvate/phosphate antiporter. *Plant Cell* **1997**, *9*, 453–462. [CrossRef]
29. Sanz, L.; Albertos, P.; Mateos, I.; Sánchez-Vicente, I.; Lechón, T.; Fernández-Marcos, M.; Lorenzo, O. Nitric oxide (NO) and phytohormones crosstalk during early plant development. *J. Exp. Bot.* **2015**, *66*, 2857–2868. [CrossRef]
30. Pető, A.; Lehotai, N.; Lozano-Juste, J.; León, J.; Tari, I.; Erdei, L.; Kolbert, Z. Involvement of nitric oxide and auxin in signal transduction of copper-induced morphological responses in *Arabidopsis* seedlings. *Ann. Bot.* **2011**, *108*, 449–457. [CrossRef] [PubMed]
31. Fu, Z.-W.; Wang, Y.-L.; Lu, Y.-T.; Yuan, T.-T. Nitric oxide is involved in stomatal development by modulating the expression of stomatal regulator genes in *Arabidopsis*. *Plant Sci.* **2016**, *252*, 282–289. [CrossRef]
32. Fernández-Marcos, M.; Sanz, L.; Lewis, D.R.; Muday, G.K.; Lorenzo, O. Nitric oxide causes root apical meristem defects and growth inhibition while reducing PIN-FORMED 1 (PIN1)-dependent acropetal auxin transport. *Proc. Natl. Acad. Sci. USA* **2011**, *108*, 18506–18511. [CrossRef]
33. Sanz, L.; Fernández-Marcos, M.; Modrego, A.; Lewis, D.R.; Muday, G.K.; Pollmann, S.; Dueñas, M.; Santos-Buelga, C.; Lorenzo, O. Nitric Oxide Plays a Role in Stem Cell Niche Homeostasis through Its Interaction with Auxin. *Plant Physiol.* **2014**, *166*, 1972–1984. [CrossRef] [PubMed]
34. Wang, J.; Wang, Y.; Lv, Q.; Wang, L.; Du, J.; Bao, F.; He, Y.-K. Nitric oxide modifies root growth by S-nitrosylation of plastidial glyceraldehyde-3-phosphate dehydrogenase. *Biochem. Biophys. Res. Commun.* **2017**, *488*, 88–94. [CrossRef]
35. Bai, S.; Li, M.; Yao, T.; Wang, H.; Zhang, Y.; Xiao, L.; Wang, J.; Zhang, Z.; Hu, Y.; Liu, W.; et al. Nitric oxide restrain root growth by DNA damage induced cell cycle arrest in *Arabidopsis thaliana*. *Nitric Oxide* **2012**, *26*, 54–60. [CrossRef]
36. Liu, W.-Z.; Kong, D.-D.; Gu, X.-X.; Gao, H.-B.; Wang, J.-Z.; Xia, M.; Gao, Q.; Tian, L.-L.; Xu, Z.-H.; Bao, F.; et al. Cytokinins can act as suppressors of nitric oxide in *Arabidopsis*. *Proc. Natl. Acad. Sci. USA* **2013**, *110*, 1548–1553. [CrossRef]
37. Yang, L.; Ji, J.; Wang, H.; Harris-Shultz, K.R.; Abd Allah, E.F.; Luo, Y.; Guan, Y.; Hu, X. Carbon Monoxide Interacts with Auxin and Nitric Oxide to Cope with Iron Deficiency in *Arabidopsis*. *Front. Plant Sci.* **2016**, *7*, 112. [CrossRef]
38. Pető, A.; Lehotai, N.; Feigl, G.; Tugyi, N.; Ördög, A.; Gémes, K.; Tari, I.; Erdei, L.; Kolbert, Z. Nitric oxide contributes to copper tolerance by influencing ROS metabolism in *Arabidopsis*. *Plant Cell Rep.* **2013**, *32*, 1913–1923. [CrossRef]
39. Yun, B.-W.; Feechan, A.; Yin, M.; Saidi, N.B.B.; Le Bihan, T.; Yu, M.; Moore, J.W.; Kang, J.-G.; Kwon, E.; Spoel, S.H.; et al. S-nitrosylation of NADPH oxidase regulates cell death in plant immunity. *Nature* **2011**, *478*, 264–268. [CrossRef]
40. Yun, B.-W.; Skelly, M.J.; Yin, M.; Yu, M.; Mun, B.-G.; Lee, S.-U.; Hussain, A.; Spoel, S.H.; Loake, G.J. Nitric oxide and S-nitrosoglutathione function additively during plant immunity. *New Phytol.* **2016**, *211*, 516–526. [CrossRef] [PubMed]
41. Kneeshaw, S.; Gelineau, S.; Tada, Y.; Loake, G.J.; Spoel, S.H. Selective Protein Denitrosylation Activity of Thioredoxin-h5 Modulates Plant Immunity. *Mol. Cell* **2014**, *56*, 153–162. [CrossRef]
42. Beligni, M.V.; Lamattina, L. Nitric oxide stimulates seed germination and de-etiolation, and inhibits hypocotyl elongation, three light-inducible responses in plants. *Planta* **2000**, *210*, 215–221. [CrossRef] [PubMed]
43. Bethke, P.C.; Libourel, I.G.L.; Reinöhl, V.; Jones, R.L. Sodium nitroprusside, cyanide, nitrite, and nitrate break *Arabidopsis* seed dormancy in a nitric oxide-dependent manner. *Planta* **2006**, *223*, 805–812. [CrossRef]
44. Bethke, P.C.; Libourel, I.G.L.; Jones, R.L. Nitric oxide reduces seed dormancy in *Arabidopsis*. *J. Exp. Bot.* **2006**, *57*, 517–526. [CrossRef]

45. Albertos, P.; Romero-Puertas, M.C.; Tatematsu, K.; Mateos, I.; Sánchez-Vicente, I.; Nambara, E.; Lorenzo, O. S-nitrosylation triggers ABI5 degradation to promote seed germination and seedling growth. *Nat. Commun.* **2015**, *6*, 8669. [CrossRef]
46. Zhao, M.-G.; Tian, Q.-Y.; Zhang, W.-H. Nitric Oxide Synthase-Dependent Nitric Oxide Production Is Associated with Salt Tolerance in *Arabidopsis*. *Plant Physiol.* **2007**, *144*, 206–217. [CrossRef]
47. Shi, H.-T.; Li, R.-J.; Cai, W.; Liu, W.; Wang, C.-L.; Lu, Y.-T. Increasing Nitric Oxide Content in *Arabidopsis thaliana* by Expressing Rat Neuronal Nitric Oxide Synthase Resulted in Enhanced Stress Tolerance. *Plant Cell Physiol.* **2012**, *53*, 344–357. [CrossRef] [PubMed]
48. Shi, H.; Liu, W.; Wei, Y.; Ye, T. Integration of auxin/indole-3-acetic acid 17 and RGA-LIKE3 confers salt stress resistance through stabilization by nitric oxide in *Arabidopsis*. *J. Exp. Bot.* **2017**, *68*, 1239–1249. [CrossRef] [PubMed]
49. Pérez-Chaca, M.V.; Rodríguez-Serrano, M.; Molina, A.S.; Pedranzani, H.E.; Zirulnik, F.; Sandalio, L.M.; Romero-Puertas, M.C. Cadmium induces two waves of reactive oxygen species in *Glycine max* (L.) roots. *Plant. Cell Environ.* **2014**, *37*, 1672–1687. [CrossRef]
50. Liu, Y.; Shi, L.; Ye, N.; Liu, R.; Jia, W.; Zhang, J. Nitric oxide-induced rapid decrease of abscisic acid concentration is required in breaking seed dormancy in *Arabidopsis*. *New Phytol.* **2009**, *183*, 1030–1042. [CrossRef]
51. Bethke, P.; Gubler, F.; Jacobsen, J.; Jones, R. Dormancy of *Arabidopsis* seeds and barley grains can be broken by nitric oxide. *Planta* **2004**, *219*, 847–855. [CrossRef]
52. Li, X.; Pan, Y.; Chang, B.; Wang, Y.; Tang, Z. NO Promotes Seed Germination and Seedling Growth Under High Salt May Depend on EIN3 Protein in *Arabidopsis*. *Front. Plant Sci.* **2015**, *6*, 1203. [CrossRef] [PubMed]
53. Corpas, F.J.; Hayashi, M.; Mano, S.; Nishimura, M.; Barroso, J.B. Peroxisomes are required for in vivo nitric oxide accumulation in the cytosol following salinity stress of *Arabidopsis* plants. *Plant Physiol.* **2009**, *151*, 2083–2094. [CrossRef]
54. Wang, Y.; Ries, A.; Wu, K.; Yang, A.; Crawford, N.M. The *Arabidopsis* prohibitin gene *PHB3* functions in nitric oxide-mediated responses and in hydrogen peroxide-induced nitric oxide accumulation. *Plant Cell* **2010**, *22*, 249–259. [CrossRef]
55. Lin, Y.; Yang, L.; Paul, M.; Zu, Y.; Tang, Z. Ethylene promotes germination of *Arabidopsis* seed under salinity by decreasing reactive oxygen species: Evidence for the involvement of nitric oxide simulated by sodium nitroprusside. *Plant Physiol. Biochem.* **2013**, *73*, 211–218. [CrossRef]
56. Zhou, S.; Jia, L.; Chu, H.; Wu, D.; Peng, X.; Liu, X.; Zhang, J.; Zhao, J.; Chen, K.; Zhao, L. *Arabidopsis* CaM1 and CaM4 promote nitric oxide production and salt resistance by inhibiting S-Nitrosoglutathione Reductase via direct binding. *PLoS Genet.* **2016**, *12*, e1006255. [CrossRef]
57. Thomson, A.J.; El-Kassaby, Y.A. Interpretation of seed-germination parameters. *New For.* **1993**, *7*, 123–132. [CrossRef]
58. Joosen, R.V.L.; Kodde, J.; Willems, L.A.J.; Ligterink, W.; van der Plas, L.H.W.; Hilhorst, H.W.M. GERMINATOR: A software package for high-throughput scoring and curve fitting of *Arabidopsis* seed germination. *Plant J.* **2010**, *62*, 148–159. [CrossRef]
59. Nambara, E.; Okamoto, M.; Tatematsu, K.; Yano, R.; Seo, M.; Kamiya, Y. Abscisic acid and the control of seed dormancy and germination. *Seed Sci. Res.* **2010**, *20*, 55–67. [CrossRef]
60. Lopez-Molina, L.; Mongrand, S.; Chua, N.H. A postgermination developmental arrest checkpoint is mediated by abscisic acid and requires the ABI5 transcription factor in *Arabidopsis*. *Proc. Natl. Acad. Sci. USA* **2001**, *98*, 4782–4787. [CrossRef]
61. Lopez-Molina, L.; Mongrand, S.; McLachlin, D.T.; Chait, B.T.; Chua, N.-H. ABI5 acts downstream of ABI3 to execute an ABA-dependent growth arrest during germination. *Plant J.* **2002**, *32*, 317–328. [CrossRef]
62. Gibbs, D.J.; Md Isa, N.; Movahedi, M.; Lozano-Juste, J.; Mendiondo, G.M.; Berckhan, S.; Marín de la Rosa, N.; Vicente Conde, J.; Sousa Correia, C.; Pearce, S.P.; et al. Nitric Oxide Sensing in Plants Is Mediated by Proteolytic Control of Group VII ERF Transcription. *Mol. Cell* **2014**, *53*, 369–379. [CrossRef]
63. Dinneny, J.R. Traversing organizational scales in plant salt-stress responses. *Curr. Opin. Plant Biol.* **2015**, *23*, 70–75. [CrossRef] [PubMed]
64. Bai, S.; Yao, T.; Li, M.; Guo, X.; Zhang, Y.; Zhu, S.; He, Y. PIF3 is involved in the primary root growth inhibition of *Arabidopsis* induced by nitric oxide in the light. *Mol. Plant* **2014**, *7*, 616–625. [CrossRef]

65. Krasylenko, Y.A.; Yemets, A.I.; Blume, Y.B. Nitric oxide synthase inhibitor L-NAME affects *Arabidopsis* root growth, morphology, and microtubule organization. *Cell Biol. Int.* **2017**, *43*, 1049–1055. [CrossRef] [PubMed]
66. Sozzani, R.; Iyer-Pascuzzi, A. Postembryonic control of root meristem growth and development. *Curr. Opin. Plant Biol.* **2014**, *17*, 7–12. [CrossRef]
67. Cederholm, H.M.; Iyer-Pascuzzi, A.S.; Benfey, P.N. Patterning the primary root in *Arabidopsis*. *Wiley Interdiscip. Rev. Dev. Biol.* **2012**, *1*, 675–691. [CrossRef]
68. Gas, E.; Flores-Pérez, U.; Sauret-Güeto, S.; Rodríguez-Concepción, M. Hunting for plant nitric oxide synthase provides new evidence of a central role for plastids in nitric oxide metabolism. *Plant Cell* **2009**, *21*, 18–23. [CrossRef]
69. Hu, W.-J.; Chen, J.; Liu, T.-W.; Liu, X.; Chen, J.; Wu, F.-H.; Wang, W.-H.; He, J.-X.; Xiao, Q.; Zheng, H.-L. Comparative proteomic analysis on wild type and nitric oxide-overproducing mutant (*nox1*) of *Arabidopsis thaliana*. *Nitric Oxide* **2014**, *36*, 19–30. [CrossRef]
70. León, J.; Costa, Á.; Castillo, M.-C. Nitric oxide triggers a transient metabolic reprogramming in *Arabidopsis*. *Sci. Rep.* **2016**, *6*, 37945. [CrossRef]
71. Zhang, Z.-W.; Luo, S.; Zhang, G.-C.; Feng, L.-Y.; Zheng, C.; Zhou, Y.-H.; Du, J.-B.; Yuan, M.; Chen, Y.-E.; Wang, C.-Q.; et al. Nitric oxide induces monosaccharide accumulation through enzyme S-nitrosylation. *Plant Cell Environ.* **2017**, *40*, 1834–1848. [CrossRef] [PubMed]
72. Lozano-Juste, J.; León, J. Enhanced abscisic acid-mediated responses in *nia1nia2noa1-2* triple mutant impaired in NIA/NR- and AtNOA1-dependent nitric oxide biosynthesis in *Arabidopsis*. *Plant Physiol.* **2010**, *152*, 891–903. [CrossRef]
73. Wimalasekera, R.; Villar, C.; Begum, T.; Scherer, G.F.E. *COPPER AMINE OXIDASE1* (*CuAO1*) of *Arabidopsis thaliana* contributes to abscisic acid- and polyamine-induced nitric oxide biosynthesis and abscisic acid signal transduction. *Mol. Plant* **2011**, *4*, 663–678. [CrossRef]
74. Ma, Z.; Marsolais, F.; Bykova, N.V.; Igamberdiev, A.U. Nitric Oxide and Reactive Oxygen Species Mediate Metabolic Changes in Barley Seed Embryo during Germination. *Front. Plant Sci.* **2016**, *7*, 138. [CrossRef]
75. Wang, P.; Zhu, J.-K.; Lang, Z. Nitric oxide suppresses the inhibitory effect of abscisic acid on seed germination by S-nitrosylation of SnRK2 proteins. *Plant Signal. Behav.* **2015**, *10*, e1031939. [CrossRef]
76. Bewley, J.D. Seed germination and dormancy. *Plant Cell* **1997**, *9*, 1055–1066. [CrossRef]
77. Dean, G.; Cao, Y.; Xiang, D.; Provart, N.J.; Ramsay, L.; Ahad, A.; White, R.; Selvaraj, G.; Datla, R.; Haughn, G. Analysis of gene expression patterns during seed coat development in *Arabidopsis*. *Mol. Plant* **2011**, *4*, 1074–1091. [CrossRef]
78. Waese, J.; Fan, J.; Pasha, A.; Yu, H.; Fucile, G.; Shi, R.; Cumming, M.; Kelley, L.A.; Sternberg, M.J.; Krishnakumar, V.; et al. ePlant: Visualizing and Exploring Multiple Levels of Data for Hypothesis Generation in Plant Biology. *Plant Cell* **2017**, *29*, 1806–1821. [CrossRef]
79. Zhao, M.; Zhao, X.; Wu, Y.; Zhang, L. Enhanced sensitivity to oxidative stress in an *Arabidopsis* nitric oxide synthase mutant. *J. Plant Physiol.* **2007**, *164*, 737–745. [CrossRef] [PubMed]
80. Du, S.-T.; Liu, Y.; Zhang, P.; Liu, H.-J.; Zhang, X.-Q.; Zhang, R.-R. Atmospheric application of trace amounts of nitric oxide enhances tolerance to salt stress and improves nutritional quality in spinach (*Spinacia oleracea* L.). *Food Chem.* **2015**, *173*, 905–911. [CrossRef]
81. Valderrama, R.; Corpas, F.J.; Carreras, A.; Fernández-Ocaña, A.; Chaki, M.; Luque, F.; Gómez-Rodríguez, M.V.; Colmenero-Varea, P.; del Río, L.A.; Barroso, J.B. Nitrosative stress in plants. *FEBS Lett.* **2007**, *581*, 453–461. [CrossRef]
82. Poór, P.; Kovács, J.; Borbély, P.; Takács, Z.; Szepesi, Á.; Tari, I. Salt stress-induced production of reactive oxygen- and nitrogen species and cell death in the ethylene receptor mutant *Never ripe* and wild type tomato roots. *Plant Physiol. Biochem.* **2015**, *97*, 313–322. [CrossRef]
83. Fancy, N.N.; Bahlmann, A.-K.; Loake, G.J. Nitric oxide function in plant abiotic stress. *Plant. Cell Environ.* **2017**, *40*, 462–472. [CrossRef] [PubMed]
84. Foresi, N.; Mayta, M.L.; Lodeyro, A.F.; Scuffi, D.; Correa-Aragunde, N.; García-Mata, C.; Casalongué, C.; Carrillo, N.; Lamattina, L. Expression of the tetrahydrofolate-dependent nitric oxide synthase from the green alga *Ostreococcus tauri* increases tolerance to abiotic stresses and influences stomatal development in *Arabidopsis*. *Plant J.* **2015**, *82*, 806–821. [CrossRef]

85. Mata-Pérez, C.; Begara-Morales, J.C.; Chaki, M.; Sánchez-Calvo, B.; Valderrama, R.; Padilla, M.N.; Corpas, F.J.; Barroso, J.B. Protein Tyrosine Nitration during Development and Abiotic Stress Response in Plants. *Front. Plant Sci.* **2016**, *7*, 1699. [CrossRef]
86. Brocard, I.M.; Lynch, T.J.; Finkelstein, R.R. Regulation and role of the Arabidopsis *Abscisic Acid-Insensitive 5* gene in abscisic acid, sugar, and stress response. *Plant Physiol.* **2002**, *129*, 1533–1543. [CrossRef] [PubMed]
87. Eastmond, P.J.; Germain, V.; Lange, P.R.; Bryce, J.H.; Smith, S.M.; Graham, I.A. Postgerminative growth and lipid catabolism in oilseeds lacking the glyoxylate cycle. *Proc. Natl. Acad. Sci. USA* **2000**, *97*, 5669–5674. [CrossRef]
88. Kircher, S.; Schopfer, P. Photosynthetic sucrose acts as cotyledon-derived long-distance signal to control root growth during early seedling development in *Arabidopsis*. *Proc. Natl. Acad. Sci. USA* **2012**, *109*, 11217–11221. [CrossRef]
89. Voll, L.; Hausler, R.E.; Hecker, R.; Weber, A.; Weissenbock, G.; Fiene, G.; Waffenschmidt, S.; Flügge, U.-I. The phenotype of the *Arabidopsis cue1* mutant is not simply caused by a general restriction of the shikimate pathway. *Plant J.* **2003**, *36*, 301–317. [CrossRef]
90. Moreau, M.; Lee, G.I.; Wang, Y.; Crane, B.R.; Klessig, D.F. AtNOS/AtNOA1 is a functional *Arabidopsis thaliana* cGTPase and not a nitric-oxide synthase. *J. Biol. Chem.* **2008**, *283*, 32957–32967. [CrossRef]
91. Guo, F.-Q.; Okamoto, M.; Crawford, N.M. Identification of a plant nitric oxide synthase gene involved in hormonal signaling. *Science* **2003**, *302*, 100–103. [CrossRef] [PubMed]
92. Van Ree, K.; Gehl, B.; Wassim Chehab, E.; Tsai, Y.-C.; Braam, J. Nitric oxide accumulation in *Arabidopsis* is independent of NOA1 in the presence of sucrose. *Plant J.* **2011**, *68*, 225–233. [CrossRef]
93. Flores-Pérez, U.; Sauret-Güeto, S.; Gas, E.; Jarvis, P.; Rodríguez-Concepción, M. A mutant impaired in the production of plastome-encoded proteins uncovers a mechanism for the homeostasis of isoprenoid biosynthetic enzymes in *Arabidopsis* plastids. *Plant Cell* **2008**, *20*, 1303–1315. [CrossRef]
94. Wilkinson, J.Q.; Crawford, N.M. Identification of the Arabidopsis *CHL3* gene as the nitrate reductase structural gene *NIA2*. *Plant Cell* **1991**, *3*, 461–471.
95. Bolle, C.; Sopory, S.; Lubberstedt, T.; Klosgen, R.B.; Herrmann, R.G.; Oelmuller, R. The Role of Plastids in the Expression of Nuclear Genes for Thylakoid Proteins Studied with Chimeric β-Glucuronidase Gene Fusions. *Plant Physiol.* **1994**, *105*, 1355–1364. [CrossRef]
96. Redei, G.P. Genetic blocks in the thiamine synthesis of the angiosperm *Arabidopsis*. *Am. J. Bot.* **1965**, *52*, 834–841. [CrossRef]
97. Murashige, T.; Skoog, F. A Revised Medium for Rapid Growth and Bio Assays with Tobacco Tissue Cultures. *Physiol. Plant.* **1962**, *15*, 473–497. [CrossRef]
98. Bradford, M.M. A rapid and sensitive method for the quantitation of microgram quantities of protein utilizing the principle of protein-dye binding. *Anal. Biochem.* **1976**, *72*, 248–254. [CrossRef]
99. Schindelin, J.; Arganda-Carreras, I.; Frise, E.; Kaynig, V.; Longair, M.; Pietzsch, T.; Preibisch, S.; Rueden, C.; Saalfeld, S.; Schmid, B.; et al. Fiji: An open-source platform for biological-image analysis. *Nat. Methods* **2012**, *9*, 676–682. [CrossRef]
100. Rymen, B.; Coppens, F.; Dhondt, S.; Fiorani, F.; Beemster, G.T.S. Kinematic Analysis of Cell Division and Expansion. In *Plant Developmental Biology: Methods and Protocols*; Hennig, L., Köhler, C., Eds.; Humana Press: Totowa, NJ, USA, 2010; pp. 203–227.
101. Nieuwland, J.; Maughan, S.; Dewitte, W.; Scofield, S.; Sanz, L.; Murray, J.A.H. The D-type cyclin CYCD4;1 modulates lateral root density in *Arabidopsis* by affecting the basal meristem region. *Proc. Natl. Acad. Sci. USA* **2009**, *106*, 22528–22533. [CrossRef]
102. French, A.P.; Wilson, M.H.; Kenobi, K.; Dietrich, D.; Voss, U.; Ubeda-Tomas, S.; Pridmore, T.P.; Wells, D.M. Identifying biological landmarks using a novel cell measuring image analysis tool: Cell-o-Tape. *Plant Methods* **2012**, *8*, 7. [CrossRef]
103. R Core Team. R: A Language Environment for Statistical Computing. R Foundation for Statistical Computing. 2020. Available online: http://www.R-project.org (accessed on 29 September 2020).
104. RStudio Team. RStudio. RStudio: Integrated Development for R. RStudio, PBC. 2020. Available online: http://www.rstudio.com (accessed on 29 September 2020).

105. Antoniou, C.; Filippou, P.; Mylona, P.; Fasoula, D.; Ioannides, I.; Polidoros, A.; Fotopoulos, V. Developmental stage- and concentration-specific sodium nitroprusside application results in nitrate reductase regulation and the modification of nitrate metabolism in leaves of *Medicago truncatula* plants. *Plant Signal. Behav.* **2013**, *8*, e25479. [CrossRef]
106. Igamberdiev, A.U.; Ratcliffe, R.G.; Gupta, K.J. Plant mitochondria: Source and target for nitric oxide. *Mitochondrion* **2014**, *19*, 329–333. [CrossRef]

Publisher's Note: MDPI stays neutral with regard to jurisdictional claims in published maps and institutional affiliations.

 © 2020 by the authors. Licensee MDPI, Basel, Switzerland. This article is an open access article distributed under the terms and conditions of the Creative Commons Attribution (CC BY) license (http://creativecommons.org/licenses/by/4.0/).

Review

Post-Translational Modification of Proteins Mediated by Nitro-Fatty Acids in Plants: Nitroalkylation

Lorena Aranda-Caño, Beatriz Sánchez-Calvo, Juan C. Begara-Morales, Mounira Chaki, Capilla Mata-Pérez, María N. Padilla, Raquel Valderrama and Juan B. Barroso *

Group of Biochemistry and Cell Signaling in Nitric Oxide, Department of Experimental Biology, Center for Advanced Studies in Olive Grove and Olive Oils, Faculty of Experimental Sciences, University Campus Las Lagunillas, University of Jaén, E-23071 Jaén, Spain; laranda@ujaen.es (L.A.-C.); sanchezcalvobeatriz@gmail.com (B.S.-C.); jbegara@ujaen.es (J.C.B.-M.); mounira@ujaen.es (M.C.); mmata@ujaen.es (C.M.-P.); npadilla@ujaen.es (M.N.P.); ravalde@ujaen.es (R.V.)
* Correspondence: jbarroso@ujaen.es; Tel.: +34-953-212764

Received: 26 February 2019; Accepted: 26 March 2019; Published: 29 March 2019

Abstract: Nitrate fatty acids (NO_2-FAs) are considered reactive lipid species derived from the non-enzymatic oxidation of polyunsaturated fatty acids by nitric oxide (NO) and related species. Nitrate fatty acids are powerful biological electrophiles which can react with biological nucleophiles such as glutathione and certain protein–amino acid residues. The adduction of NO_2-FAs to protein targets generates a reversible post-translational modification called nitroalkylation. In different animal and human systems, NO_2-FAs, such as nitro-oleic acid (NO_2-OA) and conjugated nitro-linoleic acid (NO_2-cLA), have cytoprotective and anti-inflammatory influences in a broad spectrum of pathologies by modulating various intracellular pathways. However, little knowledge on these molecules in the plant kingdom exists. The presence of NO_2-OA and NO_2-cLA in olives and extra-virgin olive oil and nitro-linolenic acid (NO_2-Ln) in *Arabidopsis thaliana* has recently been detected. Specifically, NO_2-Ln acts as a signaling molecule during seed and plant progression and beneath abiotic stress events. It can also release NO and modulate the expression of genes associated with antioxidant responses. Nevertheless, the repercussions of nitroalkylation on plant proteins are still poorly known. In this review, we demonstrate the existence of endogenous nitroalkylation and its effect on the in vitro activity of the antioxidant protein ascorbate peroxidase.

Keywords: nitro-fatty acids; nitroalkenes; nitroalkylation; electrophile; nucleophile; signaling mechanism; post-translational modification; reactive lipid species; nitro-lipid-protein adducts

1. Introduction

Reactive lipid species (RLS), or so-called lipid-derived electrophiles (LDEs), are caused by polyunsaturated fatty acids (PUFAs) peroxidation [1–4]. Reactive lipid species have been identified in sanguine fluid, plasma, urine, human tissues, and animal models using array techniques. Recently, they have also been detected in plant systems with the aid of mass spectrometry. A rise in RLS abundance under pathological and stress circumstances has been broadly reported [4–10].

Polyunsaturated fatty acids are targets of peroxidation due to their unsaturated double bonds [4,11]. The main mechanisms of PUFA peroxidation are non-enzymatic autocatalytic oxidation reactions [1,12], while enzymatic oxidation reactions involving three heme-containing metallo-enzyme families (lipoxygenases (LOXs), cyclooxigenases (COXs) [1,13], and cytochromes P450 (CYPs) [1]), as well as $NADP^+$-dependent dehydrogenases [1,14] which can also occur. Non-enzymatic mechanisms include PUFA nitration triggered by reactive nitrogen species (RNS) such as nitric oxide (NO) and its derived molecules [1,15,16]. A preferential target for lipid peroxidation is arachidonic acid, whose oxidation yields several products. The non-enzymatic oxidation reactions of PUFAs yield aldehydes such as 4-hydroxynonenal (HNE) and malondialdehyde (MDA), as well as the J- and A-series of isoprostanes [4,17]. Prostaglandins (15-deoxy-Δ12,14-prostaglandin J$_2$) and lipoxins are generated by enzymatic oxidation reactions catalyzed by COX and LOX, respectively [4,18,19]. The oxidation of arachidonic acid by NO-derived species yields 12-nitroarachidonic acid (12-NO$_2$-AA) [4,20].

The addition of aldehyde, α-β-unsaturated carbonyl, epoxide or nitroalkene substituents to PUFAs during the peroxidation process causes the formation of lipid-derived species with electrophilic properties. From a chemical perspective, electrophilic molecules contain an electron-poor moiety, which makes them chemically reactive with nucleophiles (electron-rich molecules) [1]. Nucleophiles and electrophiles can be classified according to a hard/soft acid–base (HSAB) model [21]. Hard electrophiles, whose outer layer electrons are not readily excited, are difficult to polarize. Conversely, soft electrophiles have a more diffused electron distribution or partial positive charges due to the possession of electron-withdrawing substituents such as nitro groups. Nucleophiles can be characterized in a similar manner. Hard nucleophiles are highly electronegative and difficult to polarize, in contrast to soft nucleophiles, which have empty, low-lying electron orbitals. The softest biological nucleophiles, cysteine thiols, which integrate proteins, are also present in the antioxidant tripeptide glutathione (GSH). Primary and secondary amines of lysine, arginine, and histidine residues are regarded as hard nucleophiles [22]. The reactivity of nucleophiles does not only depend on the presence of hard and soft electrophiles in their vicinity, other factors such as their microenvironment (including hydrogen bonding reactions with neighboring amino residues) can influence nucleophile ionization too. For instance, as the reactivity of thiolate anions is higher than that of protonated thiols, the decrease in cysteine pKa induced by protein conformation increases its nucleophilicity [23,24]. As a general rule, hard electrophiles preferentially react with hard nucleophiles, while soft electrophiles interact with soft nucleophiles [1,25].

The importance of RLS resides in their electrophilic reactivity, which enables them to establish covalent adducts with GSH and nucleophilic amino acid residues of proteins such as cysteine, histidine, and lysine, generating post-translational modifications (PTMs) of proteins [4,26–30]. The endogenous occurrence of electrophilic fatty acids has been detected at low concentrations in plasma and animal tissues, whose biological significance is still little known [1,31]. Due to their innate reactivity, the rapid adduction process of RLS with susceptible GSH and nucleophilic residues of proteins may be functionally significant in relation to signaling responses [1,32]. However, it should be mentioned that an equilibrium between adducted and free forms exists in the milieu [1,33].

Pathological conditions promote the enzymatic and non-enzymatic generation of endogenous RLS. In these situations, an increase in the expression of oxidases and oxygenases and in the non-enzymatic production of reactive oxygen and nitrogen species (ROS and RNS), such as reduced oxygen species and oxides of nitrogen (NO, peroxynitrite (ONOO$^-$), nitrogen dioxide ($^\cdot$NO$_2$), nitronium cation (NO$_2^+$), takes place. All these species could react with PUFAs yielding RLS. Macrophage, eosinophil, and neutrophil cells in the immune system alter lipase activation, causing the scission of fatty acids from membranes. Thus, these disengaged fatty acids may be substrates for subsequent RLS formation [1,22,34]. The electrophilic nature of RLS induces the nucleophilic attack of proteins, leading to modifications in tertiary and quaternary structures, in catalytic activities, in charge and hydrophobicity, in subcellular localization, and in protein cross-linking. The main proteins susceptible to adduction perform metabolic functions such as cytoskeletal function, transcriptional regulation, host defense, ion and macromolecule transport, and enzyme catalysis. These proteins are involved in manifold physiological processes comprising resolution of inflammation, cell death, and induction of cellular antioxidants. In this respect, the anti-inflammatory and antioxidant responses stimulated by RLS adduction suggest the existence of an equilibrium between prompting events, electrophile production, protein adduction, and adaptive cellular responses. Therefore, RLS adduction allows organisms to cope with alterations generated under conditions of metabolic stress, inflammation, and modification in cells and tissues [1,4,35–39].

In plant systems, PUFA peroxidation caused by non-enzymatic or/and enzymatic (LOX-mediated) reactions generates some products with cytotoxic effects and others with protective anti-stress effects.

The LOX pathway yields RLS related to plant defense responses to pathogen infections [40] and wounding [41], and in the regulation of hypersensitive programmed cell death [42] and senescence [43]. Non-enzymatic processes can generate both harmful products with damaging actions [44] and phytoprostanes, which have biological properties similar to those of jasmonic acid [45]. Recent knowledge has illustrated the formation of RLS that perform signaling roles and are implicated in antioxidant responses as a result of the oxidation of NO-derived molecules [9].

This review focuses on the study of reactive lipids species called nitroalkenes. Specifically, we will argue the biological properties of nitroalkenes both in animal and plant systems, as well as their signaling potential generated by a post-translational modification of proteins called nitroalkylation.

2. Nitro-Fatty Acids in Animals

The reactive lipids species resulting from the interaction of unsaturated fatty acids with NO and derived species, such as NO_2 and $ONOO^-$, are called nitro-fatty acids (NO_2-FAs), nitrolipids or nitroalkenes [46].

Although the interaction between unsaturated fatty acids and RNS has been widely studied, two distinct mechanisms have been suggested to explain the in vivo nitration of fatty acids, a process which remains unknown [47]. One mechanism involves the generation of an alkyl radical through a radical hydrogen abstraction from a bis-allylic carbon followed by a double-bond rearrangement and the incorporation of a NO_2 radical (Figure 1A) [48,49]. The other mechanism consists on the generation of a carbon-centered radical through the direct addition of NO_2, which can be further oxidized either with or without a second insertion of NO_2 in order to form the nitro-fatty acid. When the carbon-centered radical reacts with the second NO_2, an unstable nitro-nitrite or dinitro compound appears which rapidly decomposes and releases nitrous acid (HNO_2), yielding the nitro-fatty acid (Figure 1B) [49,50].

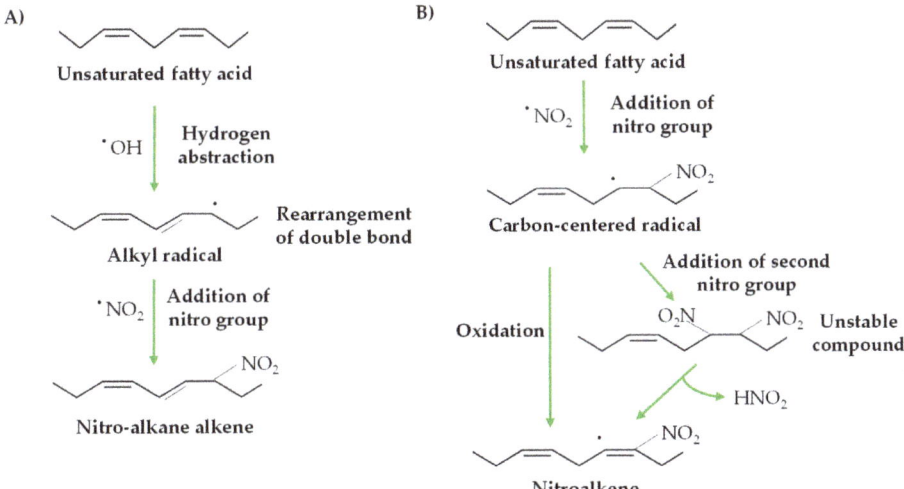

Figure 1. Possible mechanisms of nitrate fatty acid (NO_2-FA) formation. (**A**) Alkyl radical generation through a radical hydrogen abstraction from a bis-allylic carbon followed by the insertion of NO_2. (**B**) NO_2-FA formation by the direct addition of NO_2 and its oxidation (modified from Reference [49]).

In recent years, important progress in the endogenous detection of NO_2-FAs has been achieved in animal and human models. In animal systems, it is worth highlighting the detection of nitrated oleic (NO_2-OA) and linoleic acid (NO_2-LA) in the murine model of focal cardiac ischemia-reperfusion (I/R). The formation of these NO_2-FAs was due to reoxygenation-induced tissue damage which generated acidification, hypoxia, as well as ROS and RNS [49,51]. It should be mentioned that other NO_2-FAs were detected in an experimental rat model of ischemic preconditioning (IPC) [49,52]. High-resolution liquid chromatography mass spectrometry (LC-MS/MS) procedures have revealed a preferential nitration of conjugated linoleic acid (cLA) in animal systems. This fatty acid presents positional and geometric isomers of linoleic acid which have conjugated dienes in cis and trans configurations. These species have conjugated double bonds which are not separated by a methylene group [53]. Conjugated linoleic acid, which displays more reactivity with ˙NO_2 than bis-allylic fatty-acids, is the main in vivo endogenous nitration target [47,49]. The formation of nitrated cLA has been detected in activated macrophages under inflammatory conditions and in the gastric compartment following the ingest of cLA and NO_2^- [47,49,54].

Advances in chromatography mass spectrometry techniques, in vitro nitration, and animal model studies have increased our understanding of the nitration of unsaturated fatty acids in humans. Dietary products such as oils and seeds are the principal sources of unsaturated fatty acids such as oleic acid (OA), conjugated linoleic (cLA), and linolenic (cLn) acids. Pomegranates are regarded as sources of cLn, while dairy products and meat are a source of cLA. Interestingly, cLA is absorbed at much higher levels than cLn [49,55]. Dietary products such as vegetables and herbs are sources of nitrate (NO_3^-) and nitrite (NO_2^-) [49,56,57]. These NO-derived species are necessary to generate nitrated PUFAs, as nitrite is a nitrating compound derived from nitrate. However, the low level of nitrite in basal metabolic conditions is increased through the conversion of nitrate by commensal bacteria in the gastrointestinal tract [58]. As with animal models, NO_2-cLA is the principal nitroalkene generated in humans (Table 1) [47].

Table 1. Principal nitro-fatty acids detected in animal and plant systems. The lines on the middle of the double bond indicate that the nitro group could be bounded in any of the adjacent carbons. Although double bonds can generate the corresponding cis- and trans-isomers, only the cis forms are shown.

Name	Formula	Chemical Structure
Nitro-oleic acid (9-, and 10-nitro-all-cis-octadecaenoic acid)	NO_2-OA (18:1)	
Nitro-linoleic acid (9-, 10-, 12-, and 13-nitro-all-cis octadecadienoic acid)	NO_2-LA (18:2)	
Nitro-linolenic acid (9-, 10-, 12-, 13-, 15- and 16-nitro-all-cis-octadecatrienoic acid)	NO_2-Ln (18:3)	
Nitro-arachidonic acid (5-, 6-, 8-, 9-, 11-, 12-, 14- and 15-nitro-all-cis-eicosatetraenoic acid)	NO_2-AA (20:4)	

In addition to those mentioned above, other NO$_2$-FAs, such as nitro-oleic acid (NO$_2$-OA), nitro-linoleic acid (NO$_2$-LA), conjugated nitro-linoleic acid (NO$_2$-cLA), nitro-arachidonic acid (NO$_2$-AA), and cholesteryl nitrolinoleate (NO$_2$-CL) have been detected in vivo through quantitative analyses of blood and urine under both healthy and inflammatory conditions (Table 1) [59,60].

Nitrate fatty acids are endowed with a specific chemical reactivity which facilitates cellular signaling events. In addition, these molecules have potent biological properties such as a NO-releasing capacity which was observed for the first time in aqueous milieu in animal systems [15,61–63]. Two possible NO-releasing mechanisms have been proffered. The first one consists of a modified Nef reaction which generates a nitrous intermediate with an especially weak C–N bond that quickly decays to yield NO and a radical stabilized by conjugation with alkene and the OH group (Figure 2) [15,46]. The second mechanism involves the rearrangement of the nitroalkene to a nitrite ester followed by a process of homolysis to form NO and an enol group (Figure 3) [46,64,65]. Another biological property of these compounds is their hydrophobic stability in cell membranes and lipoproteins, which act as endogenous NO$_2$-FA reservoirs which can be supplied to other locations in the cell in order to act as signaling molecules [15]. An additional biological property of NO$_2$-FAs is their capacity to mediate post-translational modifications through nitroalkylation, which will be discussed below [46,51,66–68].

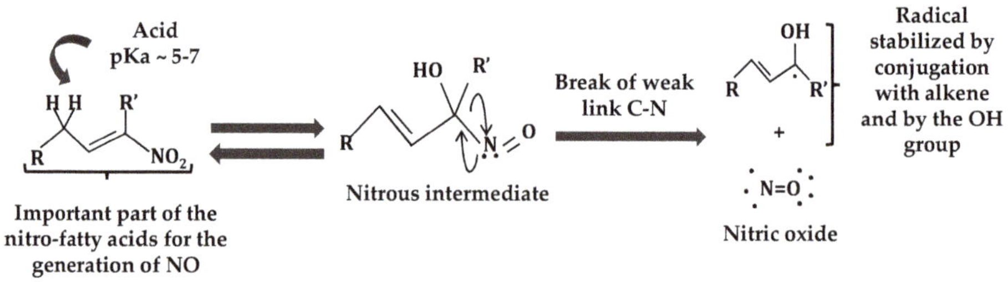

Figure 2. Mechanism of NO release through the modified Nef reaction. This mechanism consists of the generation of a nitrous intermediate which can homolyze in the aqueous medium to yield a carbon radical and nitric oxide (modified from Reference [62]).

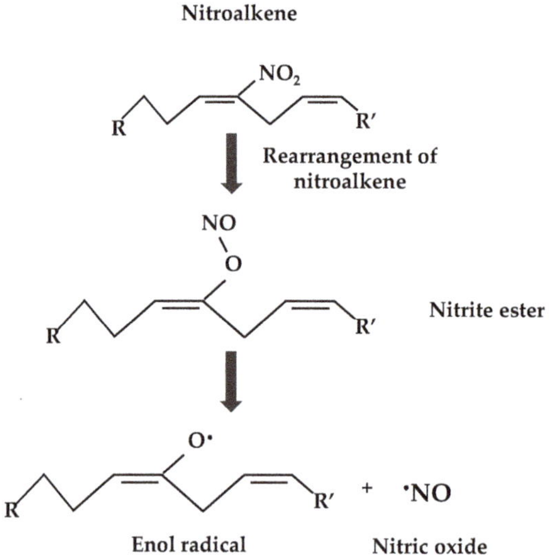

Figure 3. Release of nitric oxide from nitroalkenes through a rearrangement process. A nitrite ester is formed and homolyzed to yield NO and an enol radical (modified from Reference [15]).

Following the discovery of the presence of endogenous NO_2-FAs and their biological properties in animal and human systems, their metabolism and distribution have been examined. In this regard, NO_2-FAs have been shown to bind carrier proteins such as albumin, may be subjected to the normal lipid metabolism processes such as saturation and β-oxidation and can be esterified into complex lipids [22,49,69,70]. A recent study has shown that prostaglandin reductase leads to the reduction of NO_2-FA to electrophilic nitroalkanes and that both alkenes and nitroalkanes are subjected to β-oxidation [71]. On the other hand, gastric digestion and inflammatory conditions lead to the formation of complex lipids containing NO_2-FAs, as the formation of triglycerides (TGs) containing NO_2-FAs has been detected in adipocytes and rat plasma following the in vitro acidic gastric digestion of TGs with NO_2-OA supplementation [69]. Phospholipids containing NO_2-FAs have also been uncovered in cardiac mitochondria and cardiomyoblasts from a diabetes mellitus animal model [72]. All these studies illustrate the presence of NO_2-FAs and their metabolites in complex lipids. Lipase action can cause these NO_2-FA-containing complex lipids to release electrophilic species. In addition, free electrophilic species may travel to remote tissues to regulate cell homeostasis and tissue signaling [49].

3. Nitro-Fatty Acids in Plants

Nitrate fatty acids have been widely regarded as novel mediators of cell signaling in animal organisms. However, the knowledge about them in the plant kingdom is limited. The constitutive presence of NO_2-FAs in plant systems was initially characterized in extra-virgin olive oil (EVOO) (a basic component of the Mediterranean diet) in which oleic acid, followed by palmitic (PA), linoleic (LA), and linolenic (Ln) acids are present [73–75]. Given their properties mentioned above, the inherent occurrence of NO_2-FAs in EVOO and olives was analyzed using mass spectrometry techniques. Different endogenous NO_2-cLA isomers were detected in EVOO, while intrinsic NO_2-OA-cysteine adducts (higher levels in the olive peel) were found in olives. These reports demonstrate that both EVOO and olives are sources and endogenous reservoirs of NO_2-FAs, which could be responsible of the anti-inflammatory and anti-hypertensive properties of EVOO [10,70].

Additionally, the presence of NO_2-FAs has been recently reported in both cell–suspension cultures (ACSC) and seedlings of the model plant *Arabidopsis thaliana*. Originally, the model plant's lipid composition was analyzed, with a predominance of Ln, followed by LA and OA [10]. The biological occurrence of NO_2-Ln (Table 1) was only detected in ACSC (0.28 pmol/g FW) and seedlings (3.84 pmol/g FW) [9,10], while a modulation in NO_2-Ln levels was detected during plant growth. Seeds, 14-day-old seedlings and leaves from 30- and 45-day-old Arabidopsis plants were used. The higher NO_2-Ln content (11.18 pmol/g FW) was quantified at the seeds stage, with a continuous decline observed in the final vegetative and reproductive stages of the life cycle (0.54 pmol/g FW) [9,10]. In addition, the potential of NO_2-Ln to emit NO has been recently evidenced [9,76], and the high NO_2-Ln content in seeds could provide an additional source of NO which could favor germination and the onset of vegetative development [9,77–79].

Mass spectrometry techniques were also used to analyze the profile of NO_2-FAs in other plant species. In this sense, NO_2-Ln was detected in rice (*Oryza sativa*) leaves (0.748 pmol/g FW). The same type of NO_2-FA was identified in pea leaves (*Pisum sativum*) mitochondria (0.084 pmol/g FW) and peroxisomes (0.282 pmol/g FW) and roots (0.072 pmol/g FW). These analyses show the wide spread of NO_2-FAs in plant organisms [10,76]. Furthermore, the levels of NO_2-Ln detected in plants are similar to those found in animal systems [31], which reinforces their essential role as signaling contributors in plants [10,80].

On the other hand, the NO_2-Ln abundance was also quantified in Arabidopsis under adverse environmental conditions such as mechanical wounding, salinity, low temperature, and heavy-metal stress. Under these stress situations, a meaningful rise in NO_2-Ln content was monitored accompanied by an induction of genes associated with oxidative stress and oxygen-containing compound responses [9,80,81].

After demonstrating its relationship with plant development and plant adverse situations, a transcriptomic analysis by RNA-seq technology allowed us to analyze the signaling role played by NO$_2$-Ln. Initially, ACSC treated with increasing concentrations of NO$_2$-Ln (10 μM and 100 μM) showed this molecule's clear signaling response in terms of plant physiology and dose-dependence responses [9] previously described in animal systems [47,82]. A set of overexpressed genes related to abiotic and oxidative stress responses were detected after treatment with NO$_2$-Ln, while other genes implicated in biological procedures, such as biosynthesis of cellular metabolites, were downregulated, with a similar pattern being observed in seedlings. It is important to highlight the involvement of upregulated genes in protein folding as well as in responses to heat and H$_2$O$_2$ stress. Unexpectedly, around 40% of the genes which responded to NO$_2$-Ln were involved in heat-shock responses (HSRs) [9]. In animal systems, the treatment with NO$_2$-OA also activates a considerable number of genes related to HSRs, which reveals the presence of a conserved mechanism of response to NO$_2$-FAs in both animal and plant systems [9,82].

Among the upregulated genes which responded to reactive oxygen species (ROS) is a gene encoding for cytosolic ascorbate peroxidase 2 (APX2), which is a relevant enzyme involved in defending plants against H$_2$O$_2$. Additionally, under abiotic stress situations such as high temperatures and light intensity, interactions between APX2 and the heat shock transcription factor (HSFA2) have been detected [10,83].

Although the participation of NO$_2$-Ln in plant biology and responses to abiotic stress conditions has been previously described, the mechanisms involved in NO$_2$-Ln's defense responses to stress in plants are still little known. As with animal systems, the release of NO by NO$_2$-Ln in aqueous medium, which could be a signaling mechanism, has been demonstrated in Arabidopsis cell cultures by various in vitro experimental techniques such as ozone chemiluminescence, 4,5-diaminofluorescein (DAF-2) spectrofluorometric probes, confocal laser scanning microscopy, and the oxyhemoglobin oxidation method. Ozone chemiluminescence showed that NO-releasing from NO$_2$-FA was not propitious in acidic locations, since at neutral pH (7.4) the maximum releasing of NO was achieved. This finding may be of considerable importance inside the cells, as mitochondria, peroxisomes, and the cytosol have a basic or neutral pH [10,76]. In addition, when the leaves and roots of Arabidopsis seedlings were treated with NO$_2$-Ln, green fluorescence arose as a consequence of the increase in NO content, thus demonstrating the in vivo capability of NO$_2$-Ln to provide NO. In addition, the subsequent treatment of samples with the NO scavenger 2-(4-carboxyphenyl)-4,4,5,5-tetramethylimidazoline-1-oxyl-3-oxide (cPTIO) causes a decrease in fluorescence [9,84]. These results emphasize the important role of NO$_2$-Ln as a NO reservoir, and thus, the indirect involvement of NO$_2$-FAs in plant growth, in the response to (a)biotic stress processes and in a variety of NO-related post-translational modifications (NO-PTMs) [80,85–87].

4. Nitroalkylation

Nitro-fatty acids, which are potent electrophiles owing to the presence of electron-withdrawing nitro (-NO$_2$) substituents in the beta carbon, mainly act via post-translational modifications. For this reason, they are able to react with nucleophiles like glutathione or target amino acid residues, which affects their protein structure and eventually their function and subcellular localization [67,88]. The nitroalkylation PTM involves the establishment of a nitro-lipid-protein adduct with the cession of a couple of electrons from the nucleophile to the electrophile (NO$_2$-FA) to form a covalent bond, via a Michael adduction. This process generates lipoxidation adducts (Figure 4). Nitroalkylation provokes a chain of signaling phenomena that concludes with anti-inflammatory, anti-hypersensitive, anti-tumorigenic, cytoprotective, and antioxidant effects arbitrated by NO$_2$-FAs [46,89].

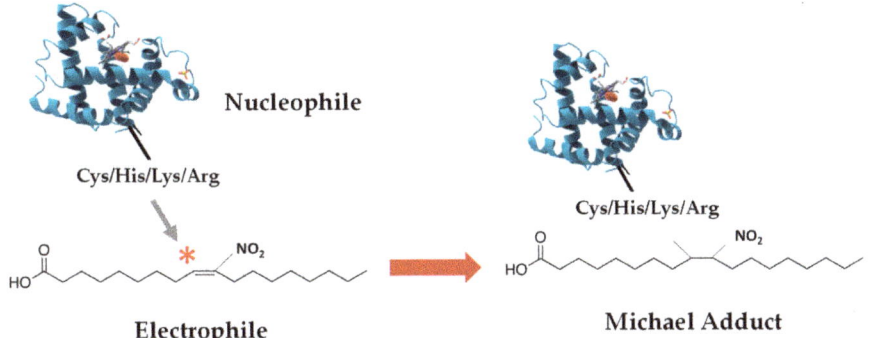

Figure 4. Nitroalkylation or formation of nitro-lipid-protein adducts. The attack of electrophilic nitro-fatty acids by nucleophilic protein residues leads to the establishment of a Michael adduct.

Diverse studies have displayed the reversible character of nitroalkylation which enables it to act as a selective signaling pathway in stressful environments. Under these conditions, the rise in the ROS and RNS levels could affect the stability of nitroalkylation. Reactive oxygen and nitrogen species (ROS and RNS) can cause the oxidation of the bond between the sulfur residues and the α-carbon of the NO_2-FAs (Michael adduct) resulting both in the generation of sulfoxides and derived species and the scission of the Michael adduct. This process results in the releasing of the nitroalkene which enables the protein to recover its initial state [22,66,81,88,90,91]. The reversible possibilities of nitroalkylation in biological processes are of considerable importance, as irreversible PTMs usually lead to permanent loss of function, and thus protein degradation [22,46,68]. Although the main nucleophiles which react with NO_2-FA are cysteine thiols (Cys-SH), and not all are able to react with electrophiles, in this sense, the deprotonated cysteine thiolate (Cys-S$^-$) is specifically the most prone to react [92,93]. Other nucleophiles are the amino substituents of lysine and arginine residues and the imidazole moiety of histidine [89].

4.1. Nitroalkylation in Animals

Nitrate fatty acids act as signaling mediators, since a scant amount of them act as powerful signal transduction cascade mediators that carry out changes in protein function through PTMs [1,66,68,94]. As mentioned above, processes such as digestion and inflammation lead to the genesis of NO_2-FAs, predominantly NO_2-cLA. In animal systems, NO_2-FAs protect against a broad cluster of diseases such as atherosclerosis, restenosis, ischemia-reperfusion, renal injury, diabetes, metabolic syndrome, endotoxemia, and triple-negative breast cancer [95,96]. Their pluripotent cell signaling capacity enables NO_2-FAs to modulate various intracellular pathways. In this line, the capacity of NO_2-FAs to release NO via the Nef reaction generates low concentrations of NO which modulates cyclic monophosphate guanosine (cGMP)-dependent cell signaling activity. Nitrate fatty acids also control the generation of NO by regulating endothelial and inducible nitric oxide synthase (eNOS and iNOS) independently of cGMP mechanisms [34,62,76].

In addition, NO_2-FAs can regulate the expression levels of differentiation-related, key inflammation, and cell proliferation genes [82,97–101]. Signaling via the Kelch-like ECH-associated protein 1 (Keap 1)-nuclear factor erythroid-derived 2-like 2 (Nrf2) pathway is a primary regulator of cellular responses to oxidative stress. The transcription factor Nrf2, which controls antioxidant protein expression, is located in the cytosol in its inactive form due to Keap1 activity which promotes Nrf2 ubiquitination and subsequent degradation by the ubiquitin–proteasome system. Keap 1 contains reactive cysteines (Cys 151, 273, and 288) which can be modified by oxidation or alkylation and used as redox state sensors. When electrophiles such as NO_2-OA, NO_2-LA, and NO_2-AA are formed, the interaction between Nrf2 and Keap1 is interrupted. This facilitates the transfer of Nrf2 to the nucleus, where it will link to specific cis targets and activate the regulation of antioxidant response element (ARE) genes [1,55,97,102–105]. The NO$_2$-FA-sensitive system involving heat-shock responses (HSRs) is a complex alliance of regulatory proteins and transcription factors which promotes cytoprotective and anti-inflammatory target gene expression [46]. Heat-shock proteins (HSPs) are chaperones whose expression is triggered by stress conditions, including heat, as well as by electrophilic and reactive species caused under inflammatory injury. Chaperones prevent the aggregation of denatured or oxidized proteins, collaborate in the transfer of these proteins to intracellular locations, and thus contribute to cellular redox homeostasis [106]. Nitro-oleic acid in human endothelial cells has been reported to activate HSF1 (Heat Shock Factor 1), the most important regulator of HSRs, followed by a remarkable induction of a large group of heat shock genes (Table 2) [82,102,107].

Nitro-fatty acid can also activate the peroxisome proliferator-activating receptor (PPAR), particularly PPARγ, which is included in the family of nuclear hormone receptors. This receptor plays a marked role in the expression of transcription factors associated with lipid generation, lipid and glucose metabolism, macrophage differentiation, and immune responses. The PPARγ regulatory domain is located in the C-terminal side which coincides with the ligand binding domain. The location of a cysteine at position 285 makes this hydrophobic region susceptible to nitroalkylation by NO$_2$-FAs such as NO$_2$-OA and NO$_2$-LA (Table 2) [1,101,108–110].

Another example is the nuclear factor kappa betta (NF-kβ) involved in transcriptional regulation under inflammatory and immune processes. The nuclear factor kappa betta is a protein complex with two subunits (p50 and p65) [1,98,111,112]. Experimental studies have shown that NF-kβ is regulated by NO$_2$-FAs at multiple levels including the inhibition of Toll-like receptor 4 (TLR4) by NO$_2$-OA. Toll-like receptor 4 is a transmembrane protein which pertains to the pattern recognition receptor (PRR) family which is able to recognize bacterial lipopolysaccharide (LPS). Its activation triggers the intracellular NF-κB signaling pathway and inflammatory cytokine production which activate the innate immune system. Thus, the inhibition of TLR4 by NO$_2$-FAs also triggers the inhibition of NF-kβ [101,113]. Another level of regulation is the inhibition of NF-kβ by nitroalkylation, specifically, the residue Cys38, placed in the DNA-binding domain of the p65 subunit, is susceptible to nitroalkylation [96,98]. The final level of regulation is the activation of PPAR by NO$_2$-FAs which causes the trans-repression of inflammatory genes such as NF-kβ (Table 2) [101,114].

In animal systems, nitroalkylation is considered to be a decisive signaling resource in anti-inflammatory processes. Nitrate fatty acids modify the anti-inflammatory response at multiple levels including gene expression, protein translation (acting on transcription factors and lipid receptors), as well as cell function, as many inflammatory proteins contain numerous nucleophilic amino acid residues which can be nitroalkylation targets. Table 2 shows a summary list of NO$_2$-FA protein targets in animal systems and how they are affected by nitroalkylation.

Table 2. NO$_2$-FA protein targets in animal systems and their effects on protein function (modified from Reference [24]).

Nitro-Fatty Acid	Protein	Nucleophile Site	Effect	References
NO$_2$-OA	GAPDH	Catalytic Cys, other Cys and His	Inhibition, increase in hydrophobicity and change in subcellular distribution	[66]
	Pro-MMP7 and Pro-MMP9	Zinc coordination Cys in the active site	Zinc release, autocatalytic cleavage of the pro-domain. MMP activation	[115]
	TRPV1 and TRPA1	Not detected	Activation of TRP channels	[116,117]

Table 2. *Cont.*

Nitro-Fatty Acid	Protein	Nucleophile Site	Effect	References
NO$_2$-OA	AT1R	Not detected	Decrease in coupling with G-protein, inhibition of downstream signaling	[118]
	PknG	Iron coordination Cys in non-catalytic domain and His	Inhibition of kinase activity	[119]
	XOR	Pterindithiolene which coordinates molybdenum	Inhibition of electron transfer reactions at the molybdenum cofactor	[120]
	HSF1	Not detected	Activation of HSFA1 and subsequent robust induction of heat shock genes	[82,107]
NO$_2$-LA	ANT1	Cys	Cardio-protection	[121]
NO$_2$-cLA	HSA	Cys and non-covalent binding		[122]
NO$_2$-AA	PGHS	Disruption of heme binding to the protein	Inhibition of PGHS-1 cyclooxygenase activity and both PGHS-1 and -2 peroxidase activity	[123]
	PKC	Probable covalent modification	Inhibitory effect on PKC activation	[124]
	NOX2	Inhibition of assembly	Inhibition of superoxide production	[125]
	PDI	Cys at active site	Inhibition of reductase and chaperone activities	[126]
NO$_2$-OA and NO$_2$-LA	NF-κB p65	DNA binding domain Cys	Inhibition of NF-κB DNA binding, abolition of pro-inflammatory responses	[98]
	PPARγ	Cys in ligand-binding domain	Agonist activation of PPARγ	[110]
NO$_2$-OA, NO$_2$-LA and NO$_2$-AA	Keap 1	Cys	Stabilization of the complex with Nrf2, newly synthesized Nrf2 translocated to the nucleus	[97,103–105]

Abbreviations: Glyceraldehyde-3-phosphate dehydrogenase (GAPDH); Pro-matrix metalloproteinases, (Pro-MMP7 and Pro-MMP9); Transient receptor potential (TRPV1, TRPA1); Angiotensin II receptor (AT1R); Protein kinase G (PknG); Xanthine oxidoreductase (XOR); Heat Shock Factor 1 (HSF1); Adenine nucleotide translocase 1 (ANT1); Human serum albumin (HSA); Prostaglandin endoperoxide H synthase (PGHS); Protein kinase C (PKC); NADPH oxidase 2 (NOX2); Protein disulfide isomerase (PDI); Nuclear factor κB subunit p65 (NF-κB p65); Peroxisome proliferator-activated receptor (PPARγ); Kelch-like ECH-associating protein 1 (Keap 1).

4.2. Nitroalkylation in Plants

Although the effects of nitroalkylation have been extensively studied in animal organisms, the impact of NO$_2$-FA action in plants, which has not been fully explored, constitutes an emerging area of interesting research work. Probably, the signaling function of NO$_2$-Ln is due to nitroalkylation processes. In this context, the endogenous presence of 37 proteins adducted with NO_2-Ln in Arabidopsis cell cultures has been identified. However, cell cultures treated with 100 μM NO_2-Ln showed an increase in the number of nitroalkylated proteins (342), belonging to different areas of cell metabolism, which included APX2 (unpublished results), whose encoding gene expression, according to the transcriptomic studies mentioned above, was induced [9].

Ascorbate peroxidase (APX2) is one of the primary antioxidant systems in plants. This enzyme belongs to the ascorbate–glutathione cycle, which detoxifies hydrogen peroxide and contains non-enzymatic antioxidants (ascorbate and glutathione) and enzymatic antioxidants such as monodehydroascorbate reductase (MDAR), glutathione reductase (GR), and dehydroascorbate reductase (DHAR), as well as the reductive coenzyme NADPH [127,128].

In this study, the APX recombinant protein from *Arabidospsis thaliana* was incubated with increasing concentrations of NO_2-Ln (1 μM and 10 μM). The enzymatic activity was spectrophotometrically monitored [129]. Furthermore, the nitroalkylation targeted residues of the treated recombinant protein were detected and characterized using LC-MS/MS. Thus, the protein was digested by trypsin and desalted by C18 columns to obtain the peptide fraction which was analyzed using an Exactive Q mass spectrometer attached to a nano-flow liquid chromatograph (nanoLC) (Thermo Fisher Scientific). The LC-MS/MS spectrum deconvolution was carried out employing Proteome Discoverer version 1.4. bioinformatics software (Thermo Fisher Scientific). The Percolator node was used to filter the peptides at a 1% false discovery rate (FDR) at the peptide-spectrum matches (PSMs).

In order to identify the position of the nitroalkylation-targeted nucleophilic residues, an in silico modeling was carried out using Raptor X bioinformatics software (http://raptorx.uchicago.edu/). The APX model was based on the structure of isoniazid (INH) bound to cytosolic soybean ascorbate peroxidase (PDB:2VCF) [130].

The treatment of recombinant APX with NO_2-Ln modulates its enzymatic activity, showing a significant decrease in the presence of 10 μM NO_2^-Ln (Figure 5). This decreased activity was associated with the post-translational modification caused by nitroalkylation, which was detected by mass spectrometry. Comparison of the spectra of control and NO_2-Ln-treated samples displayed a rise in the mass of nucleophilic residues due to treatment with NO_2-Ln. The electrophilic attack by NO_2-Ln generated the nitroalkylation of the residues showed in Figures 6 and 7A. with histidine 43 and histidine 163 being preferentially nitroalkylated. This could have functional implications (Figure 7B), as histidine 43 and histidine 163 are located at the active and metal-binding site, respectively. This fact suggests that the nitroalkylation of these residues blocks APX enzymatic activity, modulating protein function.

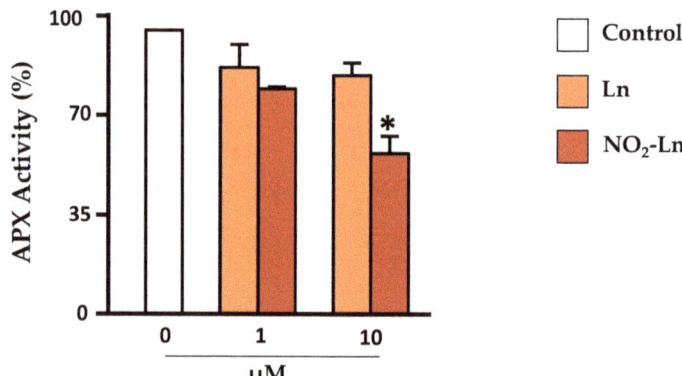

Figure 5. Modulation of the enzymatic activity of cytosolic recombinant APX following the treatment with increasing concentrations of NO_2-Ln. The negative controls methanol (NO_2-FA vehicle) and linolenic acid (non-nitrated fatty acid) were used. Vertical bars represent the mean ± standard deviation of at least three replicates. Statistically significant differences $p < 0.05$ (*) and $p < 0.01$ (**). (Ascorbate peroxidase: APX).

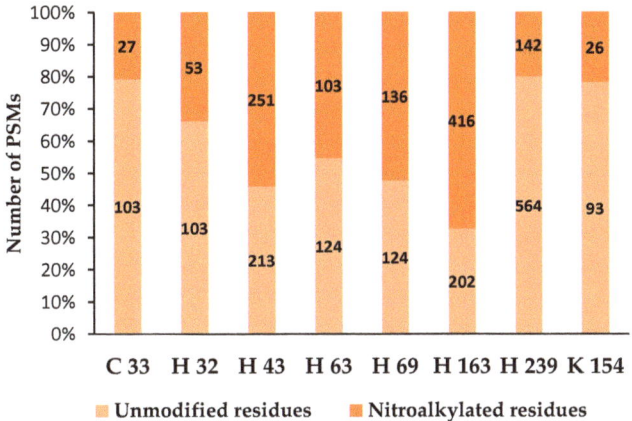

Figure 6. Detection of nitroalkylated residues in cytosolic recombinant APX by mass spectrometry (LC-MS/MS). The number on each column represents the number of PSMs of the unmodified residue related to the nitroalkylated residue. PSM: peptide-spectrum match.

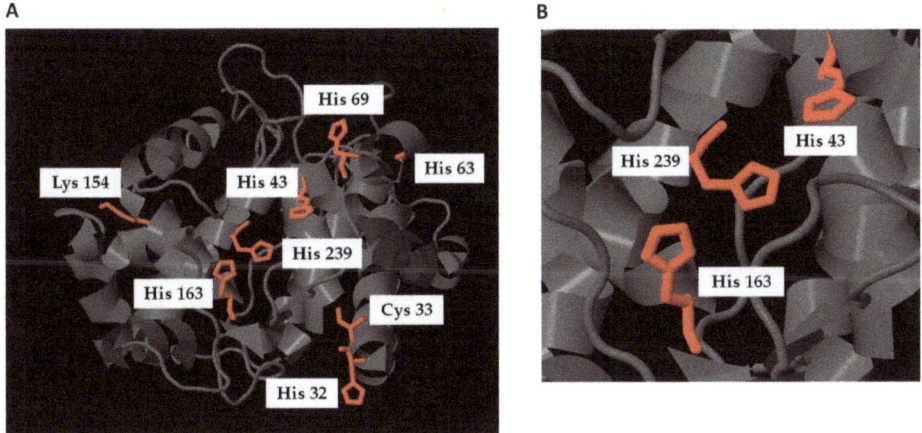

Figure 7. (**A**) In silico molecular model of cytosolic recombinant APX and localization of nitroalkylated residues. (**B**) Zoomed in illustration of the in silico molecular model where nitroalkylated histidines 43 and 163 located in the active site and in a metal-binding site, respectively, are highlighted.

Figure 8 explains the model of the nitro-lipid-protein adducts signaling mechanism in plants. Nitro-lipid-protein adducts stability can be affected by the accumulation of ROS and RNS, which could cause the oxidation of sulfhydryl substituents in proteins, and consequently the scission of the Michael adduct releasing NO_2-Ln. As was previously mentioned, the nitroalkylation of APX by NO_2-Ln generates function loss. Under nitro-oxidative conditions, the function of APX would be reactivated due to the reversibility of the nitroalkylation PTM. On the other hand, the levels of free NO_2-FA increase, being able to stimulate the expression of heat shock proteins (HSPs) and certain antioxidant systems such as APX and methionine sulfoxide reductase B (MSRB). Another possibility is that NO_2-FA could donate ·NO in the cellular aqueous environment which could act in a broad set of plant activities such as plant development, (a)biotic disorders, antioxidant responses, and NO-PTMs.

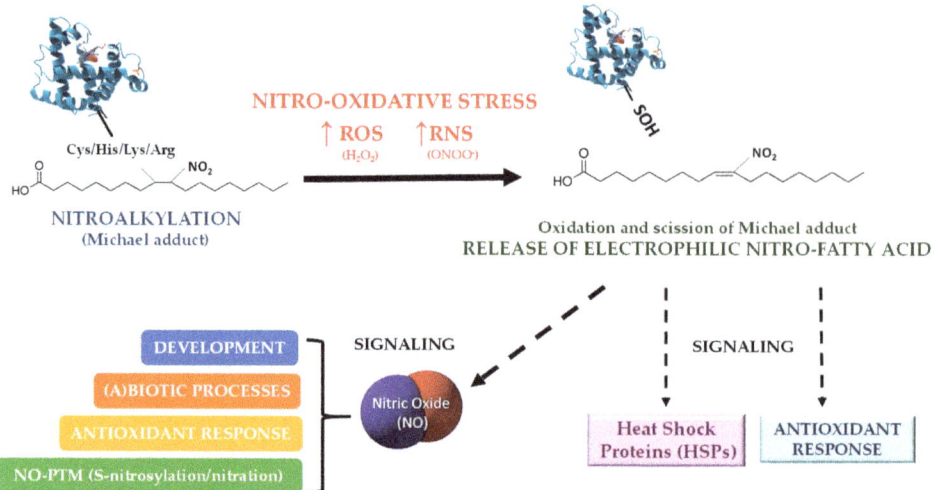

Figure 8. Model of the NO_2-FA signaling mechanism by nitro-lipid-protein adduct in plants. Nitro oxidative conditions triggers the oxidation of the protein, the subsequent scission of the Michael adduct and the releasing of the NO_2-FA. Free NO_2-FAs display signaling actions by activating the chaperone network expression and several antioxidant systems. Moreover, NO_2-FAs, which can also act as NO donors, are involved in NO signaling processes. ROS: reactive oxygen species; RNS: reactive nitrogen species; NO: nitric oxide.

The ability of NO_2-Ln to trigger pleiotropic signaling actions mainly depends on the nitroalkylation of regulatory proteins involved in plant biology and numerous types of (a)biotic-stress. Being a reversible post-translational modification, which can affect a large number of target amino acid residues (Cys, His, Lys, and Arg), together with the features outlined above, render nitroalkylation an important cell signaling mechanism mediated by NO_2-FAs.

5. Conclusions and Future Perspectives

The potent electrophilic molecules NO_2-FAs, whose electrophilicity triggers potential signaling mechanisms via nitroalkylation, were recently discovered in both animal and plant systems. This NO_2-FA-mediated PTM can be considered a NO-PTM similar to *S*-nitrosylation, because NO_2-FAs are RLS formed as a result of the oxidation of PUFA by NO-derived species. The importance of nitroalkylation resides in its reversibility and in the presence of a considerable amount of target amino acids residues that generate the formation of nitro-lipid-protein adducts, which enables this NO-PTM to trigger pleiotropic signaling actions. In animal systems, nitroalkylation is associated with signaling mechanisms in anti-inflammatory processes. However, in plant systems, this little-known NO-PTM constitutes an emerging area of research which should be developed through advances in mass spectrometry techniques.

Funding: The study was supported by an ERDF grant co-financed by the Spanish Ministry of Economy and Competitiveness (Project BIO2015-66390-P) and the Junta de Andalucía (Group BIO-286).

Acknowledgments: L.A.-C. wishes to thank the University of Jaén for funding her PhD fellowship.

Conflicts of Interest: The authors declare no conflicts of interest.

References

1. Schopfer, F.J.; Cipollina, C.; Freeman, B.A. Formation and signaling actions of electrophilic lipids. *Chem. Rev.* **2011**, *111*, 5997–6021. [CrossRef] [PubMed]

2. Beavers, W.N.; Rose, K.L.; Galligan, J.J.; Mitchener, M.M.; Rouzer, C.A.; Tallman, K.A.; Lamberson, C.R.; Wang, X.; Hill, S.; Ivanova, P.T. Protein modification by endogenously generated lipid electrophiles: Mitochondria as the source and target. *ACS Chem. Biol.* **2017**, *12*, 2062–2069. [CrossRef] [PubMed]
3. Wang, C.; Weerapana, E.; Blewett, M.M.; Cravatt, B.F. A chemoproteomic platform to quantitatively map targets of lipid-derived electrophiles. *Nat. Methods* **2013**, *11*, 79. [CrossRef] [PubMed]
4. Higdon, A.; Diers, A.R.; Oh, J.Y.; Landar, A.; Darley-Usmar, V.M. Cell signalling by reactive lipid species: New concepts and molecular mechanisms. *Biochem. J.* **2012**, *442*, 453–464. [CrossRef] [PubMed]
5. Marwah, S.; Blann, A.; Rea, C.; Phillips, J.; Wright, J.; Bareford, D. Reduced vitamin E antioxidant capacity in sickle cell disease is related to transfusion status but not to sickle crisis. *Am. J. Hematol.* **2002**, *69*, 144–146. [CrossRef] [PubMed]
6. Yin, H.; Porter, N.A. New insights regarding the autoxidation of polyunsaturated fatty acids. *Antioxid. Redox Signal.* **2005**, *7*, 170–184. [CrossRef]
7. Poon, H.F.; Calabrese, V.; Scapagnini, G.; Butterfield, D.A. Free radicals: Key to brain aging and heme oxygenase as a cellular response to oxidative stress. *J. Gerontol. A Biol. Sci. Med. Sci.* **2004**, *59*, M478–M493. [CrossRef]
8. Morrow, J.D. Quantification of isoprostanes as indices of oxidant stress and the risk of atherosclerosis in humans. *Arterioscler. Thromb. Vasc. Biol.* **2005**, *25*, 279–286. [CrossRef]
9. Mata-Pérez, C.; Sánchez-Calvo, B.; Padilla, M.N.; Begara-Morales, J.C.; Luque, F.; Melguizo, M.; Jiménez-Ruiz, J.; Fierro-Risco, J.; Peñas-Sanjuán, A.; Valderrama, R. Nitro-fatty acids in plant signaling: Nitro-linolenic acid induces the molecular chaperone network in Arabidopsis. *Plant Physiol.* **2016**, *170*, 686–701. [CrossRef]
10. Mata-Perez, C.; Sanchez-Calvo, B.; Padilla, M.N.; Begara-Morales, J.C.; Valderrama, R.; Corpas, F.J.; Barroso, J.B. Nitro-fatty acids in plant signaling: New key mediators of nitric oxide metabolism. *Redox Biol.* **2017**, *11*, 554–561. [CrossRef]
11. Porter, N.A.; Caldwell, S.E.; Mills, K.A. Mechanisms of free radical oxidation of unsaturated lipids. *Lipids* **1995**, *30*, 277–290. [CrossRef] [PubMed]
12. Addis, P. Occurrence of lipid oxidation products in foods. *Food Chem. Toxicol.* **1986**, *24*, 1021–1030. [CrossRef]
13. Hwa Lee, S.; Rangiah, K.; Williams, M.V.; Wehr, A.Y.; DuBois, R.N.; Blair, I.A. Cyclooxygenase-2-mediated metabolism of arachidonic acid to 15-oxo-eicosatetraenoic acid by rat intestinal epithelial cells. *Chem. Res. Toxicol.* **2007**, *20*, 1665–1675. [CrossRef]
14. Chiang, N.; Serhan, C.N.; Dahlén, S.-E.; Drazen, J.M.; Hay, D.W.; Rovati, G.E.; Shimizu, T.; Yokomizo, T.; Brink, C. The lipoxin receptor ALX: Potent ligand-specific and stereoselective actions in vivo. *Pharmacol. Rev.* **2006**, *58*, 463–487. [CrossRef] [PubMed]
15. Baker, P.R.; Schopfer, F.J.; O'Donnell, V.B.; Freeman, B.A. Convergence of nitric oxide and lipid signaling: Anti-inflammatory nitro-fatty acids. *Free Radic. Biol. Med.* **2009**, *46*, 989–1003. [CrossRef] [PubMed]
16. Rubbo, H.; Radi, R.; Trujillo, M.; Telleri, R.; Kalyanaraman, B.; Barnes, S.; Kirk, M.; Freeman, B.A. Nitric oxide regulation of superoxide and peroxynitrite-dependent lipid peroxidation. Formation of novel nitrogen-containing oxidized lipid derivatives. *J. Biol. Chem.* **1994**, *269*, 26066–26075. [PubMed]
17. Davies, S.S.; Amarnath, V.; Roberts, L.J., II. Isoketals: Highly reactive γ-ketoaldehydes formed from the H2-isoprostane pathway. *Chem. Phys. Lipids* **2004**, *128*, 85–99. [CrossRef]
18. Ueno, N.; Murakami, M.; Tanioka, T.; Fujimori, K.; Tanabe, T.; Urade, Y.; Kudo, I. Coupling between cyclooxygenase, terminal prostanoid synthase, and phospholipase A2. *J. Biol. Chem.* **2001**, *276*, 34918–34927. [CrossRef]
19. Prigge, S.; Boyington, J.; Faig, M.; Doctor, K.; Gaffney, B.; Amzel, L. Structure and mechanism of lipoxygenases. *Biochimie* **1997**, *79*, 629–636. [CrossRef]
20. O'donnell, V.B.; Freeman, B.A. Interactions between nitric oxide and lipid oxidation pathways: Implications for vascular disease. *Circ. Res.* **2001**, *88*, 12–21. [CrossRef]
21. Swain, C.G.; Scott, C.B. Quantitative correlation of relative rates. Comparison of hydroxide ion with other nucleophilic reagents toward alkyl halides, esters, epoxides and acyl halides1. *J. Am. Chem. Soc.* **1953**, *75*, 141–147. [CrossRef]
22. Rudolph, T.K.; Freeman, B.A. Transduction of redox signaling by electrophile-protein reactions. *Sci. Signal.* **2009**, *2*, re7. [CrossRef] [PubMed]

23. Nagahara, N.; Matsumura, T.; Okamoto, R.; Kajihara, Y. Protein cysteine modifications: (2) reactivity specificity and topics of medicinal chemistry and protein engineering. *Curr. Med. Chem.* **2009**, *16*, 4490–4501. [CrossRef]
24. Turell, L.; Steglich, M.; Alvarez, B. The chemical foundations of nitroalkene fatty acid signaling through addition reactions with thiols. *Nitric Oxide* **2018**. [CrossRef] [PubMed]
25. Pearson, R.G. Hard and soft acids and bases. *J. Am. Chem. Soc.* **1963**, *85*, 3533–3539. [CrossRef]
26. Reed, T.T. Lipid peroxidation and neurodegenerative disease. *Free Radic. Biol. Med.* **2011**, *51*, 1302–1319. [CrossRef]
27. Di Domenico, F.; Pupo, G.; Tramutola, A.; Giorgi, A.; Schininà, M.E.; Coccia, R.; Head, E.; Butterfield, D.A.; Perluigi, M. Redox proteomics analysis of HNE-modified proteins in Down syndrome brain: Clues for understanding the development of Alzheimer disease. *Free Radic. Biol. Med.* **2014**, *71*, 270–280. [CrossRef]
28. Butterfield, D.A.; Gu, L.; Domenico, F.D.; Robinson, R.A. Mass spectrometry and redox proteomics: Applications in disease. *Mass Spectrom. Rev.* **2014**, *33*, 277–301. [CrossRef]
29. Sauriasari, R.; Andrajati, R.; Saputri, D.; Muris, R.; Manfaatun, A.; Amanda, O.; Setiawan, H.; Sakano, N.; Wang, D.; Ogino, K. Marker of lipid peroxidation related to diabetic nephropathy in Indonesian type 2 diabetes mellitus patients. *Diabetes Res. Clin. Pract.* **2015**, *108*, 193–200. [CrossRef] [PubMed]
30. Uchida, K. Role of reactive aldehyde in cardiovascular diseases. *Free Radic. Biol. Med.* **2000**, *28*, 1685–1696. [CrossRef]
31. Bell-Parikh, L.C.; Ide, T.; Lawson, J.A.; McNamara, P.; Reilly, M.; FitzGerald, G.A. Biosynthesis of 15-deoxy-Δ 12,14-PGJ 2 and the ligation of PPARγ. *J. Clin. Investig.* **2003**, *112*, 945–955. [CrossRef] [PubMed]
32. Oh, J.Y.; Giles, N.; Landar, A.; Darley-Usmar, V. Accumulation of 15-deoxy-Δ12, 14-prostaglandin J2 adduct formation with Keap1 over time: Effects on potency for intracellular antioxidant defence induction. *Biochem. J.* **2008**, *411*, 297–306. [CrossRef]
33. Groeger, A.L.; Cipollina, C.; Cole, M.P.; Woodcock, S.R.; Bonacci, G.; Rudolph, T.K.; Rudolph, V.; Freeman, B.A.; Schopfer, F.J. Cyclooxygenase-2 generates anti-inflammatory mediators from omega-3 fatty acids. *Nat. Chem. Biol.* **2010**, *6*, 433–441. [CrossRef]
34. Khoo, N.K.; Freeman, B.A. Electrophilic nitro-fatty acids: Anti-inflammatory mediators in the vascular compartment. *Curr. Opin. Pharmacol.* **2010**, *10*, 179–184. [CrossRef]
35. Wong, H.L.; Liebler, D.C. Mitochondrial protein targets of thiol-reactive electrophiles. *Chem. Res. Toxicol.* **2008**, *21*, 796–804. [CrossRef]
36. Vila, A.; Tallman, K.A.; Jacobs, A.T.; Liebler, D.C.; Porter, N.A.; Marnett, L.J. Identification of protein targets of 4-hydroxynonenal using click chemistry for ex vivo biotinylation of azido and alkynyl derivatives. *Chem. Res. Toxicol.* **2008**, *21*, 432–444. [CrossRef]
37. Szapacs, M.E.; Kim, H.-Y.H.; Porter, N.A.; Liebler, D.C. Identification of proteins adducted by lipid peroxidation products in plasma and modifications of apolipoprotein A1 with a novel biotinylated phospholipid probe. *J. Proteome Res.* **2008**, *7*, 4237–4246. [CrossRef]
38. Shin, N.-Y.; Liu, Q.; Stamer, S.L.; Liebler, D.C. Protein targets of reactive electrophiles in human liver microsomes. *Chem. Res. Toxicol.* **2007**, *20*, 859–867. [CrossRef] [PubMed]
39. Dennehy, M.K.; Richards, K.A.; Wernke, G.R.; Shyr, Y.; Liebler, D.C. Cytosolic and nuclear protein targets of thiol-reactive electrophiles. *Chem. Res. Toxicol.* **2006**, *19*, 20–29. [CrossRef] [PubMed]
40. Gomi, K.; Yamamoto, H.; Akimitsu, K. Characterization of a lipoxygenase gene in rough lemon induced by *Alternaria alternata*. *J. Gen. Plant Pathol.* **2002**, *68*, 21–30. [CrossRef]
41. Kim, E.-S.; Choi, E.; Kim, Y.; Cho, K.; Lee, A.; Shim, J.; Rakwal, R.; Agrawal, G.K.; Han, O. Dual positional specificity and expression of non-traditional lipoxygenase induced by wounding and methyl jasmonate in maize seedlings. *Plant Mol. Biol.* **2003**, *52*, 1203–1213. [CrossRef] [PubMed]
42. Montillet, J.-L.; Agnel, J.-P.; Ponchet, M.; Vailleau, F.; Roby, D.; Triantaphylidès, C. Lipoxygenase-mediated production of fatty acid hydroperoxides is a specific signature of the hypersensitive reaction in plants. *Plant Physiol. Biochem.* **2002**, *40*, 633–639. [CrossRef]
43. He, Y.; Fukushige, H.; Hildebrand, D.F.; Gan, S. Evidence supporting a role of jasmonic acid in Arabidopsis leaf senescence. *Plant Physiol.* **2002**, *128*, 876–884. [CrossRef] [PubMed]
44. Uchida, K. 4-Hydroxy-2-nonenal: A product and mediator of oxidative stress. *Prog. Lipid Res.* **2003**, *42*, 318–343. [CrossRef]

45. Thoma, I.; Loeffler, C.; Sinha, A.K.; Gupta, M.; Krischke, M.; Steffan, B.; Roitsch, T.; Mueller, M.J. Cyclopentenone isoprostanes induced by reactive oxygen species trigger defense gene activation and phytoalexin accumulation in plants. *Plant J.* **2003**, *34*, 363–375. [CrossRef] [PubMed]
46. Geisler, A.C.; Rudolph, T.K. Nitroalkylation—a redox sensitive signaling pathway. *BBA-Gen. Subj.* **2012**, *1820*, 777–784. [CrossRef] [PubMed]
47. Bonacci, G.; Baker, P.R.; Salvatore, S.R.; Shores, D.; Khoo, N.K.; Koenitzer, J.R.; Vitturi, D.A.; Woodcock, S.R.; Golin-Bisello, F.; Watkins, S. Conjugated linoleic acid is a preferential substrate for fatty acid nitration. *J. Biol. Chem.* **2012**, *287*, 44071–44082. [CrossRef] [PubMed]
48. Pryor, W.A.; Lightsey, J.W.; Church, D.F. Reaction of nitrogen dioxide with alkenes and polyunsaturated fatty acids: Addition and hydrogen-abstraction mechanisms. *J. Am. Chem. Soc.* **1982**, *104*, 6685–6692. [CrossRef]
49. Buchan, G.R.; Bonacci, G.; Fazzari, M.; Salvatore, S.; Wendell, S.G. Nitro-fatty acid formation and metabolism. *Nitric Oxide* **2018**, *79*, 38–44. [CrossRef] [PubMed]
50. d'Ischia, M.; Napolitano, A.; Manini, P.; Panzella, L. Secondary targets of nitrite-derived reactive nitrogen species: Nitrosation/nitration pathways, antioxidant defense mechanisms and toxicological implications. *Chem. Res. Toxicol.* **2011**, *24*, 2071–2092. [CrossRef]
51. Rudolph, V.; Rudolph, T.K.; Schopfer, F.J.; Bonacci, G.; Woodcock, S.R.; Cole, M.P.; Baker, P.R.; Ramani, R.; Freeman, B.A. Endogenous generation and protective effects of nitro-fatty acids in a murine model of focal cardiac ischaemia and reperfusion. *Cardiovasc. Res.* **2009**, *85*, 155–166. [CrossRef] [PubMed]
52. Nadtochiy, S.M.; Baker, P.R.; Freeman, B.A.; Brookes, P.S. Mitochondrial nitroalkene formation and mild uncoupling in ischaemic preconditioning: Implications for cardioprotection. *Cardiovasc. Res.* **2008**, *82*, 333–340. [CrossRef] [PubMed]
53. Reynolds, C.; Roche, H. Conjugated linoleic acid and inflammatory cell signalling. *Prostaglandins Leukot. Essent. Fat. Acids* **2010**, *82*, 199–204. [CrossRef]
54. Vitturi, D.A.; Minarrieta, L.; Salvatore, S.R.; Postlethwait, E.M.; Fazzari, M.; Ferrer-Sueta, G.; Lancaster, J.R., Jr.; Freeman, B.A.; Schopfer, F.J. Convergence of biological nitration and nitrosation via symmetrical nitrous anhydride. *Nat. Chem. Biol.* **2015**, *11*, 504. [CrossRef] [PubMed]
55. Suzuki, T.; Yamamoto, M. Stress-sensing mechanisms and the physiological roles of the Keap1-Nrf2 system during cellular stress. *J. Biol. Chem.* **2017**, *292*, 16817–16824. [CrossRef] [PubMed]
56. Weitzberg, E.; Lundberg, J.O. Novel aspects of dietary nitrate and human health. *Annu. Rev. Nutr.* **2013**, *33*, 129–159. [CrossRef]
57. Lundberg, J.O.; Weitzberg, E. NO generation from inorganic nitrate and nitrite: Role in physiology, nutrition and therapeutics. *Arch. Pharm. Res.* **2009**, *32*, 1119–1126. [CrossRef] [PubMed]
58. Moreno-Vivián, C.; Cabello, P.; Martínez-Luque, M.; Blasco, R.; Castillo, F. Prokaryotic nitrate reduction: Molecular properties and functional distinction among bacterial nitrate reductases. *J. Bacteriol.* **1999**, *181*, 6573–6584.
59. Ferreira, A.M.; Ferrari, M.I.; Trostchansky, A.; Batthyany, C.; Souza, J.M.; Alvarez, M.N.; López, G.V.; Baker, P.R.; Schopfer, F.J.; O'Donnell, V. Macrophage activation induces formation of the anti-inflammatory lipid cholesteryl-nitrolinoleate. *Biochem. J.* **2009**, *417*, 223–238. [CrossRef]
60. Freeman, B.A.; Baker, P.R.; Schopfer, F.J.; Woodcock, S.R.; Napolitano, A.; d'Ischia, M. Nitro-fatty acid formation and signaling. *J. Biol. Chem.* **2008**, *283*, 15515–15519. [CrossRef]
61. Faine, L.A.; Cavalcanti, D.; Rudnicki, M.; Ferderbar, S.; Macedo, S.; Souza, H.P.; Farsky, S.; Boscá, L.; Abdalla, D. Bioactivity of nitrolinoleate: Effects on adhesion molecules and CD40-CD40L system. *J. Nutr. Biochem.* **2010**, *21*, 125–132. [CrossRef]
62. Schopfer, F.J.; Baker, P.R.; Giles, G.; Chumley, P.; Batthyany, C.; Crawford, J.; Patel, R.P.; Hogg, N.; Branchaud, B.P.; Lancaster, J.R. Fatty acid transduction of nitric oxide signaling Nitrolinoleic acid is a hydrophobically stabilized nitric oxide donor. *J. Biol. Chem.* **2005**, *280*, 19289–19297. [CrossRef] [PubMed]
63. Su, Y.H.; Wu, S.S.; Hu, C.H. Release of nitric oxide from nitrated fatty acids: Insights from computational chemistry. *J. Chin. Chem. Soc.* **2019**, *66*, 41–48. [CrossRef]
64. Lima, É.S.; Bonini, M.G.; Augusto, O.; Barbeiro, H.V.; Souza, H.P.; Abdalla, D.S. Nitrated lipids decompose to nitric oxide and lipid radicals and cause vasorelaxation. *Free Radic. Biol. Med.* **2005**, *39*, 532–539. [CrossRef]
65. Gorczynski, M.J.; Huang, J.; Lee, H.; King, S.B. Evaluation of nitroalkenes as nitric oxide donors. *Bioorg. Med. Chem. Lett.* **2007**, *17*, 2013–2017. [CrossRef]

66. Batthyany, C.; Schopfer, F.J.; Baker, P.R.; Durán, R.; Baker, L.M.; Huang, Y.; Cerveñansky, C.; Braunchaud, B.P.; Freeman, B.A. Reversible post-translational modification of proteins by nitrated fatty acids in vivo. *J. Biol. Chem.* **2006**, *281*, 20450–20463. [CrossRef]
67. Baker, L.M.; Baker, P.R.; Golin-Bisello, F.; Schopfer, F.J.; Fink, M.; Woodcock, S.R.; Branchaud, B.P.; Radi, R.; Freeman, B.A. Nitro-fatty acid reaction with glutathione and cysteine Kinetic analysis of thiol alkylation by a Michael addition reaction. *J. Biol. Chem.* **2007**, *282*, 31085–31093. [CrossRef] [PubMed]
68. Rubbo, H.; Radi, R. Protein and lipid nitration: Role in redox signaling and injury. *BBA-Gen. Subj.* **2008**, *1780*, 1318–1324. [CrossRef]
69. Fazzari, M.; Khoo, N.; Woodcock, S.R.; Li, L.; Freeman, B.A.; Schopfer, F.J. Generation and esterification of electrophilic fatty acid nitroalkenes in triacylglycerides. *Free Radic. Biol. Med.* **2015**, *87*, 113–124. [CrossRef]
70. Fazzari, M.; Trostchansky, A.; Schopfer, F.J.; Salvatore, S.R.; Sánchez-Calvo, B.; Vitturi, D.; Valderrama, R.; Barroso, J.B.; Radi, R.; Freeman, B.A. Olives and olive oil are sources of electrophilic fatty acid nitroalkenes. *PLoS ONE* **2014**, *9*, e84884. [CrossRef]
71. Vitturi, D.A.; Chen, C.-S.; Woodcock, S.R.; Salvatore, S.R.; Bonacci, G.; Koenitzer, J.R.; Stewart, N.A.; Wakabayashi, N.; Kensler, T.W.; Freeman, B.A. Modulation of nitro-fatty acid signaling prostaglandin reductase-1 is a nitroalkene reductase. *J. Biol. Chem.* **2013**, *288*, 25626–25637. [CrossRef] [PubMed]
72. Melo, T.N.; Domingues, P.; Ferreira, R.; Milic, I.; Fedorova, M.; Santos, S.R.M.; Segundo, M.A.; Domingues, M.R.r.M. Recent advances on mass spectrometry analysis of nitrated phospholipids. *Anal. Chem.* **2016**, *88*, 2622–2629. [CrossRef] [PubMed]
73. Catharino, R.R.; Haddad, R.; Cabrini, L.G.; Cunha, I.B.; Sawaya, A.C.; Eberlin, M.N. Characterization of vegetable oils by electrospray ionization mass spectrometry fingerprinting: Classification, quality, adulteration, and aging. *Anal. Chem.* **2005**, *77*, 7429–7433. [CrossRef]
74. Rastrelli, L.; Passi, S.; Ippolito, F.; Vacca, G.; De Simone, F. Rate of degradation of α-tocopherol, squalene, phenolics, and polyunsaturated fatty acids in olive oil during different storage conditions. *J. Agric. Food Chem.* **2002**, *50*, 5566–5570. [CrossRef] [PubMed]
75. Pérez-Camino, M.C.; Moreda, W.; Mateos, R.; Cert, A. Determination of esters of fatty acids with low molecular weight alcohols in olive oils. *J. Agric. Food Chem.* **2002**, *50*, 4721–4725. [CrossRef]
76. Mata-Pérez, C.; Sánchez-Calvo, B.; Begara-Morales, J.C.; Carreras, A.; Padilla, M.N.; Melguizo, M.; Valderrama, R.; Corpas, F.J.; Barroso, J.B. Nitro-linolenic acid is a nitric oxide donor. *Nitric Oxide* **2016**, *57*, 57–63. [CrossRef]
77. Beligni, M.V.; Lamattina, L. Nitric oxide stimulates seed germination and de-etiolation, and inhibits hypocotyl elongation, three light-inducible responses in plants. *Planta* **2000**, *210*, 215–221. [CrossRef] [PubMed]
78. Bethke, P.C.; Libourel, I.G.; Jones, R.L. Nitric oxide reduces seed dormancy in Arabidopsis. *J. Exp. Bot.* **2005**, *57*, 517–526. [CrossRef]
79. Libourel, I.G.; Bethke, P.C.; De Michele, R.; Jones, R.L. Nitric oxide gas stimulates germination of dormant Arabidopsis seeds: Use of a flow-through apparatus for delivery of nitric oxide. *Planta* **2006**, *223*, 813–820. [CrossRef] [PubMed]
80. Mata-Pérez, C.; Padilla, M.N.; Sánchez-Calvo, B.; Begara-Morales, J.C.; Valderrama, R.; Chaki, M.; Barroso, J.B. Biological properties of nitro-fatty acids in plants. *Nitric Oxide* **2018**. [CrossRef] [PubMed]
81. Padilla, M.N.; Mata-Pérez, C.; Melguizo, M.; Barroso, J.B. In vitro nitro-fatty acid release from Cys-NO2-fatty acid adducts under nitro-oxidative conditions. *Nitric Oxide* **2017**, *68*, 14–22. [CrossRef] [PubMed]
82. Kansanen, E.; Jyrkkanen, H.-K.; Volger, O.L.; Leinonen, H.; Kivela, A.M.; Hakkinen, S.-K.; Woodcock, S.R.; Schopfer, F.J.; Horrevoets, A.J.; Yla-Herttala, S. Nrf2-dependent and-independent responses to nitro-fatty acids in human endothelial cells: Identification of heat shock response as the major pathway activated by nitro-oleic acid. *J. Biol. Chem.* **2009**, *284*, 33233–33241. [CrossRef]
83. Nishizawa, A.; Yabuta, Y.; Yoshida, E.; Maruta, T.; Yoshimura, K.; Shigeoka, S. Arabidopsis heat shock transcription factor A2 as a key regulator in response to several types of environmental stress. *Plant J.* **2006**, *48*, 535–547. [CrossRef] [PubMed]
84. Sánchez-Calvo, B.; Barroso, J.B.; Corpas, F.J. Hypothesis: Nitro-fatty acids play a role in plant metabolism. *Plant Sci.* **2013**, *199*, 1–6. [CrossRef] [PubMed]
85. Astier, J.; Kulik, A.; Koen, E.; Besson-Bard, A.; Bourque, S.; Jeandroz, S.; Lamotte, O.; Wendehenne, D. Protein S-nitrosylation: What's going on in plants? *Free Radic. Biol. Med.* **2012**, *53*, 1101–1110. [CrossRef] [PubMed]

86. Begara-Morales, J.C.; Chaki, M.; Sánchez-Calvo, B.; Mata-Pérez, C.; Leterrier, M.; Palma, J.M.; Barroso, J.B.; Corpas, F.J. Protein tyrosine nitration in pea roots during development and senescence. *J. Exp. Bot.* **2013**, *64*, 1121–1134. [CrossRef] [PubMed]
87. Chaki, M.; Shekariesfahlan, A.; Ageeva, A.; Mengel, A.; von Toerne, C.; Durner, J.; Lindermayr, C. Identification of nuclear target proteins for S-nitrosylation in pathogen-treated Arabidopsis thaliana cell cultures. *Plant Sci.* **2015**, *238*, 115–126. [CrossRef]
88. Rudolph, V.; Schopfer, F.J.; Khoo, N.K.; Rudolph, T.K.; Cole, M.P.; Woodcock, S.R.; Bonacci, G.; Groeger, A.L.; Golin-Bisello, F.; Chen, C.-S. Nitro-fatty acid metabolome: Saturation, desaturation, β-oxidation, and protein adduction. *J. Biol. Chem.* **2009**, *284*, 1461–1473. [CrossRef]
89. Melo, T.; Montero-Bullón, J.-F.; Domingues, P.; Domingues, M.R. Discovery of bioactive nitrated lipids and nitro-lipid-protein adducts using mass spectrometry-based approaches. *Redox Biol.* **2019**, *2019*, 101106. [CrossRef]
90. Lin, W.-W.; Jang, Y.-J.; Wang, Y.; Liu, J.-T.; Hu, S.-R.; Wang, L.-Y.; Yao, C.-F. An improved and easy method for the preparation of 2, 2-disubstituted 1-nitroalkenes. *J. Org. Chem.* **2001**, *66*, 1984–1991. [CrossRef]
91. Jang, Y.-J.; Lin, W.-W.; Shih, Y.-K.; Liu, J.-T.; Hwang, M.-H.; Yao, C.-F. A one-pot, two step synthesis of 2, 2-disubstituted 1-nitroalkenes. *Tetrahedron* **2003**, *59*, 4979–4992. [CrossRef]
92. Sawa, T.; Arimoto, H.; Akaike, T. Regulation of redox signaling involving chemical conjugation of protein thiols by nitric oxide and electrophiles. *Bioconjugate Chem.* **2010**, *21*, 1121–1129. [CrossRef]
93. Winterbourn, C.C.; Hampton, M.B. Thiol chemistry and specificity in redox signaling. *Free Radic. Biol. Med.* **2008**, *45*, 549–561. [CrossRef] [PubMed]
94. Trostchansky, A.; Rubbo, H. Nitrated fatty acids: Mechanisms of formation, chemical characterization, and biological properties. *Free Radic. Biol. Med.* **2008**, *44*, 1887–1896. [CrossRef] [PubMed]
95. Delmastro-Greenwood, M.; Freeman, B.A.; Wendell, S.G. Redox-dependent anti-inflammatory signaling actions of unsaturated fatty acids. *Annu. Rev. Physiol.* **2014**, *76*, 79–105. [CrossRef] [PubMed]
96. Woodcock, C.-S.C.; Huang, Y.; Woodcock, S.R.; Salvatore, S.R.; Singh, B.; Golin-Bisello, F.; Davidson, N.E.; Neumann, C.A.; Freeman, B.A.; Wendell, S.G. Nitro-fatty acid inhibition of triple-negative breast cancer cell viability, migration, invasion, and tumor growth. *J. Biol. Chem.* **2018**, *293*, 1120–1137. [CrossRef]
97. Villacorta, L.; Zhang, J.; Garcia-Barrio, M.T.; Chen, X.-l.; Freeman, B.A.; Chen, Y.E.; Cui, T. Nitro-linoleic acid inhibits vascular smooth muscle cell proliferation via the Keap1/Nrf2 signaling pathway. *Am. J. Physiol. Heart Circ. Physiol.* **2007**, *293*, H770–H776. [CrossRef]
98. Cui, T.; Schopfer, F.J.; Zhang, J.; Chen, K.; Ichikawa, T.; Baker, P.R.; Batthyany, C.; Chacko, B.K.; Feng, X.; Patel, R.P. Nitrated fatty acids: Endogenous anti-inflammatory signaling mediators. *J. Biol. Chem.* **2006**, *281*, 35686–35698. [CrossRef]
99. Ambrozova, G.; Martiskova, H.; Koudelka, A.; Ravekes, T.; Rudolph, T.K.; Klinke, A.; Rudolph, V.; Freeman, B.A.; Woodcock, S.R.; Kubala, L. Nitro-oleic acid modulates classical and regulatory activation of macrophages and their involvement in pro-fibrotic responses. *Free Radic. Biol. Med.* **2016**, *90*, 252–260. [CrossRef]
100. Ambrozova, G.; Fidlerova, T.; Verescakova, H.; Koudelka, A.; Rudolph, T.K.; Woodcock, S.R.; Freeman, B.A.; Kubala, L.; Pekarova, M. Nitro-oleic acid inhibits vascular endothelial inflammatory responses and the endothelial-mesenchymal transition. *BBA-Gen. Subj.* **2016**, *1860*, 2428–2437. [CrossRef]
101. Rom, O.; Khoo, N.K.; Chen, Y.E.; Villacorta, L. Inflammatory signaling and metabolic regulation by nitro-fatty acids. *Nitric Oxide* **2018**, *78*, 140–145. [CrossRef] [PubMed]
102. Deen, A.J.; Sihvola, V.; Härkönen, J.; Patinen, T.; Adinolfi, S.; Levonen, A.-L. Regulation of stress signaling pathways by nitro-fatty acids. *Nitric Oxide* **2018**, *78*, 170–175. [CrossRef] [PubMed]
103. Kansanen, E.; Bonacci, G.; Schopfer, F.J.; Kuosmanen, S.M.; Tong, K.I.; Leinonen, H.; Woodcock, S.R.; Yamamoto, M.; Carlberg, C.; Ylä-Herttuala, S. Electrophilic nitro-fatty acids activate NRF2 by a KEAP1 cysteine 151-independent mechanism. *J. Biol. Chem.* **2011**, *286*, 14019–14027. [CrossRef]
104. Dinkova-Kostova, A.T.; Holtzclaw, W.D.; Cole, R.N.; Itoh, K.; Wakabayashi, N.; Katoh, Y.; Yamamoto, M.; Talalay, P. Direct evidence that sulfhydryl groups of Keap1 are the sensors regulating induction of phase 2 enzymes that protect against carcinogens and oxidants. *Proc. Natl. Acad. Sci. USA* **2002**, *99*, 11908–11913. [CrossRef] [PubMed]

105. Diaz-Amarilla, P.; Miquel, E.; Trostchansky, A.; Trias, E.; Ferreira, A.M.; Freeman, B.A.; Cassina, P.; Barbeito, L.; Vargas, M.R.; Rubbo, H. Electrophilic nitro-fatty acids prevent astrocyte-mediated toxicity to motor neurons in a cell model of familial amyotrophic lateral sclerosis via nuclear factor erythroid 2-related factor activation. *Free Radic. Biol. Med.* **2016**, *95*, 112–120. [CrossRef] [PubMed]
106. Benjamin, I.J.; McMillan, D.R. Stress (heat shock) proteins: Molecular chaperones in cardiovascular biology and disease. *Circ. Res.* **1998**, *83*, 117–132. [CrossRef] [PubMed]
107. Vihervaara, A.; Sistonen, L. HSF1 at a glance. *J. Cell Sci.* **2014**, *127*, 261–266. [CrossRef] [PubMed]
108. Li, Y.; Zhang, J.; Schopfer, F.J.; Martynowski, D.; Garcia-Barrio, M.T.; Kovach, A.; Suino-Powell, K.; Baker, P.R.; Freeman, B.A.; Chen, Y.E. Molecular recognition of nitrated fatty acids by PPARγ. *Nat. Struct. Mol. Biol.* **2008**, *15*, 865. [CrossRef]
109. Baker, P.; Schopfer, F.; Batthyany, C.; Groeger, A.; Branchaud, B.; Chen, Y.; Freeman, B. multiple Nitrated Unsaturated Fatty Acid Derivatives Exist In Human Blood And Urine And Serve As Endogenous Ppar Ligands: 261. *Free Radic. Biol. Med.* **2005**, *39*, S97.
110. Schopfer, F.J.; Cole, M.P.; Groeger, A.L.; Chen, C.-S.; Khoo, N.K.; Woodcock, S.R.; Golin-Bisello, F.; Motanya, U.N.; Li, Y.; Zhang, J. Covalent peroxisome proliferator-activated receptor γ adduction by nitro-fatty acids selective ligand activity and anti-diabetic signaling actions. *J. Biol. Chem.* **2010**, *285*, 12321–12333. [CrossRef] [PubMed]
111. Villacorta, L.; Minarrieta, L.; Salvatore, S.R.; Khoo, N.K.; Rom, O.; Gao, Z.; Berman, R.C.; Jobbagy, S.; Li, L.; Woodcock, S.R. In situ generation, metabolism and immunomodulatory signaling actions of nitro-conjugated linoleic acid in a murine model of inflammation. *Redox Biol.* **2018**, *15*, 522–531. [CrossRef] [PubMed]
112. Khoo, N.K.; Li, L.; Salvatore, S.R.; Schopfer, F.J.; Freeman, B.A. Electrophilic fatty acid nitroalkenes regulate Nrf2 and NF-κB signaling: A medicinal chemistry investigation of structure-function relationships. *Sci. Rep.* **2018**, *8*, 2295. [CrossRef] [PubMed]
113. Villacorta, L.; Chang, L.; Salvatore, S.R.; Ichikawa, T.; Zhang, J.; Petrovic-Djergovic, D.; Jia, L.; Carlsen, H.; Schopfer, F.J.; Freeman, B.A. Electrophilic nitro-fatty acids inhibit vascular inflammation by disrupting LPS-dependent TLR4 signalling in lipid rafts. *Cardiovasc. Res.* **2013**, *98*, 116–124. [CrossRef]
114. Ricote, M.; Glass, C.K. PPARs and molecular mechanisms of transrepression. *Biochim. Biophys. Acta Mol. Cell Biol. Lipids* **2007**, *1771*, 926–935. [CrossRef] [PubMed]
115. Bonacci, G.; Schopfer, F.J.; Batthyany, C.I.; Rudolph, T.K.; Rudolph, V.; Khoo, N.K.; Kelley, E.E.; Freeman, B.A. Electrophilic fatty acids regulate matrix metalloproteinase activity and expression. *J. Biol. Chem.* **2011**, *286*, 16074–16081. [CrossRef] [PubMed]
116. Artim, D.; Bazely, F.; Daugherty, S.; Sculptoreanu, A.; Koronowski, K.; Schopfer, F.; Woodcock, S.; Freeman, B.; de Groat, W. Nitro-oleic acid targets transient receptor potential (TRP) channels in capsaicin sensitive afferent nerves of rat urinary bladder. *Exp. Neurol.* **2011**, *232*, 90–99. [CrossRef]
117. Sculptoreanu, A.; Kullmann, F.; Artim, D.; Bazley, F.; Schopfer, F.; Woodcock, S.; Freeman, B.; De Groat, W. Nitro-oleic acid inhibits firing and activates TRPV1-and TRPA1-mediated inward currents in dorsal root ganglion neurons from adult male rats. *J. Pharmacol. Exp. Ther.* **2010**, *333*, 883–895. [CrossRef]
118. Zhang, J.; Villacorta, L.; Chang, L.; Fan, Z.; Hamblin, M.; Zhu, T.; Chen, C.S.; Cole, M.P.; Schopfer, F.J.; Deng, C.X. Nitro-oleic acid inhibits angiotensin II–induced hypertension. *Circ. Res.* **2010**, *107*, 540–548. [CrossRef] [PubMed]
119. Gil, M.; Graña, M.; Schopfer, F.J.; Wagner, T.; Denicola, A.; Freeman, B.A.; Alzari, P.M.; Batthyány, C.; Durán, R. Inhibition of Mycobacterium tuberculosis PknG by non-catalytic rubredoxin domain specific modification: Reaction of an electrophilic nitro-fatty acid with the Fe–S center. *Free Radic. Biol. Med.* **2013**, *65*, 150–161. [CrossRef]
120. Kelley, E.E.; Batthyany, C.I.; Hundley, N.J.; Woodcock, S.R.; Bonacci, G.; Del Rio, J.M.; Schopfer, F.J.; Lancaster, J.R.; Freeman, B.A.; Tarpey, M.M. Nitro-oleic acid, a novel and irreversible inhibitor of xanthine oxidoreductase. *J. Biol. Chem.* **2008**, *283*, 36176–36184. [CrossRef]
121. Nadtochiy, S.M.; Zhu, Q.; Urciuoli, W.; Rafikov, R.; Black, S.M.; Brookes, P.S. Nitroalkenes confer acute cardioprotection via adenine nucleotide translocase 1. *J. Biol. Chem.* **2012**, *287*, 3573–3580. [CrossRef] [PubMed]
122. Turell, L.; Vitturi, D.A.; Coitiño, E.L.; Lebrato, L.; Möller, M.N.; Sagasti, C.; Salvatore, S.R.; Woodcock, S.R.; Alvarez, B.; Schopfer, F.J. The chemical basis of thiol addition to nitro-conjugated linoleic acid, a protective cell-signaling lipid. *J. Biol. Chem.* **2017**, *292*, 1145–1159. [CrossRef] [PubMed]

123. Trostchansky, A.; Bonilla, L.; Thomas, C.P.; O'Donnell, V.B.; Marnett, L.J.; Radi, R.; Rubbo, H. Nitroarachidonic acid, a novel peroxidase inhibitor of prostaglandin endoperoxide H synthases 1 and 2. *J. Biol. Chem.* **2011**, *286*, 12891–12900. [CrossRef] [PubMed]
124. Bonilla, L.; O'Donnell, V.; Clark, S.; Rubbo, H.; Trostchansky, A. Regulation of protein kinase C by nitroarachidonic acid: Impact on human platelet activation. *Arch. Biochem. Biophys.* **2013**, *533*, 55–61. [CrossRef]
125. González-Perilli, L.; Álvarez, M.N.; Prolo, C.; Radi, R.; Rubbo, H.; Trostchansky, A. Nitroarachidonic acid prevents NADPH oxidase assembly and superoxide radical production in activated macrophages. *Free Radic. Biol. Med.* **2013**, *58*, 126–133. [CrossRef] [PubMed]
126. González-Perilli, L.; Mastrogiovanni, M.; de Castro Fernandes, D.; Rubbo, H.; Laurindo, F.; Trostchansky, A. Nitroarachidonic acid (NO2AA) inhibits protein disulfide isomerase (PDI) through reversible covalent adduct formation with critical cysteines. *BBA-Gen. Subj.* **2017**, *1861*, 1131–1139. [CrossRef]
127. Asada, K. Ascorbate peroxidase—A hydrogen peroxide-scavenging enzyme in plants. *Physiol. Plant.* **1992**, *85*, 235–241. [CrossRef]
128. Noctor, G.; Foyer, C.H. Ascorbate and glutathione: Keeping active oxygen under control. *Annu. Rev. Plant Biol.* **1998**, *49*, 249–279. [CrossRef]
129. Hossain, M.A.; Asada, K. Inactivation of ascorbate peroxidase in spinach chloroplasts on dark addition of hydrogen peroxide: Its protection by ascorbate. *Plant Cell Physiol.* **1984**, *25*, 1285–1295.
130. Metcalfe, C.; Macdonald, I.K.; Murphy, E.J.; Brown, K.A.; Raven, E.L.; Moody, P.C. The tuberculosis prodrug isoniazid bound to activating peroxidases. *J. Biol. Chem.* **2008**, *283*, 6193–6200. [CrossRef]

© 2019 by the authors. Licensee MDPI, Basel, Switzerland. This article is an open access article distributed under the terms and conditions of the Creative Commons Attribution (CC BY) license (http://creativecommons.org/licenses/by/4.0/).

Review

Nitrogen Dioxide at Ambient Concentrations Induces Nitration and Degradation of PYR/PYL/RCAR Receptors to Stimulate Plant Growth: A Hypothetical Model

Misa Takahashi * and Hiromichi Morikawa

Department of Mathematical and Life Sciences, Hiroshima University, Higashi-Hiroshima 739-8526, Japan
* Correspondence: misat@hiroshima-u.ac.jp; Tel.: +81-82-424-7494

Received: 20 May 2019; Accepted: 24 June 2019; Published: 30 June 2019

Abstract: Exposing *Arabidopsis thaliana* (Arabidopsis) seedlings fed with soil nitrogen to 10–50 ppb nitrogen dioxide (NO_2) for several weeks stimulated the uptake of major elements, photosynthesis, and cellular metabolisms to more than double the biomass of shoot, total leaf area and contents of N, C P, K, S, Ca and Mg per shoot relative to non-exposed control seedlings. The $^{15}N/^{14}N$ ratio analysis by mass spectrometry revealed that N derived from NO_2 (NO_2-N) comprised < 5% of the total plant N, showing that the contribution of NO_2-N as N source was minor. Moreover, histological analysis showed that leaf size and biomass were increased upon NO_2 treatment, and that these increases were attributable to leaf age-dependent enhancement of cell proliferation and enlargement. Thus, NO_2 may act as a plant growth signal rather than an N source. Exposure of Arabidopsis leaves to 40 ppm NO_2 induced virtually exclusive nitration of PsbO and PsbP proteins (a high concentration of NO_2 was used). The PMF analysis identified the ninth tyrosine residue of PsbO1 (^{9}Tyr) as a nitration site. ^{9}Tyr of PsbO1 was exclusively nitrated after incubation of the thylakoid membranes with a buffer containing NO_2 and NO_2^- or a buffer containing NO_2^- alone. Nitration was catalyzed by illumination and repressed by photosystem II (PSII) electron transport inhibitors, and decreased oxygen evolution. Thus, protein tyrosine nitration altered (downregulated) the physiological function of cellular proteins of Arabidopsis leaves. This indicates that NO_2-induced protein tyrosine nitration may stimulate plant growth. We hypothesized that atmospheric NO_2 at ambient concentrations may induce tyrosine nitration of PYR/PYL/RCAR receptors in Arabidopsis leaves, followed by degradation of PYR/PYL/RCAR, upregulation of target of rapamycin (TOR) regulatory complexes, and stimulation of plant growth.

Keywords: nitrogen dioxide; *Arabidopsis thaliana*; plant growth; cell enlargement; cell proliferation; early flowering; tyrosine nitration; PsbO

1. Introduction

Atmospheric nitrogen dioxide (NO_2) originates equally from natural sources, including soil microbes and lightning, and anthropogenic sources, including the combustion of fossil fuels [1,2]. Globally, atmospheric NO_2 is a main pollutant in urban areas and a key precursor of ozone and particulate matter (PM) [3,4]. The current World Health Organization (WHO) annual guideline value for atmospheric NO_2 is 40 µg/m^3 (21.3 ppb) [4]. It has been reported that the average NO_2 concentration in 141 countries is 50.6 µg/m^3 (~27.3 ppb) [5]; this value is clearly higher than that recommended by the WHO.

Plants emit nitric oxide (NO) and NO_2 [6–9]. Plants also absorb NO_2 and assimilate NO_2-derived nitrogen into amino acid nitrogen [10,11]. The compensation point concentration at which the emission of NO_2 from plants and absorption into plants balance is reported to be 0.3–3 ppb [12]. Therefore, at 27 ppb NO_2 (see above), plants are a sink for NO_2. These nitrogen oxides (NO and NO_2) are often considered as air pollutants [13]. For humans, NO_2 at ambient concentrations is definitely toxic [14]. In contrast, the effects of atmospheric NO and NO_2 are either toxic or non-toxic on plants contingent on their concentrations and the plant species [13,15–18]. It is noteworthy that in the nineteenth century, *von Liebig* (1827) [19] first proposed that lightning is important in the global nitrogen cycle to produce atmospheric NO and NO_2 by the oxidation of N_2, and that these nitrogen oxides serve as a natural fertilizer.

We discovered that atmospheric NO_2 at ambient concentrations (10–50 ppb) acts as a stimulant signal for plant growth. We also sought understanding of why an air-pollutant such as NO_2 can act as a stimulant factor for plant growth. Recently, based on previous research, we devised a hypothesis to answer this fundamental question, which is described below.

2. Nitrogen Dioxide at Ambient Concentrations of 10–50 ppb Acts as a Positive Plant Growth Signal in *Arabidopsis Thaliana*

Initially, we investigated the potential of plant material to mitigate atmospheric NO_2 [20–22]. We found a higher than 600-fold difference in the assimilation ability of NO_2 among 217 plant taxa [20]. We investigated hypothetical air-pollutant-philic plants [23,24] that utilize NO_2 as the sole nitrogen source. During our research, we discovered that atmospheric NO_2 at concentrations as low as 10–50 ppb positively regulates plant growth [25–32].

Arabidopsis thaliana (Arabidopsis) was grown in air without NO_2 for the first week after sowing, and then for 1–4 weeks in air with (abbreviated as $+NO_2$-treated plants) or without ($-NO_2$ control plants) NO_2 [26]. Seedlings were watered semiweekly with half-strength inorganic salts of Murashige and Skoog (M&S) medium [33] containing 19.7 mM nitrate and 10.3 mM ammonium. Plant age is expressed in weeks after sowing and corresponds to the time of harvest. The plant type utilized was accession C24 or Columbia. Their responses in terms of biomass increase and flowering time to NO_2 were very similar qualitatively, but differed quantitatively (see below) [28].

NO_2 concentration effect on the yield of shoot biomass in 4-week-old plants was first determined. Shoot biomass of $+NO_2$-treated C24 plants under 10 ± 0.2 and 50 ± 0.3 ppb NO_2 was 3.2-fold [29] and 2.5-fold greater relative to the $-NO_2$ control plants. Treatments of 100 ± 20 and 200 ± 50 ppb NO_2 produced no stimulation of growth, or somewhat repressed the growth of plants. In this study, 50 ± 0.3 ppb NO_2 treatments were used. Images of typical 4-week-old $+NO_2$ and $-NO_2$ control plants (Columbia) are shown in Figure 1.

Figure 1. Typical 4-week-old plants of *Arabidopsis thaliana* accession Columbia grown in the presence (right) or absence (left) of 50 ppb nitrogen dioxide (NO_2). Bar = 1 cm.

Increase in shoot biomass by NO_2 treatment was accompanied by increase in uptake of seven major elements, such as carbon(C), N, phosphorus (P), potttasium (K), calcium (Ca), magnesium (Mg) and sulfur (S) into shoots. The contents of these elements per shoot dry weight (DW) were virtually the same for +NO$_2$-treated plants and −NO$_2$ control plants, and the contents of these elements per shoot were two times greater in +NO$_2$-treated plants than in −NO$_2$ control plants (Table S1) [26]. These findings agreed with our previous work on *Nicotiana plumbaginifolia* [24] and other vegetable plant species [27].

Arabidopsis thaliana accession Columbia also increased shoot biomass in response to NO$_2$ treatment as in the case of accession C24 [26]. This is consistent with the report of Xu et al. (2010) [34]. Columbia grew faster than C24, and 4-week-old Columbia appeared to be close to the end of the vegetative growth. Shoot biomasses of +NO$_2$-treated plants and −NO$_2$ control plants of 4-week-old Columbia were 24.2 ± 5.5 and 14.3 ± 2.5 mg (mean ± SD, n = 5), respectively [26]. This difference in shoot biomass (1.7-fold) was smaller than that in C24 (2.5-fold, see above).

NO$_2$ treatment significantly shortened flowering time in both the C24 and Columbia accessions. In accordance with Kotchoni et al. [34], the number of days after sowing when the flower bolts became 1 cm long was a measure of flowering time.

The accession C24 exhibited a median flowering time of 41 and 42 d in +NO$_2$-treated and −NO$_2$ control plants, respectively. This difference was statistically significant by Student's *t*-test ($P < 0.05$) [25]. In the case of the accession Columbia, flowering time was remarkably shortened by NO$_2$ treatment. The median flowering time of this accession was 34 and 40 d in +NO$_2$-treated plants and −NO$_2$ control plants, respectively [31]. This was statistically significant ($P < 0.001$) by Student's *t*-test [25]. A similar flowering acceleration was observed in other plants. NO$_2$ treatment shortened the flowering time by 3.2 days and increased fruit yield by 1.4 times in tomato [31]. In addition, NO$_2$ has accelerated the flowering of mulkhiya plants [35].

NO at the same concentration as NO$_2$ increased shoot biomass in Arabidopsis C24 [25,26] and Columbia [36]. This agrees with those reports that sodium nitroprusside, a NO donor, accelerates vegetative growth of Arabidopsis [37], and that NO gas stimulates the expansion of leaf discs of pea [38] and the vegetative growth of spinach [39].

As NO$_2$ stimulate shoot biomass production, we expected a similar NO$_2$ effect on the root biomass production. Among 6 plant species we studied so far, 2 showed NO$_2$-stimulated root biomass production, but the remaining 4 showed no NO$_2$-stimulated root biomass production: Sunflower [27] and Arabidopsis C24 [25] plants that were exposed with NO2 exhibited 0.4 ± 0.04 and 4.8 ± 0.08 (g/plant) root biomass (mean of 3 or 10 plants, respectively, ± SD) which were significantly higher ($P < 0.05$ or 0.001 by Student's t test) than corresponding value of non-exposed plants (0.2 ± 0.08 and 2.5 ± 0.6). However, NO$_2$ showed no statistically significant increases in the root biomass in lettuce, cucumber, pumpkin [27] and *Nicotiana plumbaginifolia* [24]. The causes and mechanisms for this result are completely unknown, and will be an important and intriguing subject for future studies. Interestingly, NO$_2$ did stimulate the seed production of mulkhiya plants [35] although whether NO$_2$ exhibits similar effects on other plant species such as Arabidopsis is not known yet. Similarly, how NO$_2$ stimulates other aspects of whole life cycle of plants also is an important and intriguing subject of the future studies.

To investigate the physiological role of N derived from NO$_2$ (NO$_2$–N), Arabidopsis seedlings were fed with ^{15}N-labeled gaseous NO$_2$ (50 ppb) and unlabeled nitrate (19.7 mM) and ammonium (10.3 mM), and mass spectrometric N analysis [40] including the ^{15}N/^{14}N ratio analysis on the aboveground parts of plants was performed. The ^{15}N/^{14}N ratio is a measure of the content of NO$_2$–N as a relative amount of the total plant N. We found that NO$_2$–N occupied < 5% (4.05 ± 0.75%; mean ± SD, n = 3) of the total N in the +NO$_2$-treated Arabidopsis C24 shoots. Therefore, NO$_2$–N plays only a minor role as an N source, but instead plays an important role as a plant growth signal. Similar results indicating that NO$_2$ plays an important role as a plant growth signal were obtained in *Nicotiana plumbaginifolia* [24] and other vegetable plant species [27].

The increased total leaf area following NO$_2$ treatment (Table S1) indicated that NO$_2$ treatment increases the sizes of individual leaves. Therefore, the sizes of individual rosette leaves in positions 1–25 on 5-week-old +NO-treated plants and −NO$_2$ control plants, which had 28 and 25 rosette leaves, respectively, were measured. Leaves 1–11 and leaves 12–25 were in almost maturity stages and developing stages, respectively. Each of rosette leaves was separated by an angle of approximately 137° [41]. The oldest leaf located at the bottom (root side) was numbered as leaf 1, and progressively leaves were numbered as leaf 2, 3, 4 etc. to the youngest one located at the tip of the stem (close to apical meristem) as leaf 25 or 28. Leaves 1–11 and leaves 12–25 or 28 were in almost maturity stages and developing stages, respectively.

Microscopic study was performed according to Tsuge et al. (1996) [42]. The leaves of Arabidopsis C24 plants were fixed with a FAA solution (formaldehyde-acetic acid-ethanol), and microscopic observations were carried using a stereo microscope and a differential interference microscope. Microphotographs were taken to measure leaf area, cell number, and cell size (Figure 2). Leaves 1 (the oldest) to 25 (the youngest) in +NO_2-treated plants had 1.3–8.4 times greater leaf areas compared with –NO_2 control plants in the corresponding leaf positions (Figure 2A). The observed differences were significant statistically at all positions according to the Student's t-test (Figure 2A) [25].

It is known that determinants of organ size are cell number and cell size [43–45]. Therefore, we investigated whether the increases in leaf areas following NO_2 treatment were ascribable to increases in cell numbers or cell sizes, or both. In both the +NO_2-treated and –NO_2 control plants of Arabidopsis, palisade cells in the adaxial sub-epidermal layer were neatly aligned in the paradermal plane throughout leaf development [42], as reported previously [25]. Thus, the sizes and numbers of palisade cells in leaves of positions 1 to 25 in 5-week-old +NO_2-treated plants and –NO_2 control plants of Arabidopsis C24 were determined. (Figure 2B,C).

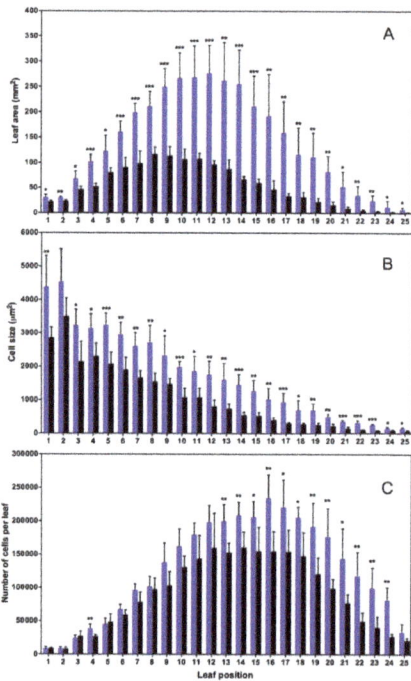

Figure 2. Area of leaves (**A**), size of cells (**B**), and number of cells (**C**) in 5-week-old *Arabidopsis thaliana* C24 plants as a function of leaf position. Plants were grown in the presence (+NO_2-treated plants, blue columns) or absence (–NO_2 control plants, black columns) of NO_2. Values are expressed as means ± SD; n = 5. Statistical significance assessed by Student's t-test (* $P < 0.05$; ** $P < 0.01$; *** $P < 0.001$).

Leaf area as a function of leaf position was more or less asymmetric in both –NO$_2$ control (black columns) and +NO$_2$-treated plants (blue columns) (Figure 2A); –NO$_2$ control plants exhibited a positively-skewed bell-shaped pattern, while +NO$_2$-treated plants exhibited less skewed and less asymmetric pattern [25]. The ratio of leaf area of +NO$_2$-treated plants to that of –NO$_2$ control plants at the corresponding leaf positions (designated RLA) was calculated [25]. The RLA varied from 1.3–2.5 in leaves 1–11, while varied from 2.9–8.4 in leaves 12–25. This difference in RLA between leaves 1–11 and 12–25 was significant ($P < 0.01$), as assessed by Mann-Whitney U test [25].

The size of the cells decreased as a linear function of leaf position in plants with or without NO$_2$ treatment (Figure 2B). This suggests that increase in cell size is a linear function of leaf age. Presence or absence of NO$_2$ did not affect this result. Student's t test showed that these results in the cell size at all positions, except leaf 2, were statistically significant (Figure 2B) [25].

The ratio of the cell size of +NO$_2$-treated plants to that of the –NO$_2$ control plants at the corresponding leaf position (designated RCS) was calculated. The RCS of leaves 12–25 (2.0–3.2) was larger than that of leaves 1–11 (1.3–1.9). This difference in RCS between leaves 12–25 and 1–11 was significant statistically ($P < 0.01$) by Mann-Whitney U test. This finding is in line with our previous observation that NO$_2$ exerts a greater effect on leaf expansion in younger leaves than in older leaves [24].

Cell number as a function of leaf position exhibited a normal distribution for both the +NO$_2$-treated plants and–NO$_2$ control plants (Figure 2C). Plants with or without NO$_2$ treatment did not show significant differences in cell numbers in almost all maturing leaves (positions 1–11, except leaf 4). This suggest that the NO$_2$ treatment did not affect cell numbers in the leaves in maturity. Almost all younger leaves (positions 13–24 of leaves 12–25) showed statistically significant differences in cell number, suggesting that NO$_2$ did increase cell numbers in developing leaves.

The ratio of the cell number of the +NO$_2$-treated plants to that of the –NO$_2$ control plants at the corresponding leaf position (designated RCN) was calculated. Leaves 12–25 had larger RCN (1.2–3.1) than that of leaves 1–11 (0.9–1.5). This difference in RCN between leaves 12–25 and 1–11 was found to be significant ($P < 0.01$) by the Mann-Whitney U test [25]. This observation indicated that the effect of NO$_2$ on cell proliferation changed depending on developmental stage of the leaves, and was greater in younger leaves than in older leaves.

To investigate the correlations between ratio in leaf area and the ratio in cell size (or cell number), log(RLA), log(RCS) and log(RCN) values were calculated and analyzed using Pearson's correlation analysis and Bonferroni's correction (Table S2). The correlation between leaf area and cell size was found to be high and significant in leaves 1–25 ($R = 0.9$, $P < 0.001$). Interestingly enough, the correlation between leaf areas and cell size was found to be stronger in older leaves than in younger leaves; $R = 0.7$, $P < 0.05$ for 1–11 leaves, while $R = 0.3$, $P > 0.5$ for 12–25 leaves (Table S2). This means that the correlation between NO$_2$-induced leaf expansion and cell size expansion was higher in older leaves than in younger leaves. Leaf area and cell number in leaves 1–25 were found to have a significantly high correlation ($R = 0.9$, $P < 0.001$). The same was found to be true when developing (12–25) ($R = 0.9$, $P < 0.001$) and maturing (1–11) ($R = 0.7$, $P < 0.05$) leaves were separately analyzed (Table S2) [25].

Thus, NO$_2$-induced leaf expansion correlated well with cell proliferation in both younger and older leaves. It is concluded that NO$_2$-mediated leaf expansion can largely be ascribed to cell proliferation in younger leaves, while the NO$_2$ effect can be ascribed to both cell proliferation and enlargement in older leaves [25].

3. NO$_2$ Selectively Nitrates Specific Cellular Proteins in Arabidopsis Leaves

Nitration of protein tyrosine is the addition of a nitro group on the carbon-3 of tyrosine residues of proteins to produce 3-nitrotyrosine (3-NT), which accompanies a drastic decrease (from 10.0 to 7.2) in the pKa of the tyrosine hydroxy group. Protein tyrosine nitration is an important post-translational modification in cell physiology, including cellular signaling [46,47]. According to a free radical mechanism [46–48], prior to their nitration, tyrosine residues are oxidized to tyrosyl radicals by an oxidation mechanism. Tyrosyl radicals undergo rapid radical-radical combination with NO$_2$ radicals that exist in the close vicinity of the tyrosyl radicals to produce 3-NT. Nonetheless, biological protein nitration is not a simple chemical process, but is instead a characteristic selective process in which only a restricted number of proteins are nitrated [46–48].

Selectivity of protein nitration is central for protein nitration to play a vital role in signal transduction that reflects the cellular redox state [46–50]. Selectivity of protein tyrosine nitration has been investigated mainly in mammals [45–47,50,51]. Although a number (12–127 kinds) of plant proteins are reported to be nitratable [47,51–53], experimental substantiation on this issue in plant protein nitration is rather scarce. NO_2 is a potent nitrating agent that nitrates tyrosine residues on proteins to yield NT [54,55] (see Section 4). Furthermore, NO_2 is a hydrophobic molecule (less hydrophobic than NO but more so than carbon dioxide), and thus is almost freely permeable to cell membranes [56]. In addition to its signaling role in plant growth, NO_2 is an in vivo intermediate involved in biological protein tyrosine nitration in animals [51] and plants [49]. Therefore, we used NO_2 as a nitrating agent; for the sake of facilitating nitrated protein and nitration site identification, plants were exposed to high (4–40 ppm) concentrations of NO_2 [57].

Arabidopsis (accession C24) plants were exposed to air containing or not containing 40 ppm NO_2 for 8 h under illumination. Proteins were extracted from whole leaves (abbreviated as whole leaf protein). Alternatively, chloroplasts were isolated and fractionated into soluble (stromal and lumenal) and insoluble (thylakoid membrane) fractions, and proteins were extracted from each fraction (abbreviated as chloroplast protein) [57]. Proteins were analyzed using two-dimensional polyacrylamide gel electrophoresis (2D PAGE), followed by staining with SYPRO Ruby stain and Western blotting using a 3-NT-specific antibody.

The 2D PAGE images of whole leaf proteins and chloroplast proteins are shown in Figures 3 and 4, respectively. The relative intensities of spots on Western blots (abbreviated as RISI), and those of the spots on SYPRO Ruby gels (abbreviated as RISS) were determined. Nitrated proteins identified in chloroplast protein fractions and their electrophoretic and proteomic characteristics are summarized in Table S3. Proteins that showed a high RISI and/or a high RISI/RISS were concluded to be selectively nitrated [57]. Seven 3-NT-positive spots were detected on a Western blot of whole leaf proteins from exposed leaves (Figure 3), all of which were identified as PsbO1, PsbO2 or PsbP1 by peptide mass fingerprinting (PMF) [57].

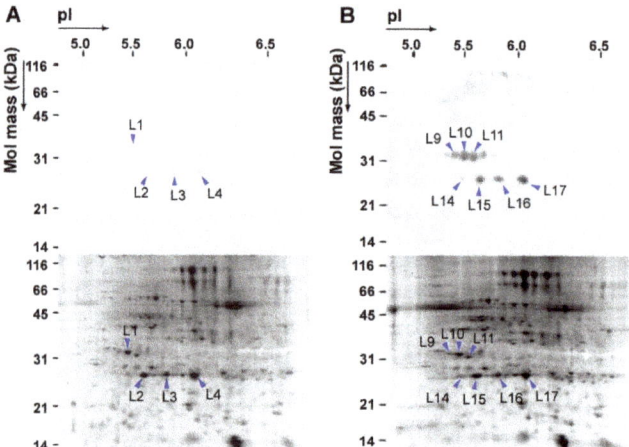

Figure 3. 2D PAGE gel patterns of Arabidopsis whole leaf proteins extracted from +NO_2-treated (right) and −NO_2 control (left) plants. Western blots detected using 3-NT-specific antibody (upper panels) and gels stained with SYPRO Ruby (lower panels). Each gel was loaded with 100 µg protein.

PsbO and PsbP are external proteins localized on the stromal side of the thylakoid membrane in PSII. PsbO and PsbP stabilize the oxygen-evolving complex (OEC) of PSII together with other external proteins, including PsbQ and PsbR [58–60]. No nitration of PsbQ or PsbR was detected. Thus, nitration was specific to PsbO and PsbP, while their RISI/RISS ratio was low (≤ 1.5) (Table S3). Non-exposed control plants exhibited very faint 3-NT-positive spots.

The number of 3-NT-positive spots was markedly increased in purified and fractionated chloroplast proteins (Figure 4, Table S3) [57]. Distinct 3-NT-positive protein spots were lined at 32 kDa (SL7–12), and distinct but clearly visible spots were lined at 27 kDa (SL13–18) on the Western blot of the soluble (stromal and lumenal) chloroplast protein fraction from $+NO_2$-treated plants (Figure 4A, upper panel). Lined spots of less in number at 32 kDa (IS7–10) were detected on the Western blot of the thylakoid membrane protein fraction from $+NO_2$-treated plants (Figure 4B, upper panel). PsbO and PsbP accounted for > 80% of the total RISI values [Table S3], and high RISI/RISS ratios (2.5–6.6) were exhibited by four non-PSII proteins such as peroxiredoxin II E (PRXII E) (spot SL21), thylakoid lumenal protein (SL22), RuBisCO activase (RCA, SL31), and the delta subunit of chloroplast ATP synthase (SL19) [Table S3]. Thus, PsbO, PsbP and these four non-PSII proteins are concluded to be selectively nitrated.

Despite that use of purified/fractionated chloroplast proteins markedly increased the number of 3-NT positive spots on Western blots (Figure 4), no 3-NT-positive spots attributable to other extrinsic (such as PsbQ and PsbR) or intrinsic (such as D1 and D2) proteins of PSII were detected, and nor were 3-NT-positive spots attributable to RuBisCO subunits (Figure 4, Table S3). Thus, NO_2 selectively nitrates two PSII and four non-PSII proteins in Arabidopsis. PMF analysis using MALDI-TOFMS provided evidence that the ninth tyrosine residue (^{9}Tyr) of PsbO1 is a nitration site [57].

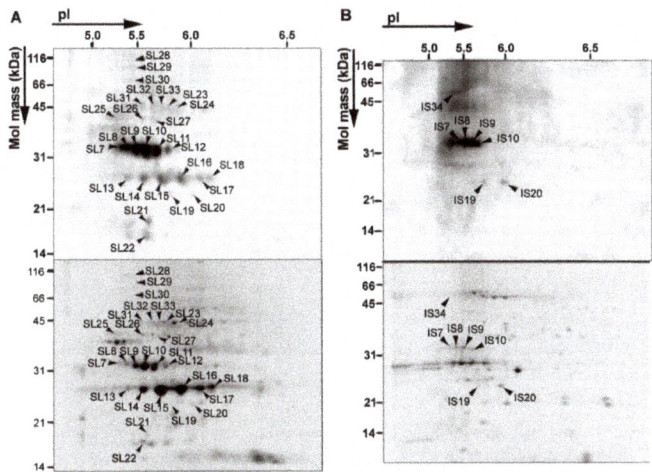

Figure 4. 2D PAGE gel patterns of chloroplast proteins extracted from $+NO_2$-treated plants. The stromal and lumenal protein fraction (**A**), and the thylakoid membranous protein fraction (**B**) are shown. Spots on gel (**A**) and (**B**) were numbered as SLn and ISn, respectively. Upper and lower panels of (**A**) and (**B**) correspond to Western blots detected using 3-NT-specific antibody and gels stained with SYPRO Ruby, respectively. Each lane of stromal and lumenal protein fraction and thylakoid membranous protein fraction was loaded with 40 and 20 μg protein, respectively.

4. PsbO1 May Function as an Electron Element Like Yz in PSII Electron Transport Chain

To investigate the physiological significance of protein nitration, thylakoid membranes were isolated from Arabidopsis leaves and incubated in a buffer solution bubbled with NO_2 gas or a buffer solution of potassium nitrite (KNO_2). The former buffer contains NO_2 and nitrite (NO_2^-), while the latter contains NO_2- alone [61]. NO_2 dissociates in water as shown in reaction 1 [62], as described previously [61]. Concentrations of NO_2 in the buffer were quantified by numerical solution of kinetic Equations (1)–(3). Nitrite (NO_2^-) concentrations in the buffer were quantified by capillary electrophoresis [63].

$$NO_2 \underset{k_2}{\overset{k_1}{\rightleftarrows}} N_2O_4 \overset{k_3}{\to} NO^{2-} + NO^{3-} \qquad \text{(Reaction 1)}$$

$$\frac{d[NO_2]}{dt} = -2k[NO_2]^2 + 2k_2[N_2O_4] \qquad (1)$$

$$\frac{d[N_2O_4]}{dt} = k_1[NO_2]^2 - k_2[N_2O_4] - k_3[N_2O_4] \qquad (2)$$

$$\frac{d[NO_2^-]}{dt} = k_3[N_2O_4] \qquad (3)$$

where k_1, k_2 and k_3 are rate constants 4.5×10^8 mol^{-1} s^{-1}, 6.4×10^3 s^{-1}, and 10^3 s^{-1}, respectively [62].

A distinct 3-NT-positive band of 32.5 kDa was detected on a Western blot of proteins extracted from thylakoid membranes that were incubated in a buffer containing NO_2 and NO_2^- under illumination (Figure 5A). This band was assigned to PsbO1 by liquid chromatography/mass spectrometry (LC/MS), followed by a Mascot search analysis [64]. On the other hand, no such band was detected following incubation thylakoid membranes in the same buffer in the dark at all concentrations of NO_2 and NO_2^- (Figure 5). Thus, illumination is essential in NO_2/NO_2^--mediated protein nitration. The intensities of the PsbO1 band on the Western blots were quantified using PDQuest software (ver. 7.0; Bio-Rad, Hercules, CA, USA) [64]. The intensity of the 3-NT-positive PsbO1 band after incubation in a buffer containing NO_2 and NO_2^- was divided by the intensity of the 3-NT-positive PsbO1 band before incubation in the buffer. This value was designated fold-change in the PsbO1 band intensity, and plotted against the concentrations of NO_2 and NO_2^- (Figure 5B). Incubation in the dark resulted null intensity of PsbO1 band at all concentrations of NO_2 and NO_2–except 44.4 µM NO_2 and 6.52 mM NO_2^- (Figure 5). This confirms that illumination is essential in NO_2/NO_2^--mediated protein nitration of PsbO1 in Arabidopsis thylakoid membranes.

Redox-active tyrosines play a key role in the photosynthetic electron in PSII. Yz (161Tyr of the D1 protein) in PSII is the most well-studied redox-active tyrosine residue in plants. Under illumination, it donates an electron to the PSII electron transport chain and itself is oxidized to tyrosyl radical [65,66]. It is reduced back to tyrosine by an electron derived from oxidation of water at the OEC. Thus, Y_z functions as an electron relay element between P680 and OEC Mn_4 cluster (Mn_4Ca) through photosynthetic electron transfer [67]. Another tyrosine that has a similar function, Y_D (161Tyr of the D2 protein), is also known [65,66].

In light of our finding of the illumination-triggered nitration of ^{9}Tyr of PsbO, it is conceivable that this tyrosine residue of PsbO1 is also redox-active, and that the photosynthetic electron transport chain can oxidize, upon illumination, this tyrosine residue to tyrosyl radical that is highly sensitive to nitration. The formed tyrosyl radical may rapidly react with NO_2 to yield 3-NT. Therefore, we hypothesized a nitration mechanism that prior to nitration PSII photosynthetic electron transport, in response to illumination, oxidizes the nitratable tyrosine residue of PsbO1 to tyrosyl radical to react with NO_2 to yield 3-NT [64].

Thylakoid membranes were incubated in a buffer containing NO_2 and NO_2^- in the presence or absence of electron transport inhibitors such as 3-(3,4-dichlorophenyl)-1,1-dimethylurea (DCMU), sodium azide and 1,5-diphenylcarbazide (DPC). Proteins were extracted from the treated thylakoid membranes, nitration of PsbO1 was determined by quantification of intensity of PsbO1 band. The results are shown in Figure 6. Fold-change in PsbO1 band intensity is given by (intensity of PSBO1 band after incubation in a buffer containing NO_2 and NO_2^-)/(intensity of PSBO1 band before incubation in the buffer). DCMU inhibits the photosynthetic electron transport by inhibiting binding of plastoquinone [68], and decreased the fold-change in PsbO1 band intensity to about one-fifth of the control value (Figure 6). Azide inhibits the photosynthetic electron transport by inhibiting a variety of reactions, including oxidation of water [69]. Azide also decreased the fold-change in PsbO1 band intensity to one-tenth of the control value (Figure 6). DPC inhibits the photosynthetic electron transport by inhibiting photosynthetic electron flow [70]. DPC decreased the fold-change in PsbO1 band intensity to one-tenth of the control value (Figure 6). Our present findings that nitration of PsbO1 was substantially inhibited by photosynthetic electron transport inhibitors substantiate our postulated nitration mechanism, whereby nitratable tyrosine residue of PsbO1 undergoes one-electron oxidation to tyrosyl radical that is highly reactive with NO_2 under illumination through PSII photosynthetic electron transport.

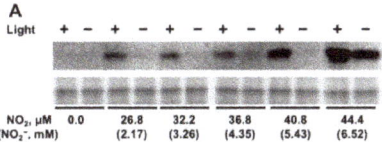

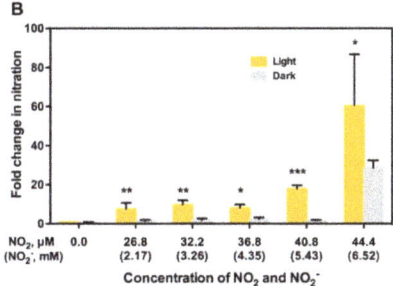

Figure 5. Demonstration that light is essential to induce nitration of PsbO1. (**A**) Arabidopsis thylakoid membranes were incubated in a buffer containing NO_2 and NO_2^- with or without illumination. Upper and lower panel show 3-NT-positive band and SYPRO-Ruby-stained band of PSBO1, respectively. (**B**) Fold-change in the PsbO1 band intensity (FCPSBO1) as a function of NO_2 and NO_2^- concentrations in a buffer solution bubbled with NO_2 gas. FCPSBO1 = (PsbO1 band intensity following incubation in a buffer bubbled with NO_2 gas)/(PsbO1 band intensity following incubation in buffer without NO_2 or NO_2^-). Data represent means of 3 independent experiments ± SD. *, $P < 0.05$; ***, $P < 0.001$. Student's t-test was done using GraphPad Prism 6.0 (GraphPad Software, La Jolla, CA, USA).

We next investigated oxygen evolution from isolated thylakoid membranes that had been treated or not treated with a buffer containing NO_2 and NO_2^- [71]. This buffer contained NO_2 and NO_2^- as nitrating agent [61]. As it is reported that nitrite anion inhibits PSII to decrease oxygen evolution [72–74], it is necessary to separately evaluate these two effects of nitrite on the oxygen evolution. Thylakoid membranes isolated from Arabidopsis leaves were incubated in a buffer containing NO_2 and NO_2^- or a buffer containing NO_2^- alone in the light or in the dark [71]. After incubation, each of the treated thylakoid membrane samples was divided into two portions. The first portion was analyzed for nitration of PsbO1 by Western blotting using 3-NT-specific antibody. The intensity of the 3-NT-positive PsbO1 band was quantified. Using the second portion, oxygen evolution was quantified [71]. Results are shown in Figure 7.

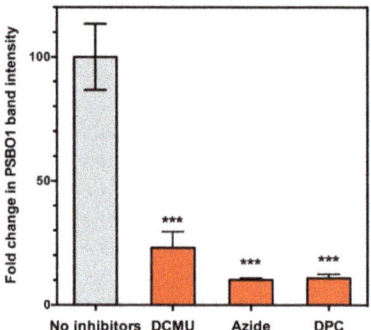

Figure 6. Demonstration that PSII electron transport inhibitors inhibit the nitration of PsbO1. Thylakoid membranes were incubated in a buffer bubbled with NO_2 gas (containing 36.8 µM NO_2 and 4.35 mM NO_2^-). Inhibitors such as 30 µM 3-(3,4-dichlorophenyl)-1,1-dimethylurea (DCMU), 10 mM sodium azide, or 1 mM 1,5-diphenylcarbazide (DPC) were added or not added to the buffer. Proteins were extracted, electrophoresed and Western blotted using a 3-NT-specific antibody followed by quantification of the PSBO1 band intensity. See text for details. Fold-change in PsbO1 band intensity = (intensity of PSBO1 band following incubation in a buffer containing NO_2 and NO_2^-)/(intensity of PSBO1 band before incubation in the buffer). Mean ± SD of three independent experiments. One-way ANOVA with Tukey's multiple comparison test was used to assess statistical significance: ***, $P < 0.001$.

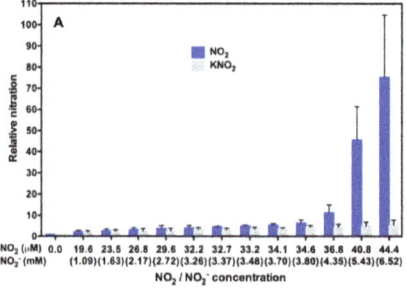

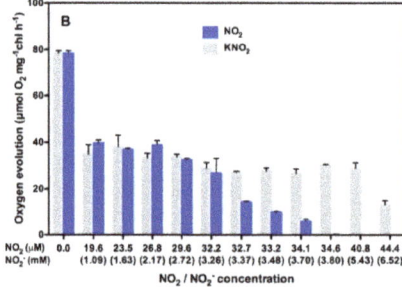

Figure 7. Demonstration that nitration of PsbO1 inhibits oxygen evolution. Arabidopsis thylakoid membranes were incubated in light in a buffer containing NO_2 and NO_2^- or a buffer containing NO_2^- alone. Incubated thylakoid membranes were divided into equal parts: first one for Western blot analysis and second one for oxygen evolution analysis. (**A**) Relative nitration of PsbO1 as a function of concentrations of NO_2 or NO_2^-. (**B**) Oxygen evolution as a function of concentrations of NO_2 or NO_2. The data represent the mean ± SD of three independent experiments.

Incubation of thylakoid membranes in a buffer containing NO_2^- alone at concentrations higher than 3.80 mM NO_2^- did not decrease oxygen evolution to null, but decreased it to one-third to half of the initial value (Figure 7B). On the other hand, oxygen evolution was decreased to almost null when co-existing NO_2 concentration exceeded 34.6 µM (Figure 7B). This indicates that the effect of NO_2 higher than 34.6 µM exceeds the effect of NO_2^- to inhibit oxygen evolution when thylakoid membranes were incubated in a buffer containing NO_2 and NO_2^-. This decrease in oxygen evolution is primarily ascribable to nitration of PsbO1 by NO_2. This substantiates our hypothesis [71,75] that PsbO1 functions as an electron element, like Y_z in photosynthetic electron transport.

In light of the present findings regarding the nitration characteristics of ^{9}Tyr of PsbO1, selectivity, light dependence, inhibitor-inhibitable and inhibiting oxygen evolution [61,71], and the widely accepted free radical mechanism of tyrosine nitration [45,46], we suggest that illumination induces selective and preferential photo-oxidation of ^{9}Tyr of PsbO1, similar to Y_z. ^{9}Tyr may act as an electronic element, similarly to Y_z in PSII electron transport chain.

The 3D structure for plant PSII from pea [76] is the only currently available crystal structure of higher-plant PSII. Using this structure for plant PSII from pea [76] and a molecular graphics software (PyMOL Molecular Graphics System Software, ver. 2.0.7; Schrödinger, New York, NY, USA), ^{9}Tyr of PsbO1 and the OEC were calculated to be 36.1Å apart. This is approximately five times greater than the distance between Y_z and the OEC Mn_4 cluster (Mn_4Ca) (7.5–8.0 Å) [77,78], making it too large for direct interactions [77,79] between the ^{9}Tyr of PsbO1 and OEC. However, electron transfer via peptide bonds as distant as more than 40 Å is reported [80]. Furthermore, a 134-Å electron transfer through the helical peptide was also reported [81]. In these cases, the amide groups reportedly act as quantum mechanical hopping sites for electron transfer. Long-range inter-protein electron transfer such as from cytochrome c to cytochrome c peroxidase has also been reported [82]. Moreover, electron transfer between the photosynthetic reaction center and cytochrome c across in *Rhodobacter sphaeroides* has been reported [83]. Taken together, inter-protein electron transfer plays a vital role in cellular metabolism including photosynthesis [82,83]. It is therefore postulated that long-range intra- and inter-protein electron transfer from PSII Mn cluster→^{9}Tyr of PsbO1→P680+ (PSII primary electron donor) could support a hypothesis that ^{9}Tyr of PsbO1 functions as an electronic element, like Y_z, in PSII electron transport (Figure 8) [75].

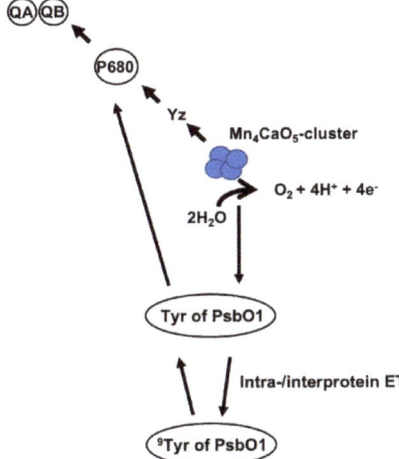

Figure 8. A model to hypothesize a novel role of the ninth tyrosine residue of PsbO1 (^{9}Tyr) in photosynthetic electron transport in PSII. Hypothetical long range inter- and intra-molecular electron transfer from manganese cluster to P680+ via ^{9}Tyr of PsbO1 supported the ^{9}Tyr as a novel electronic element, like Y_z, in the PSII electron transport.

5. NO$_2$ May Induce Tyrosine Nitration of PYR/PYL/RCAR ABA Receptors Leading to Degradation of the Receptors and Upregulation of TOR, to Stimulate Plant Growth

Our finding that NO$_2$-induced nitration of PsbO1 results in reduced oxygen evolution from Arabidopsis thylakoid membranes shows that protein tyrosine nitration alters (downregulates) the physiological function cellular proteins of Arabidopsis leaves. This finding indicates that NO$_2$-induced protein tyrosine nitration may be involved in NO$_2$-stimulated plant growth. However, as the concentration of NO$_2$ used in the study of plant growth (10–50 ppb) was about 800-4000 times lower than that used in the study of protein nitration (40 ppm), further investigations are required to clarify the physiological significance of the NO$_2$-mediated nitration of cellular proteins.

Protein nitration always inhibits protein function in plants [49,84]. In mammalian cells, protein nitration also usually inhibits protein function [45–47,50,51], as in plants, but rarely results in gain-of-function of proteins [51]. It remains unknown how protein nitration, a negative regulator, stimulates plant growth. This question is similar to the fundamental and long-standing question as to why an air-pollutant and toxic compounds, such as NO$_2$, act as a positive signal for plant growth. Inhibition of negative factors should induce plant growth. Figure 9 depicts a hypothetical model of how NO$_2$-induced protein nitration stimulates plant growth. The rationale for this model is as follows:

Plant growth requires the orchestration of a variety of cellular processes, which are controlled by regulatory proteins such as the serine/threonine protein kinase target of rapamycin (TOR), which forms complexes with regulatory proteins [82,85,86]. TOR plays a central role in auxin signal transduction in Arabidopsis [87]. TOR is downregulated by the plant hormone abscisic acid (ABA). ABA detection and signaling are mediated by the pyrabactin resistance1/PYR1-like/regulatory components of the ABA receptor (PYR/PYL/RCAR) family [83,88,89]. Tyrosine nitration of PYR/PYL/RCAR proteins reportedly results in polyubiquitylation and proteasome-mediated degradation [89]. Thus, the degradation of PYR/PYL/RCAR receptor proteins eventually results in upregulation of TOR and stimulation of plant growth [85]. Therefore, it is conceivable that NO$_2$ may induce tyrosine nitration of PYR/PYL/RCAR proteins, to degrade these proteins and upregulate TOR regulatory complexes to stimulate plant growth (Figure 9).

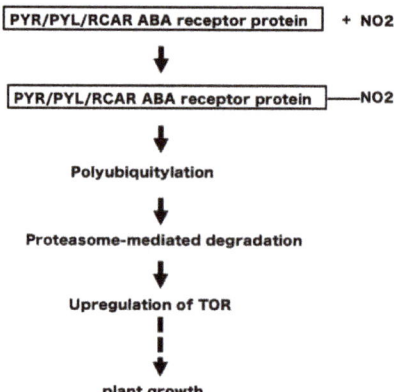

Figure 9. NO$_2$ may induce tyrosine nitration of PYR/PYL/RCAR to degrade PYR/PYL/RCAR, and upregulate target of rapamycin (TOR) to stimulate plant growth.

6. Future Perspectives

In future studies, antibody-assisted proteomic analysis is needed of nitratable proteins in Arabidopsis leaves that are exposed to low concentrations of NO$_2$ (10–50 ppb), to ascertain whether PYR/PYL/RCAR proteins from Arabidopsis leaves are nitratable at such low concentrations of NO$_2$. Future studies should also investigate whether auxin signal transduction in Arabidopsis leaves [90] is increased following exposure to ambient concentrations of NO_2. In both cases, special care is needed to ensure that the samples are always isolated from ambient air that contains 10–50 ppb NO_2. If NO_2-mediated nitration of PYR/PYL/RCAR proteins cannot be detected, or the involvement of TOR/ABA in NO_2-mediated plant growth stimulation cannot be ascertained, other target proteins of NO_2, such as hexokinase-like (HKL) proteins [91], a negative effector of plant growth in Arabidopsis, should be investigated in Arabidopsis leaves.

Supplementary Materials: The following are available online at http://www.mdpi.com/2223-7747/8/7/198/s1, Table S1: Shoot biomass (mg), total leaf area (mm^2), and content (µg/shoot) of carbon (C), nitrogen (N), phosphorus (P), potassium (K), calcium (Ca), magnesium (Mg), and sulfur (S) in 5-week-old *Arabidopsis thaliana* C24 plants grown with (+NO_2-treated plants) and without (–NO_2 control plants) NO_2 treatment, Table S2: Correlation analysis between leaf area (RLA) and cell size (RCS), and that between leaf area (RLA) and cell number RCN), Table S3: Identified nitrated proteins and their electrophoretic and mass spectrometric characteristics in chloroplast proteins extracted from *Arabidopsis thaliana* leaves exposed to NO_2.

Author Contributions: Writing—original draft preparation, M.T.; editing, H.M.

Funding: This work was supported by a grant from the Nippon Life Insurance Foundation (to MT), a grant from the Nissan Science Foundation (to MT), a Grant-in-Aid for Creative Scientific Research from the Japan Science and Technology Agency (no. 13GS0023 to HM), a Grant-in-Aid for Scientific Research from the Japan Society for the Promotion of Science (no. 15710149 to MT) and a grant from the Naito Foundation (to MT).

Acknowledgments: We appreciate financial supports from the Nippon Life Insurance Foundation (to MT), a grant from the Nissan Science Foundation (to MT), a Grant-in-Aid for Creative Scientific Research from the Japan Science and Technology Agency (no. 13GS0023 to HM), a Grant-in-Aid for Scientific Research from the Japan Society for the Promotion of Science (no. 15710149 to MT) and a grant from the Naito Foundation (to MT).

Conflicts of Interest: The authors declare no conflict of interest.

References

1. Aneja, V.P.; Roelle, P.A.; Murray, G.C.; Southerland, J.; Erisman, J.W.; Fowler, D.; Asman, W.A.H.; Naveen Patni, H. Atmospheric nitrogen compounds II: Emissions, transport, transformation, deposition and assessment. *Atmos. Environ.* **2001**, *35*, 1903–1911. [CrossRef]
2. Oswald, R.; Behrendt, T.; Ermel, M.; Wu, D.; Su, H.; Cheng, Y.; Breuninger, C.; Moravek, A.; Mougin, E.; Delon, C.; et al. HONO emissions from soil bacteria as a major source of atmospheric reactive nitrogen. *Science* **2013**, *6151*, 1233–1235. [CrossRef] [PubMed]
3. United States Environmental Protection Agency Home Page Integrated Science Assessment for Oxides of Nitrogen—Health Criteria (2016 Final Report). Available online: https://cfpub.epa.gov/ncea/isa/recordisplay.cfm?deid=310879 (accessed on 1 January 2019).
4. World Health Organization. Home Page. Available online: http://www.who.int/mediacentre/factsheets/fs313/en/ (accessed on 1 January 2019).
5. NationMaster Home Page. Available online: http://www.nationmaster.com/index.php (accessed on 1 January 2019).
6. Klepper, L. Nitric oxide (NO) and nitrogen dioxide (NO_2) emissions from herbicide-treated soybean plants. *Atmos. Environ.* **1979**, *13*, 537–542. [CrossRef]
7. Dean, J.V.; Harper, J.E. Nitric oxide and nitrous oxide production by soybean and winged bean during the in vivo nitrate reductase assay. *Plant Physiol.* **1986**, *82*, 718–723. [CrossRef] [PubMed]
8. Wildt, J.; Kley, D.; Rockel, A.; Rockel, P.; Segschneider, H.J. Emission of NO from several higher plant species. *J. Geochem. Res.* **1997**, *102*, 5919–5927. [CrossRef]
9. Hari, P.; Raivonen, M.; Vesala, T.; Munger, J.W.; Pilegaard, K.; Kulmala, M. Atmospheric science: Ultraviolet light and leaf emission of NO(x). *Nature* **2003**, *422*, 134. [CrossRef] [PubMed]
10. Yoneyama, T.; Sasakawa, H. Transformation of atmospheric NO_2 absorbed in spinach leaves. *Plant Cell Physiol.* **1979**, *20*, 263–266.
11. Yoneyama, T.; Sasakawa, H.; Ishizuka, S.; Totsuka, T. Absorption of atmospheric NO_2 by plants and soils. *Soil Sci. Plant Nutr.* **1979**, *25*, 255–265.
12. Breuninger, C.; Meixner, F.X.; Kesselmeier, J. Field investigations of nitrogen dioxide (NO_2) exchange between plants and the atmosphere. *Atmos. Chem. Phys.* **2013**, *13*, 773–790. [CrossRef]
13. Wellburn, A.R. Why are atmospheric oxides of nitrogen usually phytotoxic and not alternative fertilizers? *New Phytol.* **1990**, *115*, 395–429. [CrossRef]

14. Chauhan, A.J.; Inskip, H.M.; Linaker, C.H.; Smith, S.; Schreiber, J.; Johnston, S.L.; Holgate, S.T. Personal exposure to nitrogen dioxide (NO_2) and the severity of virus-induced asthma in children. *Lancet* **2003**, *9373*, 1939–1944. [CrossRef]
15. Capron, T.M.; Mansfield, T.M. Inhibition of growth in tomato by air polluted with nitrogen oxides. *J. Exp. Bot.* **1977**, *28*, 112–116. [CrossRef]
16. Sandhu, R.; Gupta, G. Effects of nitrogen dioxide on growth and yield of black turtle bean (Phaseolus vulgaris L.) cv. 'Domino'. *Environ. Pollut.* **1989**, *59*, 337–344. [CrossRef]
17. Saxe, H. Relative sensitivity of greenhouse pot plants to long-term exposures of NO- and NO_2-containing air. *Environ. Pollut.* **1994**, *85*, 283–290. [CrossRef]
18. Morikawa, H.; Higaki, A.; Nohno, M.; Takahashi, M.; Kamada, M.; Nakata, M.; Toyohara, G.; Okamura, Y.; Matsui, K.; Kitani, S.; et al. More than a 600-fold variation in nitrogen dioxide assimilation among 217 plant taxa. *Plant Cell Environ.* **1998**, *21*, 180–190. [CrossRef]
19. von Liebig, J. Extrait d'une note sur la nitrification. *Ann. Chem. Phys.* **1827**, *35*, 329–333.
20. Takahashi, M.; Higaki, A.; Nohno, M.; Kamada, M.; Okamura, Y.; Matsui, K.; Kitani, S.; Morikawa, H. Differential assimilation of nitrogen dioxide by 70 taxa of roadside trees at an urban pollution level. *Chemosphere* **2005**, *61*, 633–639. [CrossRef]
21. Takahashi, M.; Sasaki, Y.; Ida, S.; Morikawa, H. Enrichment of nitrite reductase gene improves the ability of *Arabidopsis thaliana* plants to assimilate nitrogen dioxide. *Plant Physiol.* **2001**, *126*, 731–741. [CrossRef] [PubMed]
22. Morikawa, H.; Takahashi, M.; Kawamura, Y. Metabolism and genetics of atmospheric nitrogen dioxide control using pollutant-philic plants. In *Phytoremediation Transformation and Control of Contaminants*; McCutcheon, S.C., Schnoor, J.L., Eds.; John Wiley & Sons: Hoboken, NJ, USA, 2003; pp. 765–786. ISBN 0-471-39435-1.
23. Takahashi, M.; Morikawa, H. Air-pollutant-philic plants for air remediation. *J. Environ. Prot.* **2012**, *3*, 1346–1352. [CrossRef]
24. Takahashi, M.; Nakagawa, M.; Sakamoto, A.; Ohsumi, C.; Matsubara, T.; Morikawa, H. Atmospheric nitrogen dioxide gas is a plant vitalization signal to increase plant size and the contents of cell constituents. *New Phytol.* **2005**, *168*, 149–154. [CrossRef]
25. Takahashi, M.; Furuhashi, T.; Ishikawa, N.; Horiguchi, G.; Sakamoto, A.; Tsukaya, H.; Morikawa, H. Nitrogen dioxide regulates organ growth by controlling cell proliferation and enlargement in Arabidopsis. *New Phytol.* **2014**, *201*, 1304–1315. [CrossRef] [PubMed]
26. Takahashi, M.; Morikawa, H. Nitrogen dioxide is a positive regulator of plant growth. *Plant Signal. Behav.* **2014**, *9*, e28033. [CrossRef] [PubMed]
27. Adam, S.E.H.; Shigeto, J.; Sakamoto, A.; Takahashi, M.; Morikawa, H. Atmospheric nitrogen dioxide at ambient levels stimulates growth and development of horticultural plants. *Botany* **2008**, *86*, 213–217. [CrossRef]
28. Takahashi, M.; Morikawa, H. Differential responses of *Arabidopsis thaliana* accessions to atmospheric nitrogen dioxide at ambient concentrations. *Plant Signal. Behav.* **2014**, *9*, e28563. [CrossRef] [PubMed]
29. Takahashi, M.; Morikawa, H. Kinematic evidence that atmospheric nitrogen dioxide increases the rates of cell proliferation and enlargement to stimulate leaf expansion in Arabidopsis. *Plant Signal. Behav.* **2015**, *10*, e1022011. [CrossRef] [PubMed]
30. Takahashi, M.; Morikawa, H. Nitrogen dioxide accelerates flowering without changing the number of leaves at flowering in *Arabidopsis thaliana*. *Plant Signal. Behav.* **2014**, *9*, e970433. [CrossRef] [PubMed]
31. Takahashi, M.; Sakamoto, A.; Ezura, H.; Morikawa, H. Prolonged exposure to atmospheric nitrogen dioxide increases fruit yield of tomato plants. *Plant Biotechnol.* **2011**, *8*, 485–487. [CrossRef]
32. Murashige, T.; Skoog, F. A revised medium for rapid growth and bioassays with tobacco cultures. *Physiol. Plant.* **1962**, *15*, 473–497. [CrossRef]
33. Xu, Q.; Zhou, B.; Ma, C.; Xu, X.; Xu, J.; Jiang, Y.; Liu, C.; Li, G.; Herbert, S.J.; Hao, L. Salicylic acid-altering Arabidopsis mutants response to NO_2 exposure. *Environ. Contam. Toxicol.* **2010**, *84*, 106–111. [CrossRef]
34. Kotchoni, S.O.; Larrimore, K.E.; Mukherjee, M.; Kempinski, C.F.; Barth, C. Alterations in the endogenous ascorbic acid content affect flowering time in Arabidopsis. *Plant Physiol.* **2009**, *149*, 803–815. [CrossRef]
35. Adam, S.E.H.; Abdel-Banat, B.M.A.; Sakamoto, A.; Takahashi, M.; Morikawa, H. Effect of Atmospheric Nitrogen Dioxide on Mulukhiya (Corchorus olitorius) Growth and Flowering. *Am. J. Plant Physiol.* **2008**, *3*, 180–184. [CrossRef]

36. He, Y.; Tang, R.H.; Hao, Y.; Stevens, R.D.; Cook, C.W.; Ahn, S.M.; Jing, L.; Yang, Z.; Chen, L.; Guo, F.; et al. Nitric oxide represses the Arabidopsis floral transition. *Science* **2004**, *305*, 1968–1971. [CrossRef] [PubMed]
37. Leshem, Y.Y.; Haramaty, E. The characterization and contrasting effects of the nitric oxide free radical in vegetative stress and senescence of Pisum sativum Linn. foliage. *J. Plant Physiol.* **1996**, *148*, 258–263. [CrossRef]
38. Jin, C.W.; Du, S.T.; Zhang, Y.S.; Tang, C.X.; Lin, X.Y. Atmospheric nitric oxide stimulates plant growth and improves the quality of spinach (Spinaciaoleracea). *Ann. Appl. Biol.* **2009**, *155*, 113–120. [CrossRef]
39. Morikawa, H.; Takahashi, M.; Sakamoto, A.; Matsubara, T.; Arimura, G.-I.; Kawamura, Y.; Fukunaga, K.; Fujita, K.; Sakurai, N.; Hirata, T.; et al. Formation of unidentified nitrogen in plants: An implication for a novel nitrogen metabolism. *Planta* **2004**, *219*, 14–22. [CrossRef] [PubMed]
40. Tsuge, T.; Tsukaya, H.; Uchimiya, H. Two independent and polarized processes of cell elongation regulate leaf blade. *Development* **1996**, *122*, 1589–1600. [PubMed]
41. Callos, J.D.; Medford, J.I. Organ positions and pattern formation in the shoot apex. *Plant J.* **1994**, *6*, 1–7. [CrossRef]
42. Potters, G.; Pasternak, T.P.; Guisez, Y.; Palme, K.J.; Jansen, M.A. Stress-induced morphogenic responses: Growing out of trouble? *Trends Plant Sci.* **2007**, *3*, 98–105. [CrossRef] [PubMed]
43. Granier, C.; Tardieu, F. Multi-scale phenotyping of leaf expansion in response to environmental changes: The whole is more than the sum of parts. *Plant Cell Environ.* **2009**, *32*, 1175–1184. [CrossRef] [PubMed]
44. Kawade, K.; Horiguchi, G.; Tsukaya, H. Non-cell-autonomously coordinated organ size regulation in leaf development. *Development* **2010**, *137*, 4221–4227. [CrossRef]
45. Ischiropoulos, H. Protein tyrosine nitration: An update. *Arch. Biochem. Biophys.* **2009**, *484*, 117–121. [CrossRef] [PubMed]
46. Radi, R. Nitric oxide, oxidants, and protein tyrosine nitration. *Proc. Nat. Acad. Sci. USA* **2004**, *101*, 4003–4008. [CrossRef] [PubMed]
47. Rubbo, H.; Radi, R. Protein and lipid nitration: Role in redox signaling and injury. *Biochim. Biophys. Acta* **2008**, *1780*, 1318–1324. [CrossRef] [PubMed]
48. Corpas, F.J.; Chaki, M.; Leterrier, M.; Barroso, J.B. Protein tyrosine nitration: A new challenge in plants. *Plant Signal. Behav.* **2009**, *4*, 920–923. [CrossRef] [PubMed]
49. Yadav, S.; David, A.; Baluška, F.; Bhatla, S.C. Rapid auxin-induced nitric oxide accumulation and subsequent tyrosine nitration of proteins during adventitious root formation in sunflower hypocotyls. *Plant Signal. Behav.* **2013**, *8*, e23196. [CrossRef] [PubMed]
50. Ara, J.; Przedborski, S.; Naini, A.B.; Jackson-Lewis, V.; Trifiletti, R.R.; Horwitz, J.; Ischiropoulos, H. Inactivation of tyrosine hydroxylase by nitration following exposure to peroxynitrite and 1-methyl-4-phenyl-1,2,3,6-tetrahydropyridine (MPTP). *Proc. Natl. Acad. Sci. USA* **1998**, *95*, 7659–7663. [CrossRef]
51. Souza, J.M.; Peluffo, G.; Radi, R. Protein tyrosine nitration–Functional alteration or just a biomarker? *Free Radic. Biol. Med.* **2008**, *45*, 357–366. [CrossRef]
52. Álvarez, C.; Lozano-Juste, J.; Romero, L.C.; García, I.; Gotor, C.; León, J. Inhibition of Arabidopsis O-acetylserine(thiol)lyase A1 by tyrosine nitration. *J. Biol. Chem.* **2011**, *286*, 578–586. [CrossRef]
53. Abello, N.; Kerstjens, H.A.M.; Postma, D.S.; Bischoff, R. Protein tyrosine nitration: Selectivity, physicochemical and biological consequences, denitration, and proteomics methods for the identification of tyrosine-nitrated proteins. *J. Proteome Res.* **2009**, *8*, 3222–3238. [CrossRef]
54. Koeck, T.; Fu, X.; Hazen, S.L.; Crabb, J.W.; Stuehr, D.J.; Aulark, K.S. Rapid and selective oxygen-regulated protein tyrosine denitration and nitration in mitochondria. *J. Biol. Chem.* **2004**, *279*, 27257–27262. [CrossRef]
55. Morot-Gaudry-Talarmain, Y.; Rockel, P.; Moureaux, T.; Quilleré, I.; Leydecker, M.T.; Kaiser, W.M.; Morot-Gaudry, J.F. Nitrite accumulation and nitric oxide emission in relation to cellular signaling in nitrite reductase antisense tobacco. *Planta* **2002**, *215*, 708–715. [PubMed]
56. Cecconi, D.; Orzetti, S.; Vandelle, E.; Rinalducci, S.; Zolla, L.; Delledonne, M. Protein nitration during defense response in *Arabidopsis thaliana*. *Electrophoresis* **2009**, *30*, 2460–2468. [CrossRef]
57. Takahashi, M.; Shigeto, J.; Sakamoto, A.; Izumi, S.; Asada, K.; Morikawa, H. Dual selective nitration in Arabidopsis: Almost exclusive nitration of PsbO and PsbP, and highly susceptible nitration of four non-PSII proteins, including peroxiredoxin II E. *Electrophoresis* **2015**, *36*, 2569–2578. [CrossRef] [PubMed]

58. Nelson, N.; Yocum, C.F. Structure and function of photosystems I and II. *Annu. Rev. Plant Biol.* **2006**, *57*, 521–565. [CrossRef] [PubMed]
59. Anderson, J.M.; Chow, W.S.; De Las Rivas, J. Dynamic flexibility in the structure and function of photosystem II in higher plant thylakoid membranes: The grana enigma. *Photosynth. Res.* **2008**, *98*, 575–587. [CrossRef] [PubMed]
60. Suorsa, M.; Sirpiö, S.; Allahverdiyeva, Y.; Paakkarinen, V.; Mamedov, F.; Styring, S.; Aro, E.M. PsbR, a missing link in the assembly of the oxygen-evolving complex of plant photosystem II. *J. Biol. Chem.* **2006**, *281*, 145–150. [CrossRef] [PubMed]
61. Takahashi, M.; Shigeto, J.; Shibata, T.; Sakamoto, A.; Izumi, S.; Morikawa, H. Differential abilities of nitrogen dioxide and nitrite to nitrate proteins in thylakoid membranes isolated from Arabidopsis leaves. *Plant Signal. Behav.* **2016**, *11*, e1237329. [CrossRef] [PubMed]
62. Huie, R.E. The reaction kinetics of NO_2. *Toxicology* **1994**, *89*, 193–216. [CrossRef]
63. Kawamura, Y.; Takahashi, M.; Arimura, G.; Isayama, T.; Irifune, K.; Goshima, N.; Morikawa, H. Determination of levels of NO_3^-, NO_2^- and NH_4^+ ions in leaves of various plants by capillary electrophoresis. *Plant Cell Physiol.* **1996**, *37*, 878–880. [CrossRef]
64. Takahashi, M.; Shigeto, J.; Sakamoto, A.; Morikawa, H. Light-triggered selective nitration of PsbO1 in isolated Arabidopsis thylakoid membranes is inhibited by photosynthetic electron transport inhibitors. *Plant Signal. Behav.* **2016**, *11*, e1263413. [CrossRef]
65. Peltier, J.B.; Friso, G.; Kalume, D.E.; Roepstorff, P.; Nilsson, F.; Adamska, I.; Van Wijk, K.J. Proteomics of the chloroplast: Systematic identification and targeting analysis of lumenal and peripheral thylakoid proteins. *Plant Cell* **2000**, *12*, 319–341. [CrossRef] [PubMed]
66. Nakamura, S.; Noguchi, T. Infrared detection of a proton released from tyrosine Y_D to the bulk upon its photo-oxidation in photosystem II. *Biochemistry* **2015**, *54*, 5045–5053. [CrossRef] [PubMed]
67. Fork, D.C.; Urbach, W. Evidence for the localization of plastocyanin in the electron-transport chain of photosynthesis. *Proc. Natl. Acad. Sci. USA* **1965**, *53*, 1307–1315. [CrossRef] [PubMed]
68. Kawamoto, K.; Mano, J.; Asada, K. Photoproduction of the azidyl radical from the azide anion on the oxidizing side of photosystem II and suppression of photooxidation of tyrosine Z by the azidyl radical. *Plant Cell Physiol.* **1995**, *36*, 121–129. [CrossRef]
69. Kovacs, L.; Hegde, U.; Padhye, S.; Bernat, G.; Demeter, S. Effect of potassium-(picrate)-(18-crown-6) on the photosynthetic electron transport. *Z. Naturforsch.* **1996**, *51*, 539–547. [CrossRef]
70. Takahashi, M.; Shigeto, J.; Sakamoto, A.; Morikawa, H. Selective nitration of PsbO1 inhibits oxygen evolution from isolated Arabidopsis thylakoid membranes. *Plant Signal. Behav.* **2017**, *12*, e1304342. [CrossRef] [PubMed]
71. Sinclair, J. Changes in spinach thylakoid activity due to nitrite ions. *Photosynth. Res.* **1987**, *12*, 255–263. [CrossRef]
72. Wincencjusz, H.; Yocum, C.F.; Van Gorkom, H.J. Activating anions that replace Cl^- in the O_2-evolving complex of photosystem II slow the kinetics of the terminal step in water oxidation and destabilize the S2 and S3 states. *Biochemistry* **1999**, *38*, 3719–3725. [CrossRef]
73. Pokhrel, R.; Brudvig, G.W. Investigation of the inhibitory effect of nitrite on Photosystem II. *Biochemistry* **2013**, *52*, 3781–3789. [CrossRef]
74. Takahashi, M.; Morikawa, H. A novel role for PsbO1 in photosynthetic electron transport as suggested by its light-triggered selective nitration in *Arabidopsis thaliana*. *Plant Signal. Behav.* **2018**, *13*, e1513298. [CrossRef]
75. Su, X.; Ma, J.; Wei, X.; Cao, P.; Zhu, D.; Chang, W.; Liu, Z.; Zhang, X.; Li, M. Structure and assembly mechanism of plant C2S2M2-type PSII-LHCII supercomplex. *Science* **2017**, *357*, 815–820. [CrossRef] [PubMed]
76. Umena, Y.; Kawakami, K.; Shen, J.-R.; Kamiya, N. Crystal structure of oxygen-evolving photosystem II at a resolution of 1.9 Å. *Nature* **2011**, *473*, 55–60. [CrossRef] [PubMed]
77. Lakshmi, K.V.; Brudvi, G.W. Electron paramagnetic resonance distance measurements in photosynthetic reaction centers. In *Distance Measurements in Biological Systems by EPR*; Berliner, L.J., Eaton, G.R., Eaton, S.S., Eds.; Springer: London, UK, 2000; Volume 19, pp. 513–567. ISBN 978-1-4757-0575-1.
78. Pigolev, A.V.; Klimov, V.V. The green alga Chlamydomonas reinhardtii as a tool for in vivo study of site-directed mutations in PsbO protein of photosystem II. *Biochemistry (Moscow)* **2015**, *80*, 662–673. [CrossRef] [PubMed]

79. Morita, T.; Kimura, S. Long-range electron transfer over 4 nm governed by an inelastic hopping mechanism in self-assembled monolayers of helical peptides. *J. Am. Chem. Soc.* **2003**, *125*, 8732–8733. [CrossRef] [PubMed]
80. Arikuma, Y.; Nakayama, H.; Morita, T.; Kimura, S. Ultra-long-range electron transfer through a self-assembled monolayer on gold composed of 120-Å-long α-helices. *Langmuir* **2011**, *27*, 1530–1535. [CrossRef] [PubMed]
81. Mclendon, G.; Hake, R. Interprotein electron transfer. *Chem. Rev.* **1992**, *92*, 481–490. [CrossRef]
82. Miyashita, O.; Okamura, M.Y.; Onuchic, J.N. Interprotein electron transfer from cytochrome c2 to photosynthetic reaction center: Tunneling across an aqueous interface. *Proc. Natl. Acad. Sci. USA* **2005**, *102*, 3558–3563. [CrossRef] [PubMed]
83. Kolbert, Z.; Feigl, G.; Bordé, Á.; Molnár, Á.; Erdei, L. Protein tyrosine nitration in plants: Present knowledge, computational prediction and future perspectives. *Plant Physiol. Biochem.* **2017**, *113*, 56–63. [CrossRef]
84. Salem, M.A.; Li, Y.; Bajdzienko, K.; Fisahn, J.; Watanabe, M.; Hoefgen, R.; Schöttler, M.A.; Giavalisco, P. RAPTOR controls developmental growth transitions by altering the hormonal and metabolic balance. *Plant Physiol.* **2018**, *177*, 565–593. [CrossRef]
85. Chen, G.H.; Liu, M.J.; Xiong, Y.; Sheen, J.; Wu, S.H. TOR and RPS6 transmit light signals to enhance protein translation in deetiolating Arabidopsis seedlings. *Proc. Natl. Acad. Sci. USA* **2018**, *115*, 12823–12828. [CrossRef]
86. Schepetilnikov, M.; Ryabova, L.A. Recent discoveries on the role of TOR (Target of Rapamycin) signaling in translation in plants. *Plant Physiol.* **2018**, *176*, 1095–1105. [CrossRef] [PubMed]
87. Wang, P.; Zhao, Y.; Li, Z.; Hsu, C.C.; Liu, X.; Fu, L.; Hou, Y.J.; Du, Y.; Xie, S.; Zhang, C.; et al. Reciprocal regulation of the TOR kinase and ABA receptor balances plant growth and stress response. *Mol. Cell.* **2018**, *69*, 100–112.e6. [CrossRef] [PubMed]
88. Castillo, M.C.; Lozano-Juste, J.; González-Guzmán, M.; Rodriguez, L.; Rodriguez, P.L.; León, J. Inactivation of PYR/PYL/RCAR ABA receptors by tyrosine nitration may enable rapid inhibition of ABA signaling by nitric oxide in plants. *Sci. Signal.* **2015**, *8*, 89. [CrossRef] [PubMed]
89. Karve, A.; Brandon, D.; Moore, B.D. Function of Arabidopsis hexokinase-like1 as a negative regulator of plant growth. *J. Exp. Bot.* **2009**, *60*, 4137–4149. [CrossRef] [PubMed]
90. John, F.; Roffler, S.; Wicker, T.; Ringli, C. Plant TOR signaling components. *Plant Signal. Behav.* **2011**, *6*, 1700–1705. [CrossRef]
91. Antoniuk-Pablant, A.; Sherman, B.D.; Kodis, G.; Gervaldo, M.; Moore, T.A.; Moore, A.L.; Gust, D.; Megiatto, J.D., Jr. Mimicking the electron transfer chain in photosystem II with a molecular triad thermodynamically capable of water oxidation. *Proc. Natl. Acad. Sci. USA* **2012**, *109*, 15578–15583.

© 2019 by the authors. Licensee MDPI, Basel, Switzerland. This article is an open access article distributed under the terms and conditions of the Creative Commons Attribution (CC BY) license (http://creativecommons.org/licenses/by/4.0/).

Review

S-Nitrosoglutathione Reductase—The Master Regulator of Protein S-Nitrosation in Plant NO Signaling

Jana Jahnová, Lenka Luhová and Marek Petřivalský *

Department of Biochemistry, Faculty of Science, Palacky University, Šlechtitelů 11, 78371 Olomouc, Czech Republic; jana.jahnova@upol.cz (J.J.); lenka.luhova@upol.cz (L.L.)
* Correspondence: marek.petrivalsky@upol.cz

Received: 29 January 2019; Accepted: 13 February 2019; Published: 21 February 2019

Abstract: S-nitrosation has been recognized as an important mechanism of protein posttranslational regulations, based on the attachment of a nitroso group to cysteine thiols. Reversible S-nitrosation, similarly to other redox-base modifications of protein thiols, has a profound effect on protein structure and activity and is considered as a convergence of signaling pathways of reactive nitrogen and oxygen species. In plant, S-nitrosation is involved in a wide array of cellular processes during normal development and stress responses. This review summarizes current knowledge on S-nitrosoglutathione reductase (GSNOR), a key enzyme which regulates intracellular levels of S-nitrosoglutathione (GSNO) and indirectly also of protein S-nitrosothiols. GSNOR functions are mediated by its enzymatic activity, which catalyzes irreversible GSNO conversion to oxidized glutathione within the cellular catabolism of nitric oxide. GSNOR is involved in the maintenance of balanced levels of reactive nitrogen species and in the control of cellular redox state. Multiple functions of GSNOR in plant development via NO-dependent and -independent signaling mechanisms and in plant defense responses to abiotic and biotic stress conditions have been uncovered. Extensive studies of plants with down- and upregulated GSNOR, together with application of transcriptomics and proteomics approaches, seem promising for new insights into plant S-nitrosothiol metabolism and its regulation.

Keywords: S-nitrosation; S-nitrosothiols; nitric oxide; S-nitrosoglutathione reductase; S-(hydroxymethyl)glutathione

1. Introduction

Nitric oxide (NO) is an important messenger included in many physiological processes. It is an uncharged, relatively stable free radical with unpaired electrons allowing diverse chemistry. Rapid reactions with other radicals including reactive oxygen species (ROSs) [1] lead to the formation of reactive nitrogen species (RNSs), substances with versatile chemical properties triggering specific physiological responses. NO is involved in the regulation of plant growth and development, immunity and environmental interactions with the inclusion of signaling cascades of responses to stress conditions [2,3].

In general, the biosynthesis of NO in plants can proceed by pathways either oxidative or reductive, either enzymatic or non-enzymatic reactions, anyway depending on the site and the nature of stimulus for NO production [3–5]. The nitrate reductase (NR; EC 1.6.6.1) pathway, localized in the cytosol, is the best-characterized production pathway of NO in plants [4]. Another well-described way of NO production is the nitrite reduction in electron transport chains of mitochondria or chloroplasts [5]. In mammals, constitutive and inducible isoforms of nitric oxide synthase (NOS, 1.14.13.39) are the major enzyme sources of NO. NOS-like enzymatic activities were described in plants, however neither gene with significant homology nor protein with similarity to bacterial or animal NOS have been found [6–9]. It is assumed that NOS-like activity in plants is carried out by several enzymes, which can together generate NO from L-arginine and have the same cofactors requirements as the NOS in mammals and bacteria [6,7]. However, a recent analysis of 1087 land plant transcriptomes confirmed the absence of evolutionarily conserved NOS sequences within the plant kingdom [9].

S-nitrosothiols (SNOs) represent relatively a stable reserve and transport form of NO in vivo [10,11]. They are formed by S-nitrosation, a selective and reversible covalent addition of nitric oxide moiety to the sulfur atom of cysteines both in low-molecular weight thiols and proteins. S-nitrosation is considered to be an important redox-based post-translational protein modification, an integral part of signaling pathways of NO and RNSs [12]. It is supposed to be implicated in the regulation of a variety of protein functions and cell activities—programmed cell death, metabolism, control of redox balance, iron homeostasis, control of protein quality, and gene transcription [13]. Importantly, S-nitrosothiols are considered key elements of the interplay between RNSs and ROSs, both under physiological and stress conditions leading to various scenarios of oxidative, nitrosative, or nitro-oxidative stress [14].

The most abundant low-molecular weight S-nitrosothiol is suggested to be S-nitrosoglutathione (GSNO), generated by an O_2-dependent reaction of NO-derived RNSs and the major antioxidant tripeptide glutathione (GSH; γ-Glu-Cys-Gly). GSNO is regarded to be an intracellular reservoir of NO bioactivity and a transport form of NO as well, even though NO and GSNO do not always interact with the same target proteins [15]. Reactions including S-nitrosation, transnitrosation, when nitroso group is transferred from SNO to the thiol group of another molecule, S-glutathionylation are involved in its metabolism [16,17]. Acting as a buffer for NO, GSNO could maintain the level of protein S-nitrosation [18]. However, more detailed knowledge on the distribution, intracellular levels, and modulation of GSNO under natural and stress conditions is needed [17].

2. S-Nitrosoglutathione Reductase: Key Enzyme of the Regulation of S-Nitrosation and Formaldehyde Detoxification

GSNOR is an evolutionarily conserved, cytosolic enzyme that catalyzes the NADH-dependent reduction of GSNO, leading to the formation of glutathione disulfide (GSSG) and ammonium [18]. Sakamoto et al. [19] demonstrated for the first time in plants that GSNOR is glutathione-dependent formaldehyde dehydrogenase (FALDH; EC 1.2.1.1). The proper substrate for FALDH is the hemithioacetal S-hydroxymethylglutathione (HMGSH), formed nonenzymatically from formaldehyde and glutathione [20]. HMGSH is oxidized to S-formylglutathione using NAD^+ as a coenzyme (Figure 1A). After the elucidation of an exact reaction mechanism, the enzyme was reclassified as S-(hydroxymethyl)glutathione dehydrogenase (EC 1.1.1.284). Formerly, Koivusalo et al. [21] reported evidence that FALDH and ADH3 are identical enzymes. Thus, in accordance with the formal enzyme classification, GSNOR is a Zn-dependent medium-chain class III alcohol dehydrogenase (ADH3; EC 1.1.1.1). Since the GSNO has been uncovered as among the most effective substrates for this enzyme [18,22,23], the designation as GSNOR is currently widely extended within the scientific literature. However, the denomination has not been accepted by IUBMB nomenclature commission up to the present day.

Via removing GSNO, GSNOR plays a critical role in the metabolism of RNSs, in the homeostasis of intracellular levels of NO and in control of the trans-nitrosation equilibrium between S-nitrosylated proteins and GSNO, the most common low-molecular weight S-nitrosothiol [15,18,22,24]. In trans-nitrosation reactions, the nitroso group is transferred among thiols on proteins and low-molecular weight peptides. GSNO reduction by GSNOR is an irreversible reaction, and the products can no longer nitrosate cellular proteins.

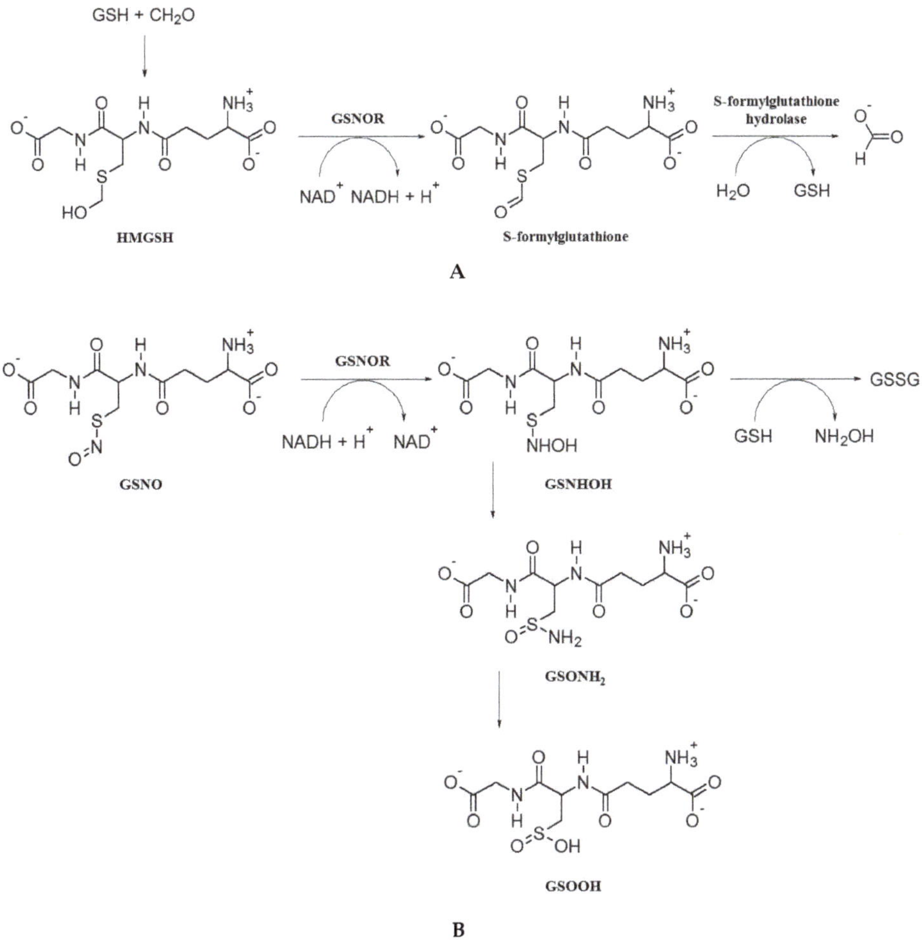

Figure 1. Reaction mechanisms of alcohol dehydrogenase/S-nitrosoglutathione reductase (ADH3/GSNOR) in formaldehyde and S-nitrosoglutathione catabolism. (**A**) In the dehydrogenase mode, GSNOR using NAD$^+$ as a coenzyme catalyzes the oxidation of S-hydroxymethylglutathione (HMGSH), spontaneously formed from formaldehyde and glutathione to S-formylglutathione, which was further hydrolyzed to glutathione and formate by S-formylglutathione hydrolase. (**B**) In the reductase mode, GSNOR catalyzes the reduction of S-nitrosoglutathione (GSNO) using NADH to an unstable intermediate N-hydroxysulfinamide (GSNHOH). Depending on the local concentration of GSH, GSNHOH is either decomposed to glutathione disulfide (GSSG) and hydroxylamine at high GSH levels, or at low GSH levels spontaneously converts to glutathione sulfinamid (GSONH2), which can be hydrolyzed to glutathione sulfinic acid (GSOOH) and ammonia.

Through the GSNO reductase activity, GSNO is reduced to an unstable intermediate N-hydroxysulfinamide (GSNHOH) in the first reaction step, using NADH as a specific co-substrate (Figure 1B). Different final products are produced in the next reaction step, depending on the local concentration of GSH. Thus, the cellular redox potential in terms of NADH and GSH levels is an important factor in control of the product formation [25]. Common cellular concentrations of GSH are found in a millimolar range, which favors a reaction shift from GSNHOH to the formation of glutathione disulfide (GSSG) accompanied with the release of hydroxylamine [23,25]. However, the cellular levels of GSH are widely fluctuating under different biotic and abiotic stress conditions.

In vitro studies have demonstrated that, at low levels of GSH, GSNHOH spontaneously converts to glutathione sulfinamide ($GSONH_2$). $GSONH_2$ is further hydrolyzed to glutathione sulfinic acid (GSO_2H), which can be oxidized even to glutathione sulfonic acid (GSO_3H) under oxidative stress induced by various stress conditions [25]. The latter metabolites inhibit glutathione transferases, enzymes with an important role in the glutathione-dependent detoxification of xenobiotics by their conjugation with GSH [26].

Another factor involved in the regulation of GSNO turnover is the accessibility of NADH, a co-substrate in the reduction of GSNO. The cellular ratio of free oxidized and reduced form of dinucleotides (NAD^+/NADH) is high under physiological conditions, which is not favorable for reductive pathways [27]. The $NADP^+$/NADPH ratio is much lower, which enables NADPH to be used in biosynthetic reductive pathways [28]. Since GSNOR cannot use NADPH in the reduction of GSNO, it is controlled by NADH availability and increasing levels of NADH are proposed to trigger the GSNO reduction. GSNOR enzymes themselves produce NADH in the process of the oxidation of formaldehyde; formaldehyde likely triggers the reduction of GSNO [25].

In plants, formaldehyde can originate from various processes. Among them, the major sources of formaldehyde include the dissociation of 5,10-methylene-tetrahydrofolate and the oxidation of methanol formed by demethylation of pectin. Formaldehyde can also be formed by oxidative demethylation reactions, decarboxylation of glyoxylate, and P450-dependent oxidation of xenobiotics [29,30]. The compound is highly reactive because of the polarized carbonyl group and can participate in a nucleophile as well as an electrophile addition and substitution reactions. The carbonyl group can react with DNA and proteins producing stable carboxylated products. GSNOR, by another name FALDH, is the main enzyme metabolizing formaldehyde [23]. This is implicated in this process by oxidation of HMGSH, spontaneously formed from formaldehyde and glutathione. Emergent S-formylglutathione is decomposed by S-formylglutathione hydrolase (EC 3.1.2.12) to glutathione and formate [29].

3. Molecular Properties of S-Nitrosoglutathione Reductase

3.1. GSNOR Structure

A few studies on kinetic and structural analysis of plant GSNOR enzymes indicate a high similarity between the plant and human homologues [31–35]. GSNOR described in tomato (*Solanum lycopersicum*; SlGSNOR) plants is a homodimeric enzyme consisting of two 40 kDa subunits containing a big catalytic and a small coenzyme-binding domain with an active site localized in a cleft between them [34]. Non-catalytic domain includes a binding site for NAD^+ coenzyme; six beta-strands of each coenzyme-binding domain form 12 pseudo-continuous beta-sheets. Each catalytic domain includes two zinc atoms. One of them is involved in the catalytic mechanism by activating the hydroxyl and carbonyl groups of substrates for transfer of hydride, and is bonded to Cys47, Cys177, His69, and either Glu70 or a water molecule. The second zinc atom is considered to have purely a structural role and is coordinated to four cysteine residues, Cys99, Cys102, Cys105, and Cys113 [34].

Crystal structures of SlGSNOR apoenzyme, binary complex with NAD^+ and a structure crystallized in the presence of NADH and GSH were described to understand the role of specific residues in the active site and the structural changes occurring during the catalytic cycle of GSNOR activity [34]. Catalytic domains of the apoenzyme and of the binary complex with NAD^+ are both in the semi-open conformation. The catalytic zinc atoms in the apoenzyme are in a tetrahedral configuration, H-bonded to Cys47, Cys177, His69 and coordinated to the molecule of water in the active site. The coenzyme binding is associated with the catalytic zinc atoms movement towards Glu70 in the catalytic domain in a hydrogen-bonding interaction with the carboxylate oxygen of Glu70. Zinc atoms are in a tetrahedral configuration coordinated with Cys47, Cys177, His69, and Glu70, and they are no longer coordinated with the water molecule. In the SlGSNOR structure crystallized with NADH and GSH, the enzyme appears in closed conformation; rotation of the catalytic domains by approximately

3° towards the coenzyme-binding domains was observed [34]. This structure is highly similar to the complex of human GSNOR (hGSNOR) with NADH and HMGSH, where a catalytic domain moves towards the coenzyme-binding domain during the formation of the ternary complex [32,33,36]. In the hGSNOR, the domain closure brings one molecule of water close to 2′-hydroxyl of nicotinamide riboside moiety, suggesting that the proton from the substrate is transferred to the solvent directly from the coenzyme. Similarly, in SlGSNOR, the domain closure brings Thr49 and His48 closer to the 2′-hydroxyl of nicotinamide riboside moiety, which might facilitate the proton transfer [34]. In the hGSNOR, the HMGSH substrate is directly coordinated to active site zinc atoms and interacts with highly conserved residues Arg114, Asp55, Glu57, and Thr46, and the zinc atom is in a tetrahedral configuration coordinated with Cys44, Cys177, His66, and HMGSH [32,33].

Eukaryotic GSNORs are highly conserved and unusually cysteine-rich proteins [35]. Most of the cysteines are inaccessible to the solvent, having usually only a structural function [37]. Three positionally conserved cysteines accessible to the solvent are predicted to be the site of post-translational modifications, e.g., S-nitrosation or glutathionylation [35]. Regulation of GSNOR activity through S-nitrosation of that conserved cysteines was observed in *A. thaliana* plants [15,38]. In vitro studies showed susceptibility of the enzymatic activity to NO donors and its subsequent restoration after treatment with dithiothreitol (DTT), a reducing agent [38]. Mono-, di-, and trinitrosation, which were confirmed by mass spectrometry, lead to subtle changes in enzyme conformation. GSNOR monomers within the same dimer interact with each other and the substrate binding cleft alters the shape. Thus, GSNOR activity might be regulated by high levels of NO donors.

3.2. GSNOR Substrate Specificity and Inhibition

Enzymes from the alcohol dehydrogenase class I (ADH1) and class III (ADH3) families have a very similar tertiary structure, but despite this fact their substrate specificity and kinetic mechanism are very different [39,40]. GSNOR, belonging to the class III family, can work in two modes catalyzing a conversion of plenty of substrates, including long-chain primary alcohols, aldehydes, and ω-hydroxyfatty acids. In the dehydrogenase mode, it catalyzes oxidation in the presence of NAD^+, whereas in the reductase mode it catalyzes reduction in the presence of NADH. $NADP^+$ and NADPH are very poor coenzymes reaching negligible reaction rates compared to those with NAD^+ and NADH [34].

Several studies on hGSNOR showed that an anion binding pocket, containing Gln111, Arg114, and Lys283, is presented in the active site of hGSNOR, and the positive charge of Arg114 enables correct orientation of negatively charged substrates, HMGSH and GSNO [26,32,33,36]. The plant GSNOR enzyme exhibits significant difference in the anion-binding pocket of the active site, which is composed of only two residues, Arg117 and Lys287, while the glutamine (Gln111 in hGSNOR) is missing and replaced by Gly114. Since the Gln111 in hGSNOR forms a hydrogen bond with carboxyl oxygen atoms of substrate, the different composition of the anion binding pocket of plant GSNOR results in the reduced affinity for the carboxyl group of ω-hydroxyfatty acids [34].

Plant GSNOR catalyzes the oxidation of HMGSH, geraniol, cinnamyl alcohol, ω-hydroxyfatty acids, and aliphatic alcohols with chains longer than four carbons, to corresponding aldehydes using NAD^+ as a coenzyme. Short-chain alcohols, e.g., ethanol and propanol, are not enzyme substrates. In the reductase mode, plant GSNOR preferentially catalyzes the reduction of GSNO, while reactions with either aliphatic or aromatic aldehydes are insignificant. The observed K_m values for various plausible substrates of SlGSNOR were in the same range as those for AtGSNO, which indicates that the substrate preferences of plant GSNOR are similar [34,41]. SlGSNOR shows similar K_m values for HMGSH and GSNO, 58 and 57 μM, respectively, while GSNO is reduced with 15–20 times higher catalytic efficiency compared to the oxidation of HMGSH [34]. Similarly, higher reaction rates of GSNO reduction compared to HMGSH oxidation were described in hGSNOR [42].

Fatty acids with medium chains (e.g., dodecanoic, decanoic, and octanoic acid), glutathione, and its derivatives (e.g., S-methylglutathione) were described as non-competitive inhibitors of plant GSNOR. Lacking an S-nitrosyl or S-hydroxymethyl group that binds to the active site zinc atom, the affinity of inhibitors GSH and S-methylglutathione is reduced by 2–3 orders of magnitude compared to GSNO and HMGSH. N6022, a pyrolle-based compound, was found to be a significantly stronger non-competitive inhibitor compared to fatty acids, inhibiting SlGSNOR at nanomolar concentrations [34].

4. The GSNOR Role in Plants

Biochemical and genetic characterizations of plant GSNOR enzyme, previously named either glutathione-dependent formaldehyde dehydrogenase (FALDH) or class III alcohol dehydrogenase (ADH3), have been well described in several reports [43–45]. Sakamoto et al. [19] identified FALDH in *Arabidopsis thaliana* as GSNOR, an enzyme able to catalyze GSNO reduction and thus regulate intracellular levels of protein S-nitrosation. GSNOR activity has been demonstrated in many plant species, e.g., *A. thaliana*, lettuce, maize, pea, rice, sunflower, and tomato [34,43–50]. Available data indicate that GSNOR is involved in numerous developmental processes and metabolic programs in plants via regulation of NO homeostasis. The enzyme is highly evolutionarily conserved [18]. Most sequenced green plant genomes encode a single copy of a GSNOR protein, predicted to be localized in cytosol [35]. The presence of multiple gene copies has only been reported in several plant species.

GSNOR is found throughout the plant suggesting the regulation of GSNO concentration in all plant cell types [51]. Experimental evidence suggests localization in the cytosol, nucleus (excluding nucleolus), and peroxisomes of *A. thaliana* [35]. Since GSNOR lacks a nuclear targeting signal, a transportation step in association with another protein is supposed. Studies on pea leaves cells showed GSNOR localized identically with *A. thaliana* in cytosol and peroxisomes and in chloroplasts and mitochondria [52]. Mitochondrial targeting peptide was predicted for *Physcomitrella* GSNOR paralog [35]. Modulation of the mitochondrial functionality by GSNOR, using cell suspension cultures with both higher and lower GSNOR levels, was demonstrated in *A. thaliana* plants [53]. Changes in GSNOR levels have an influence on the activities of mitochondrial complex I, external NADH dehydrogenase, alternative oxidase and uncoupling protein. GSNOR modulates the activity of the mitochondrial respiratory chain through controlling NO/SNO homeostasis under physiological conditions and under nutritional stress. In addition to its role in the reduction of GSNO, it may control the redox state of cells by affecting to intracellular levels of NADH and GSH.

Similarly to other organisms, plant GSNOR regulates levels of S-nitrosothiols through an irreversible NADH-dependent degradation of S-nitrosoglutathione, and it plays an important regulatory role in overall NO metabolism. Modulations of GSNOR both on the transcriptional and post-translational level can therefore contribute to a fine-tuning of NO signaling pathways in plants (Figure 2). Interestingly, reversible oxidative modification of GSNOR cysteine residues are known to inhibit its enzyme activity in vitro, suggesting a potential direct crosstalk of RNSs and ROSs signaling at this point [51,54]. Moreover, negative regulation of GSNOR activity by nitrosative modifications might present another important mechanism to control GSNO levels, a critical mediator of the downstream signaling effects of NO [38], as well as for formaldehyde detoxification in the enzyme dehydrogenase reaction mode.

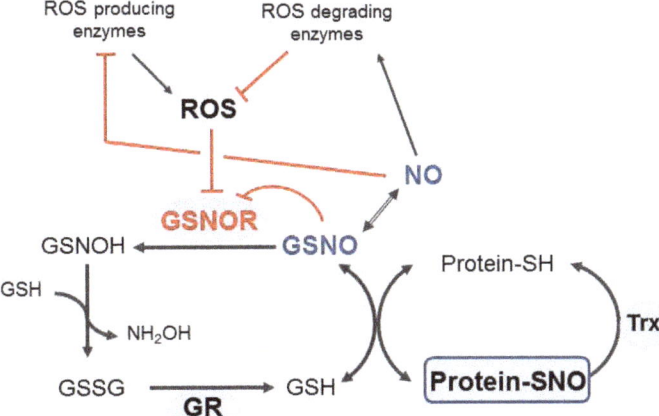

Figure 2. Regulatory mechanisms of GSNOR in protein denitrosation on the intersection of signaling pathways of ROSs and RNSs. Trans-nitrosation reactions of S-nitrosated proteins and reduced glutathione (GSH) can be shifted by the GSNOR activity through irreversible NADH-dependent reduction of S-nitrosoglutathione. GSH can be eventually regenerated by an NADPH-dependent reduction of GSSG catalyzed by glutathione reductase (GR). GSNOR activity can be inhibited by oxidative modification, resulting in GSNO accumulation and hence increased NO bioactivity, which can in turn regulate activities of enzymes of ROS metabolism. GSNOR activity can be also inhibited by S-nitrosation, to enable transient accumulations of its substrate GSNO and eventually to influence the cellular status of protein S-nitrosation.

4.1. GSNOR in Plant Growth and Development

Nitric oxide is well-known to be involved in regulation of a broad spectrum of activities during plant growth and development. Its action is supposed to be mediated via formation of S-nitrosothiols and trans-nitrosation reactions. Relative stable S-nitrosothiols enable signal transfer at large distances, S-nitrosation and denitrosation reactions are strongly controlled by the GSNOR. Although a constitutive GSNOR expression was suggested through the plant, different expression in organs of *A. thaliana* was found using histochemical activity staining and immunolocalization [30,55]. Higher levels of GSNOR were observed in the roots and leaves from the first stages of development. In transgenic *A. thaliana* plants, both up- and down regulation of GSNOR levels resulted in noticeable changes in the phenotype, namely a shortening of root length [30]. Experiments with *A. thaliana* HOT5 (sensitive to hot temperatures) mutants demonstrated that GSNOR function was necessary for normal plant growth, fertility, and plant acclimation to high temperatures [56]. Mutant plants failed to grow on nutrient plates and showed increased reproductive shoots and reduced fertility. Both *hot5* missense and null mutations showed increased NO species, supporting the statement that GSNOR regulates NO homeostasis. Furthermore, *A. thaliana* null mutants exhibit defects in stem and trichome branching [35]. The ubiquitous expression throughout the plant was confirmed using GFP-tagged GSNOR, with especially high fluorescent signal in the root tip, apical meristem, and flowers. Additional experiments [24,57,58] demonstrate that GSNOR has an influence on shoot branching, hypocotyl growth, seed yield and flowering time, decreased stature or loss of apical dominance, and fewer rosette leaves. Defective growth and development of the *gsnor1-3* mutant of *A. thaliana* with reduced GSNOR activity result from impaired, but not completely abolished, auxin signaling, auxin polar transport, and auxin distribution [58]. The processes mentioned here might be regulated by S-nitrosation of components in auxin signaling and transport, e.g., integral membrane proteins transporting auxin, intracellular receptor TIR1 (transport inhibitor response 1), and ubiquitin-conjugating enzyme E2.

Abscisic acid (ABA) is another phytohormone important for plant growth, development, and adaptation to stress conditions. ABA signaling in guard cells is impaired in *gsnor1–3* plants via S-nitrosation of sucrose nonfermenting 1 (SNF1)-related protein kinase 2.6 (SnRK2.6), which is one of the central components of the ABA signaling pathway, at cysteine 137, a residue close to kinase catalytic site [59]. Frungillo et al. [15] described the influence of GSNOR on the assimilation of nitrogen, which is a major nutrient in plant growth and development. *A. thaliana* plants overexpressing the GSNOR gene exhibit increased nitrate reductase (NR) activity; conversely, GSNOR mutant plants show a significant decrease in NR activity. Simultaneously, GSNOR enzymatic activity, but not gene expression, is inhibited by the nitrogen assimilatory pathway via post-transcriptional S-nitrosation, preventing any scavenging of GSNO. These data indicate that NO and S-nitrosothiols control their own generation and scavenging via modulation of GSNOR activity and nitrate assimilation [15]. Taken together, acquired data show that GSNOR is essential for normal growth and development of *A. thaliana*.

The spatial distribution of GSNOR activity and gene expression in pepper plants (*Capsicum annuum*) [60] was found to be in agreement with the data from *A. thaliana* [30]. At the early stages of development up to 14 days after germination, the highest activity of GSNOR was found in roots in comparison to hypocotyls and cotyledons. The activity of the enzyme decreased with age in roots and, on the contrary, increased in hypocotyls and cotyledons; however, no relevant changes in the gene expression were observed [60]. Different GSNOR gene expression was observed in organs of tomato (*Solanum lycopersicum*), with a contradictory trend during plant ageing [34]. At the early stage of development, both GSNOR gene expression and activity were found to be higher in cotyledons compared to roots, whereas the expression is higher in roots and stem compared to leaves and shoot apex at later stages. The GSNOR gene is highly expressed in stamens and pistil and in fruits during ripening. Similar to phenotypes of *A. thaliana* mutants, GSNOR overexpression in tomato plant had little effect on growth and development, whereas GSNOR downregulated plants are significantly smaller, suggesting a role for NO and S-nitrosothiol signaling [61].

4.2. GSNOR in Plant Responses to Abiotic Stress

Accumulated experimental evidence has delineated the importance of GSNOR in plant responses to diverse abiotic stress conditions (reviewed in [3,62]). GSNOR gene expression and enzymatic activity are altered by plant exposure to abiotic stress stimuli, e.g., low and high temperatures, wounding, continuous light and darkness and exposure to heavy metals [63–68]. Here we present an updated overview including recent advances and reports on the modulation of GSNOR gene expression and enzymatic activity by plant exposure to abiotic stress stimuli, e.g., low and high temperatures, wounding, continuous light and darkness, and exposure to heavy metals.

4.2.1. Mechanical Injury and Wounding

GSNOR gene expression is downregulated in Arabidopsis after wounding; moreover, both GSNOR mRNA and protein levels are decreased in tobacco plants after treatment with jasmonic acid, the hormone implicated in the wounding signal transduction [69]. Another study with *A. thaliana* plants described the role of GSNOR in modulating levels of GSNO and its consequence for wound response [68]. Using wild-type and GSNOR-antisense plants, the data showed wounding-induced GSNO accumulation controlled by GSNOR. The rapid increase of GSNO was observed in the injured leaves, whereas it was detected later in vascular tissues and parenchyma of systemic leaves, suggesting the role of GSNO in the wound signal transmission through vascular tissue. In addition, GSNO accumulation was required to activate the jasmonic acid-dependent wound responses at local and systemic levels [68]. GSNOR is downregulated, at the level of gene and protein expression and enzymatic activity, in mechanically damaged sunflower (*Helianthus annuus*) seedlings, which in turn leads to an accumulation of S-nitrosothiols, specifically GSNO [64]. An increase in GSNOR activity in roots, stems, and leaves was observed in two genotypes of *Cucumis* spp., *C. sativus* and *C. melo*, and pea (*Pisum sativum*) exposed to mechanical damage of stem and leaf [67]. GSNOR activity was generally higher, but any unequivocal tendency in changes in the activity in the time of experiment relevant for all studied plants was found.

A potential role of GSNOR in plant resistance to herbivory *Manduca sexta* was examined in tobacco (*N. attenuata*) plants using a virus-induced silencing of GSNOR [70]. GSNOR-silenced plants were more susceptible to herbivore attack and decreased the herbivore-induced accumulation of phytohormones jasmonic acid (JA) and ethylene and activity of trypsin proteinase inhibitors. Moreover, it was found that GSNOR mediates some jasmonate-dependent responses, e.g., the accumulation of defense secondary metabolites.

4.2.2. Thermotolerance

GSNOR is involved in plant responses to cold and heat stress. Enzymatic activity of GSNOR is essential for the acclimation of *A. thaliana* plants to high temperature, since HOT5 mutants, plants with defect GSNOR gene, are more sensitive to high temperature as a consequence of disturbed homeostasis of S-nitrosothiols and NO-derived ROS signaling pathways [56]. GSNOR is expressed constitutively during plant development, and any significant regulation at the transcriptional level or at the level of protein induced by heat was observed. Posttranscriptional redox regulation, possibly by cysteine modifications, might be a mechanism by which the enzymatic activity is controlled [56]. NO and GSNO, as S-nitrosating agents, and GSNOR were found to be involved in the programmed cell death (PCD) induced by heat shock or H_2O_2 in tobacco (*Nicotiana tabacum*) bright yellow-2 cells [71]. NO increased in both experimentally induced PCDs, and GSNO level increased in H_2O_2-treated cells and decreased in cells exposed to heat shock, which is in accordance with lower GSNOR expression and activity observed in H_2O_2-treated cells and with higher GSNOR expression and activity in heat-shocked cells.

Low and high temperatures induce nitrosative stress in pea plants, since higher levels of NO, SNOs, and protein tyrosine nitration, markers of nitrosative stress, were detected, together with increased GSNOR activity [63]. Similarly, GSNOR activity is induced by cold stress in leaves of pepper (*Capsicum annuum* L.) plants, while NO content is lower [65]. Coincidently with previous experiments, Kubienová et al. [67] described the same trend, an increase in activity of GSNOR in pea, *Cucumis sativus*, and *Cucumis melo* plants, with differences among the studied plants and their organs and in general with stronger changes induced by cold stress in comparison with heat stress. GSNOR regulates germination of recalcitrant *Baccaurea ramiflora* seeds under chilling stress probably by modulating the total RNSs content, as enzyme inhibitors dodecanoic acid and 5-chloro-3-(2-[(4-ethoxyphenyl)ethylamino]-2-oxoethyl)-1H-indole 2-carboxylicacid caused a significant increase in total RNSs and reduced germination [72]. Chilling stress enhanced the GSNOR activity and increased the level of S-nitrosothiols, while exogenous NO and CO treatment suppressed the chilling-induced accumulation of S-nitrosothiols and induced GSNOR activity. Similarly, an increase in both S-nitrosothiols and non-protein thiols was observed in plants of *Brassica juncea* under cold-stress, suggesting that S-nitrosation might regulate redox and stress-related proteins in apoplasts [73].

The role of GSNOR in the nitrosative responses was examined in citrus plants exposed to various types of abiotic stresses. GSNOR was considerably downregulated at the level of mRNA by continuous light, salinity, and especially cold, and together the enzymatic activity was decreased in plants exposed to continuous light or dark and cold [66]. A possible role for GSNOR in regulating of cytosolic redox status and SNOs content during chilling stress was also suggested in poplar (*Populus trichocarpa*), a fast growing woody plant. NO and SNO content as well as GSNOR protein and enzymatic activity were increased in poplar leaves after chilling treatment [74].

4.2.3. Toxic Metals

Heavy metals have a toxic effect on plants, e.g., an induction of oxidative and nitrosative stress, leading to a severe growth inhibition, decreased photosynthesis, transpiration and chlorophyll content.

The relation between ROSs and RNSs and the role of NO and enzymes affecting the NO level were examined in several plant species. Peto et al. [75] described the behavior in wild-type *Nox1* and *Gsnor1-3* mutant *A. thaliana* plants during copper stress. *Nox1* is an NO overproducing plant with higher levels of L-arginine and L-citrulline, and *Gsnor1-3* is a plant with reduced GSNOR activity with higher levels of NO, S-nitrosothiols, and nitrate [24,55,56]. The strength of the stress determines the role of NO [75]. A high NO level, due to the reduced GSNOR activity, increases sensitivity under mild stress conditions, while it supports tolerance under severe stress conditions. A forty percent increase in GSNOR activity was observed in *A. thaliana* plants grown in the presence of 0.5 mM arsenate, accompanied with a significant reduction of GSNO content and a significant increase in NO content [76]. *Gsnor1-3* mutant *A. thaliana* plants with a high S-nitrosothiols level show an increased selenite tolerance [77].

GSNOR modulates NO-induced nitrosative stress in rice plants grown under aluminum stress, which leads to accumulation of both ROSs and RNSs. GSNOR gene expression and enzymatic activity were slightly higher and the enzymatic activity was significantly increased by NO treatment in rice plants grown under aluminum stress [78]. A fast increase in S-nitrosothiol content and a reduction of the leaf photosynthesis ratio is a result of suppressed GSNOR activity with specific inhibitors. In potato plants exposed to aluminum, GSNOR activity is not affected in roots and it is increased by about 20 and 45% in leaves and stems, respectively [79]. A contrary trend in the regulation of GSNOR during heavy metal stress was observed in pea leaves treated with 50 µM cadmium, where GSNOR expression and activity were decreased by about 30% [46].

4.2.4. Soil Salinity and Alkalinity

Salinity and alkalinity of soil are significant factors limiting plant growth, where $NaHCO_3$ and Na_2CO_3 are the main contributors, leading to the creation of osmotic stress, high soil pH, and excess Na^+. Metabolic regulation of NO and S-nitrosothiols was examined in tomato plants grown under alkaline stress [61]. GSNOR expression as well as protein is significantly inhibited in response to alkaline stress with levels fluctuating during the alkaline treatment. Plants overexpressing GSNOR are alkaline-tolerant, while under-expressing plants are alkaline-sensitive. During the alkaline treatment, overexpressing plants exhibit significantly increased efficiency of ROS scavenging, while under-expressing plants accumulate both ROSs and RNSs, thus leading to oxidative stress and programmed cell death. GSNOR may regulate tolerance of tomato plant to alkaline stress, having a role in regulating redox balance [61]. Similarly, a decrease in the GSNOR enzymatic activity was observed in roots of tomato (*Solanum lycopersicum*) plants treated with 120 mM NaCl [80]. Salinity caused an overall decrease in the content of redox molecules nicotinamide adenine dinucleotide phosphate (NADPH) and reduced glutathione (GSH), in contrast to increased NO levels. Salt stress upregulates GSNOR in citrus plants, and the GSNOR function is controlled by polyamines, substances involved in plant responses to abiotic stress. Significant suppression of GSNOR gene expression and enzymatic activity by polyamines in salinized citrus plants was reported, suggesting the role of GSNOR in modulating of nitrosative signaling [81].

Actions of NO, calmodulins (CaM), and GSNOR in *A. thaliana* plants in response to salt stress were described by Zhou et al. [82]. CaM is a significant Ca^{2+} sensor protein in plants acting as signaling molecule mediating reactions against various stresses; two isoforms *AtCaM1* and *AtCaM4*, encoding the same protein, are induced by salinity. Both AtCaM1 and AtCaM4 proteins bind directly to GSNOR. The protein–protein interaction inhibits GSNOR enzymatic activity and results in an increased NO level. Moreover, AtCaM4–GSNOR interaction regulates the ion balance, so it increases plant resistance to saline stress [82].

An important role of NO, GSNOR, and S-nitrosation in response to salt stress was described in a unicellular green alga *Chlamydomonas reinhardtii* [83]. NO production via increased nitrate reductase, but not NOS-like enzyme, activity was induced by salt stress to trigger the defense response. Induction or inactivation of antioxidant enzymes and GSNOR varied in connection with the duration of salt stress. Short-term stress caused the enzymes to scavenge ROSs and RNSs and balance cellular redox status. Long-term stress inactivated them significantly by RNS-induced protein S-nitrosation, resulting in oxidative damage and reduced cell viability. Salt stress induced the accumulation of S-nitrosothiols and S-nitrosation of GSNOR, glutathione S-transferase, and ubiquitin-like protein; S-nitrosation was reduced by thioredoxin-h5 (TRXh5), while it was enhanced by GSNOR inhibitor DA, suggesting the important role of GSNOR and S-nitrosation in adaptation of *C. reinhardtii* to salt stress [83].

4.2.5. Other Abiotic Stresses

GSNOR activity is modulated in response to altered light conditions, as described for the first time in *A. thaliana* HOT5 plants grown in the dark [56]. A significant increase was observed in leaves of pea plants exposed to continuous light and continuous dark [63]. A distinct trend was observed in plants of pea, *Cucumis sativus* and *Cucumis melo*, grown either in continuous darkness (etiolated plants) or transferred to a normal light regime after 7 days in the dark (de-etiolated plants) [67]. GSNOR activity in roots decreased in time in all studied plants and was not affected by different light conditions. Continuous darkness led to a significant decrease in GSNOR activity in etiolated hypocotyls, which did not recover to values of control green plants until 168 h after de-etiolation.

Water stress, a problem for plant growth and productivity, in *Lotus japonicus* leads to both oxidative and nitrosative stress. Among others, cellular NO and S-nitrosothiol content are increased, GSNOR activity is reduced, and protein tyrosine nitration is stimulated [84]. The role of GSNOR in plants of *Lamiophlomis rotata* at high altitude was described by Ma et al. [85]. The GSNOR protein level and enzymatic activity increases in connection with a rising altitude. Since GSNOR is supposed to scavenge excess RNSs, the enzyme restricts RNA damage by decomposition of RNSs.

A direct crosstalk between ROS- and NO-dependent signaling pathways was described in *A. thaliana* plants. GSNOR activity is inhibited both by H_2O_2 in vitro and by oxidative stress induced by paraquat treatment in vivo, which leads to enhanced S-nitrosothiol and nitrite levels. The loss of enzymatic activity is caused by the release of one Zn^{2+} per subunit, probably that one from the active center of the protein [86]. *A. thaliana* GSNOR1/HOT5 mutant was identified to be identical to the *paraquat resistant2-1* (par2-1) mutant, showing an anti-cell death phenotype, supporting the role of GSNOR in regulation of cell death in plants via modulation of intracellular level of NO [48]. A higher NO level was found in paraquat-resistant mutant plants and, in a similar way, wild-type plants treated with an NO donor were also resistant to paraquat. The protein level of GSNOR was increased by paraquat and decreased by NO donors, while the mRNA level was not influenced.

The role of GSNOR and the regulation of intracellular SNO levels were studied in an NO accumulation mutant (*nitric oxide excess1, noe1*) in rice (*Oryza sativa*) [87]. NOE1 was identified as a rice catalase, and an increased level of H_2O_2 was a result of its mutation, promoting the nitrate reductase-dependent induction of NO production. The overexpression of GSNOR gene reduces intracellular S-nitrosothiol content and alleviates cell death in the leaves of *noe1* plants. GSNOR is supposed to have a role during the desiccation of seeds of recalcitrant *Antiaris toxicaria*. Desiccation induces ROS accumulation leading to oxidative stress, enhances carbonylation, and reduces the S-nitrosation of antioxidant enzymes, antioxidant enzyme activities, and the seed germination rate. Treatment with GSNOR inhibitors dodecanoic acid or 3-[1-(4-acetylphenyl)-5-phenyl-1H-pyrrol-2-yl]propanoic acid further increases the level of antioxidant enzymes S-nitrosation and reverses seed germination inhibited by desiccation [88]. Exposure to both of these GSNOR inhibitors prior to NO gas, which is a well-known inducer of seed germination, leads to enhanced S-nitrosation and the activity of antioxidant enzymes ascorbate peroxidase, dehydroascorbate reductase, and glutathione reductase [88].

4.3. GSNOR in Plant Responses to Biotic Stress

NO and S-nitrosylated proteins are important signal molecules activating an immune response of plants to microbial pathogens. Thus, plant defense response to the pathogen is expected to be controlled by GSNOR, which manages NO/S-nitrosothiol homeostasis (reviewed in [89]). Díaz et al. [69] demonstrated for the first time that GSNOR gene expression is transcriptionally regulated in response to signals associated with plant defense in *A. thaliana* and tobacco. The gene expression in both mentioned plant species is induced by salicylic acid (SA), a mediator of biotic stress. The following experiments with *A. thaliana* transgenic plants [55] described GSNOR as an important component of resistance protein signaling networks. Transgenic plants with decreased GSNOR gene expression, achieved by antisense strategy, show increased content of intracellular S-nitrosothiols, constitutive activation of the pathogenesis-related (PR-1) gene, and enhanced basal resistance against biotrophic pathogen *Peronospora parasitica*. Systemic acquired resistance (SAR) was demonstrated to be enhanced in plants under-expressing GSNOR and decreased in overexpressing plants; in agreement with the rate of GSNOR expression, the level of S-nitrosothiols changed in local and systemic leaves. The SAR signal transmission might be regulated by GSNOR through the vascular system, as the enzyme was found to be localized in the phloem. Taken together with the previous published data [69], downregulation of GSNOR accompanied by an increase in the level of S-nitrosothiols is suggested to result in enhanced plant immunity [55]. The hypothesis that GSNOR is a key regulator of systemic defense responses in pathogenesis was supported by another study [68], where GSNO was found to act synergistically with salicylic acid in SAR.

Rather opposing results, when GSNOR was proposed to be a positive regulator of plant immune responses suppressing pathogen growth early in the infection process, were described earlier by Feechan et al. [24] in *A. thaliana* plants exposed to diverse microbial pathogens, e.g., bacteria *Pseudomonas syringae*, powdery mildew *Blumeria graminis*, and downy mildew *Hyaloperonospora parasitica*. Basal disease resistance was strongly reduced in the absence of GSNOR, accompanied with reduced and delayed expression of SA-dependent genes, while non-host resistance was increased in *A. thaliana* mutants overexpressing GSNOR, accompanied with accelerated expression of SA-dependent genes [24]. The content of SNOs and GSNOR activity was studied in two types of sunflower (*Helianthus annuus* L.) cultivars with different sensitivity to the pathogen *Plasmopara halstedii*, susceptible and resistant ones [47]. After infection, enzymatic activity slightly increased in a susceptible cultivar, while more explicit increase was observed in a resistant cultivar. As for the level of S-nitrosothiols after infection, it was enhanced 3.5-fold and reduced 1.5-fold in susceptible and resistant cultivars, respectively. Different spatial localization of S-nitrosothiols in hypocotyls depending on the susceptibility was observed. Different GSNOR activity was found under normal conditions in leaves of two genotypes of *Cucumis* spp. varying in susceptibility to biotrophic pathogen *Golovinomyces cichoracearum* [67]. Significantly higher enzymatic activity was found in leaves of the susceptible one compared to the moderately resistant one.

GSNOR might have an important role in NO-mediated biochemical modifications that subsequently lead to the more effective defense responses of potato plants to an attack of *Phytophthora infestans*. Potato leaf treatment with SAR inducers β-aminobutyric acid (BABA) and laminarin decreased GSNOR activity and provoked accumulation of NO and ROSs. Pre-treatment with mentioned SAR inducers before inoculation with *P. infestans* by contrast increased GSNOR activity significantly, the S-nitrosothiol pool was depleted, and potato defense responses to the pathogen were enhanced, while non-inducer pre-treated plants showed unaltered enzymatic activity, a high level of S-nitrosothiols, and lower defense responses [90]. Interestingly, no significant changes in the activity of GSNOR and S-nitrosothiols levels were observed in plants of *Medicago truncatula* after infection with *Aphanomyces euteiches* [91]. Data in that study show that resistance of *M. truncatula* against *A. euteiches* is connected with NO homeostasis, which is closely related with N nutrition.

In the plant immune response, oxidoreductase TRXh5 was found to be an effective protein-SNO reductase, providing reversibility and specificity to signaling via protein-SNO [92]. The data indicate that TRXh5 and GSNOR, enzymes exhibiting similar subcellular localization, might have partially distinct groups of protein-SNO substrates, thus regulating different immune signaling pathways. Significantly enhanced transcription of TRXh and GSNOR genes was found in transgenic plants of *Nicotiana tabacum* overexpressing γ-glutamylcysteine synthetase with higher glutathione levels, which have increased tolerance to biotrophic *Pseudomonas syringae* pv. *tabaci* [93]. GSNOR, the TRXh gene, and other genes of SA-mediated pathway dependent on non-expressor of pathogenesis-related gene 1, a transcriptional coactivator, were upregulated in tobacco BY-2 cells treated with exogenous GSH. Accumulation of NPR1 was induced by GSNO together with an enhanced SA concentration and subsequent activation of pathogenesis-related genes, leading to enhanced resistance of *A. thaliana* plants to *Pseudomonas syringae* pv. *tomato* [94]. NO induced an increase in GSH, which is indispensable to SA accumulation and non-expressor of pathogenesis-related gene 1-dependent activation of defense response. Interestingly, both SA synthesis and signaling are decreased in *Nox1*, a NO-overproducing mutant of *A. thaliana* [95]. Those plants show disabled basal resistance and resistance gene-mediated protection. Moreover, using different double mutant plants *Nox1* and *atgsnor1-1* (plant overexpressing GSNOR) or *atgsnor1-3* (plant under-expressing GSNOR), the authors suggest that NO and GSNO control cellular processes in different ways via distinct or overlapping molecular targets.

Understanding of GSNOR, thioredoxin (TRX), and their roles during biosynthesis of phenylpropanoid-derived styrylpyrone polyphenols, components inhibiting tumor proliferation and reducing hypertension and various neurodegenerative disorders [96], in co-cultured *Inonotus obliquus* and *Phellinus morii* might be, in future, employed for medicinal applications. Zhao et al. [97] described the interplay between GSNOR and TRX and their regulation via S-nitrosation/denitrosation and an impact to styrylpyrone biosynthesis. S-nitrosation of the key enzymes in the phenylpropanoid biosynthesis decreases their activity, which can be restored by TRX-mediated denitrosation. Moreover, TRX acts as a trans-nitrosylase leading to the S-nitrosation of GSNOR via a protein–protein interaction and thus a decrease in its enzymatic activity.

5. Conclusions

S-nitrosation has emerged among the key components of redox-based NO signaling that regulate the structure and activity of proteins through reversible post-translational modification of cysteine thiols. Despite the important advances in the understanding of the functions of S-nitrosation and S-nitrosothiols in plant metabolism and stress responses, major gaps in the picture of S-nitrosation on the intersection of signaling pathways of NO and ROSs still remain. Among them, the identification of NO sources and their localization contributing to S-nitrosation reactions in distinct tissues, cells, and subcellular compartments continues to be crucial in plant NO research in general. On the other hand, the mechanisms of in vivo regulations regarding the activity of GSNOR, and potentially of other more specific denitrosylases acting directly on proteins S-nitrosothiol, are still poorly described. Moreover, the development of highly specific and sensitive analytical tools to evaluate levels of both low molecular- and protein S-nitrosothiols will certainly contribute to advancement in the plant S-nitrosation field. Finally, the transfer of the knowledge obtained with model plants such as *A. thaliana* to important agricultural crops is expected to be exploited through genetic manipulation of GSNOR levels to eventually achieve desired improvements in crop yields and stress tolerance.

Author Contributions: Conceptualization, L.L. and M.P.; writing—original draft preparation, J.J.; writing—review, and editing, J.J., L.L., and M.P.

Funding: This research was funded by Palacký University in Olomouc (IGA_2018_033).

Conflicts of Interest: The authors declare no conflict of interest.

References

1. Hill, B.G.; Dranka, B.P.; Bailey, S.M.; Lancaster, J.R., Jr.; Darley-Usmar, V.M. What part of NO don't you under-stand? Some answers to the cardinal questions in nitric oxide biology. *J. Biol. Chem.* **2010**, *285*, 19699–19704. [CrossRef]
2. Groß, F.; Durner, J.; Gaupels, F. Nitric oxide, antioxidants and prooxidants in plant defense responses. *Front. Plant Sci.* **2013**, *4*, 419. [CrossRef] [PubMed]
3. Yu, M.; Lamattina, L.; Spoel, S.H.; Loake, G.J. Nitric oxide function in plant biology: A redox cue in deconvolution. *New Phytol.* **2014**, *202*, 1142–1156. [CrossRef] [PubMed]
4. Sakihama, Y.; Nakamura, S.; Yamasaki, H. Nitric oxide production mediated by nitrate reductase in the green alga *Chlamydomonas reinhardtii*: An alternative NO production pathway in photosynthetic organisms. *Plant Cell Physiol.* **2002**, *43*, 290–297. [CrossRef] [PubMed]
5. Mur, L.A.J.; Mandon, J.; Persijn, S.; Cristescu, S.M.; Moshkov, I.E.; Novikova, G.V.; Hall, M.A.; Harren, F.J.M.; Hebelstrup, K.H.; Gupta, K.J. Nitric oxide in plants: An assessment of the current state of knowledge. *AoB Plants* **2013**, *5*, pls052. [CrossRef]
6. Corpas, F.J.; Palma, J.M.; del Río, L.A.; Barroso, J.B. Evidence supporting the existence of L-arginine-dependent nitric oxide synthase activity in plants. *New Phytol.* **2009**, *184*, 9–14. [CrossRef]
7. Corpas, F.J.; Barroso, J.B. Peroxisomal plant nitric oxide synthase (NOS) protein is imported by peroxisomal targeting signal type 2 (PTS2) in a process that depends on the cytosolic receptor PEX7 and calmodulin. *FEBS Lett.* **2014**, *588*, 2049–2054. [CrossRef] [PubMed]
8. Foresi, N.; Correa-Aragunde, N.; Parisi, G.; Caló, G.; Salerno, G.; Lamattina, L. Characterization of a nitric oxide synthase from the plant kingdom: NO generation from the green alga *Ostreococcus tauri* is light irradiance and growth phase dependent. *Plant Cell* **2010**, *22*, 3816–3830. [CrossRef]
9. Jeandroz, S.; Wipf, D.; Stuehr, D.J.; Lamattina, L.; Melkonian, M.; Tian, Z.; Zhu, Y.; Carpenter, E.J.; Wong, G.K.; Wendehenne, D. Occurrence, structure, and evolution of nitric oxide synthase–like proteins in the plant kingdom. *Sci. Signal.* **2016**, *9*, re2. [CrossRef]
10. Gaston, B. Nitric oxide and thiol groups. *Biochim. Biophys. Acta* **1999**, *1411*, 323–333. [CrossRef]
11. Handy, D.E.; Loscalzo, J. Nitric Oxide and Posttranslational Modification of the Vascular Proteome. *Arterioscler. Thromb. Vasc. Biol.* **2006**, *26*, 1207–1214. [CrossRef] [PubMed]
12. Hess, D.T.; Stamler, J.S. Regulation by S-nitrosylation of Protein Posttranslational Modification. *J. Biol. Chem.* **2011**, *287*, 4411–4418. [CrossRef]
13. Lamotte, O.; Bertoldo, J.B.; Besson-Bard, A.; Rosnoblet, C.; Aimé, S.; Hichami, S.; Terenzi, H.; Wendehenne, D. Protein S-nitrosylation: Specificity and identification strategies in plants. *Front. Chem.* **2015**, *2*, 114. [CrossRef] [PubMed]
14. Corpas, F.J.; Barroso, J.B. Nitro-oxidative stress vs. oxidative or nitrosative stress in higher plants. *New Phytol.* **2013**, *199*, 633–635. [CrossRef] [PubMed]
15. Frungillo, L.; Skelly, M.J.; Loake, G.J.; Spoel, S.H.; Salgado, I. S-nitrosothiols regulate nitric oxide production and storage in plants through the nitrogen assimilation pathway. *Nat. Commun.* **2014**, *5*, 5401. [CrossRef] [PubMed]
16. Martínez-Ruiz, A.; Lamas, S. S-nitrosylation: A potential new paradigm in signal transduction. *Cardiovasc. Res.* **2004**, *62*, 43–52. [CrossRef] [PubMed]
17. Corpas, F.J.; Alché, J.D.; Barroso, J.B. Current overview of S-nitrosoglutathione (GSNO) in higher plants. *Front. Plant Sci.* **2013**, *4*, 126. [CrossRef]
18. Liu, L.; Hausladen, A.; Zeng, M.; Que, L.; Heitman, J.; Stamler, J.S. A metabolic enzyme for S-nitrosothiol conserved from bacteria to humans. *Nature* **2001**, *410*, 490–494. [CrossRef]
19. Sakamoto, A.; Ueda, M.; Morikawa, H. Arabidopsis glutathione-dependent formaldehyde dehydrogenase is an S-nitrosoglutathione reductase. *FEBS Lett.* **2002**, *515*, 20–24. [CrossRef]
20. Uotila, L.; Mannervik, B. A steady-state-kinetic model for formaldehyde dehydrogenase from human liver. A mechanism involving NAD+ and the hemimercaptal adduct of glutathione and formaldehyde as substrates and free glutathione as an allosteric activator of the enzyme. *Biochem. J.* **1979**, *177*, 869–878. [CrossRef]
21. Koivusalo, M.; Baumann, M.; Uotila, L. Evidence for the identity of glutathione-dependent formaldehyde dehydrogenase and class III alcohol dehydrogenase. *FEBS Lett.* **1989**, *257*, 105–109. [CrossRef]

22. Jensen, D.; Belka, G.; Du Bois, G. S-nitrosoglutathione is a substrate for rat alcohol dehydrogenase class III isoenzyme. *Biochem. J.* **1998**, *331*, 659–668. [CrossRef] [PubMed]
23. Staab, C.A.; Hellgren, M.; Höög, J.O. Medium- and short-chain dehydrogenase/reductase gene and protein families. *Cell. Mol. Life Sci.* **2008**, *65*, 3950–3960. [CrossRef] [PubMed]
24. Feechan, A.; Kwon, E.; Yun, B.W.; Wang, Y.; Pallas, J.A.; Loake, G.J. A central role for S-nitrosothiols in plant disease resistance. *Proc. Natl. Acad. Sci. USA* **2005**, *102*, 8054–8059. [CrossRef] [PubMed]
25. Staab, C.A.; Ålander, J.; Brandt, M.; Lengqvist, J.; Morgenstern, R.; Grafström, R.C.; Höög, J.O. Reduction of S-nitrosoglutathione by alcohol dehydrogenase 3 is facilitated by substrate alcohols via direct cofactor recycling and leads to GSH-controlled formation of glutathione transferase inhibitors. *Biochem. J.* **2008**, *413*, 493–504. [CrossRef] [PubMed]
26. Staab, C.A.; Ålander, J.; Morgenstern, R.; Grafström, R.C.; Höög, J.O. The Janus face of alcohol dehydrogenase 3. *Chem. Biol. Interact.* **2009**, *178*, 29–35. [CrossRef] [PubMed]
27. Williamson, D.H.; Lund, P.; Krebs, H.A. The redox state of free nicotinamide-adenine dinucleotide in the cytoplasm and mitochondria of rat liver. *Biochem. J.* **1967**, *103*, 514–527. [CrossRef]
28. Veech, R.L.; Guynn, R.; Veloso, D. The Time-Course of the Effects of Ethanol on the Redox and Phosphorylation States of Rat Liver. *Biochem. J.* **1972**, *127*, 387–397. [CrossRef] [PubMed]
29. Hanson, A.D.; Gage, D.A.; Shachar-Hill, Y. Plant one-carbon metabolism and its engineering. *Trend Plant Sci.* **2000**, *5*, 206–213. [CrossRef]
30. Espunya, M.C.; Díaz, M.; Moreno-Romero, J.; Martínez, M.C. Modification of intracellular levels of glutathione-dependent formaldehyde dehydrogenase alters glutathione homeostasis and root development. *Plant Cell Environ.* **2006**, *29*, 1002–1011. [CrossRef] [PubMed]
31. Engeland, K.; Höög, J.O.; Holmquist, B.; Estonius, M.; Jörnvall, H.; Vallee, B.L. Mutation of Arg-115 of human class III alcohol dehydrogenase: A binding site required for formaldehyde dehydrogenase activity and fatty acid activation. *Proc. Natl. Acad. Sci. USA* **1993**, *90*, 2491–2494. [CrossRef] [PubMed]
32. Sanghani, P.C.; Bosron, W.F.; Hurley, T.D. Human Glutathione-Dependent Formaldehyde Dehydrogenase. Structural Changes Associated with Ternary Complex Formation. *Biochemistry* **2002**, *41*, 15189–15194. [CrossRef] [PubMed]
33. Sanghani, P.C.; Robinson, H.; Bosron, W.F.; Hurley, T.D. Human Glutathione-Dependent Formaldehyde Dehydrogenase. Structures of Apo, Binary, and Inhibitory Ternary Complexes. *Biochemistry* **2002**, *41*, 10778–10786. [CrossRef] [PubMed]
34. Kubienová, L.; Kopečný, D.; Tylichová, M.; Briozzo, P.; Skopalová, J.; Šebela, M.; Navrátil, M.; Tâche, R.; Luhová, L.; Barroso, J.B.; et al. Structural and functional characterization of a plant S-nitrosoglutathione reductase from *Solanum lycopersicum*. *Biochimie* **2013**, *95*, 889–902. [CrossRef] [PubMed]
35. Xu, S.; Guerra, D.; Lee, U.; Vierling, E. S-nitrosoglutathione reductases are low-copy number, cysteine-rich proteins in plants that control multiple developmental and defense responses in Arabidopsis. *Front. Plant Sci.* **2013**, *4*, 430. [CrossRef] [PubMed]
36. Sanghani, P.C.; Robinson, H.; Bennett-Lovsey, R.; Hurley, T.D.; Bosron, W.F. Structure–function relationships in human Class III alcohol dehydrogenase (formaldehyde dehydrogenase). *Chem. Biol. Interact.* **2003**, *143–144*, 195–200. [CrossRef]
37. Crotty, J. Crystal Structures and Kinetics of S-Nitrosoglutathione Reductase from *Arabidopsis thaliana* and Human. Ph.D. Thesis, The University of Arizona, Tuscon, AZ, USA, 2009.
38. Guerra, D.; Ballard, K.; Truebridge, I.; Vierling, E. S-nitrosation of Conserved Cysteines Modulates Activity and Stability of S-nitrosoglutathione Reductase (GSNOR). *Biochemistry* **2016**, *55*, 2452–2464. [CrossRef]
39. Moulis, J.M.; Holmquist, B.; Vallee, B.L. Hydrophobic anion activation of human liver chi chi alcohol dehydrogenase. *Biochemistry* **1991**, *30*, 5743–5749. [CrossRef]
40. Wagner, F.W.; Burger, A.R.; Vallee, B.L. Kinetic properties of human liver alcohol dehydrogenase: Oxidation of alcohols by class I isoenzymes. *Biochemistry* **1983**, *22*, 1857–1863. [CrossRef]
41. Achkor, H.; Díaz, M.; Fernández, M.R.; Biosca, J.A.; Parés, X.; Martínez, M.C. Enhanced Formaldehyde Detoxification by Overexpression of Glutathione-Dependent Formaldehyde Dehydrogenase from Arabidopsis. *Plant Physiol.* **2003**, *132*, 2248–2255. [CrossRef]
42. Hedberg, J.J.; Griffiths, W.J.; Nilsson, S.J.F.; Höög, J.O. Reduction of S-nitrosoglutathione by human alcohol dehydrogenase 3 is an irreversible reaction as analysed by electrospray mass spectrometry. *Eur. J. Biochem.* **2003**, *270*, 1249–1256. [CrossRef] [PubMed]

43. Uotila, L.; Koivusalo, M. Purification of formaldehyde and formate dehydrogenases from pea seeds by affinity chromatography and *S*-formylglutathione as the intermediate of formaldehyde metabolism. *Arch. Biochem. Biophys.* **1979**, *196*, 33–45. [CrossRef]
44. Martínez, M.C.; Achkor, H.; Persson, B.; Fernández, M.R.; Shafqat, J.; Farrés, J.; Jornvall, H.; Parés, X. Arabidopsis Formaldehyde Dehydrogenase. *Eur. J. Biochem.* **1996**, *241*, 849–857. [PubMed]
45. Dolferus, R.; Osterman, J.C.; Peacock, W.J.; Dennis, E.S. Cloning of the Arabidopsis and Rice Formaldehyde Dehydrogenase Genes: Implications for the Origin of Plant ADH Enzymes. *Genetics* **1997**, *146*, 1131–1141. [PubMed]
46. Barroso, J.B.; Corpas, F.J.; Carreras, A.; Rodríguez-Serrano, M.; Esteban, F.J.; Fernández-Ocaña, A.; Chaki, M.; Romero-Puertas, M.C.; Valderrama, R.; Sandalio, L.M.; et al. Localization of *S*-nitrosoglutathione and expression of *S*-nitrosoglutathione reductase in pea plants under cadmium stress. *J. Exp. Bot.* **2006**, *57*, 1785–1793. [CrossRef] [PubMed]
47. Chaki, M.; Fernández-Ocaña, A.M.; Valderrama, R.; Carreras, A.; Esteban, F.J.; Luque, F.; Gómez-Rodríguez, M.V.; Begara-Morales, J.C.; Corpas, F.J.; Barroso, J.B. Involvement of reactive nitrogen and oxygen species (RNS and ROS) in sunflower-mildew interaction. *Plant Cell Physiol.* **2009**, *50*, 265–279. [CrossRef] [PubMed]
48. Chen, R.; Sun, S.; Wang, C.; Li, Y.; Liang, Y.; An, F.; Li, C.; Dong, H.; Yang, X.; Zhang, J.; et al. The Arabidopsis PARAQUAT RESISTANT2 gene encodes an *S*-nitrosoglutathione reductase that is a key regulator of cell death. *Cell Res.* **2009**, *19*, 1377–1387. [CrossRef] [PubMed]
49. Airaki, M.; Sánchez-Moreno, L.; Leterrier, M.; Barroso, J.B.; Palma, J.M.; Corpas, F.J. Detection and quantification of S-nitrosoglutathione (GSNO) in pepper (*Capsicum annuum* L.) plant organs by LC-ES/MS. *Plant Cell Physiol.* **2011**, *52*, 2006–2015. [CrossRef]
50. Tichá, T.; Činčalová, L.; Kopečný, D.; Sedlářová, M.; Kopečná, M.; Luhová, L.; Petřivalský, M. Characterization of *S*-nitrosoglutathione reductase from *Brassica* and *Lactuca* spp. and its modulation during plant development. *Nitric Oxide* **2017**, *68*, 68–76. [CrossRef]
51. Lindermayr, C. Crosstalk between reactive oxygen species and nitric oxide in plants: Key role of *S*-nitrosoglutathione reductase. *Free Radic. Biol. Med.* **2018**, *122*, 110–115. [CrossRef]
52. Barroso, J.B.; Valderrama, R.; Corpas, F.J. Immunolocalization of *S*-nitrosoglutathione, *S*-nitrosoglutathione reductase and tyrosine nitration in pea leaf organelles. *Acta Physiol. Plant.* **2013**, *35*, 2635–2640. [CrossRef]
53. Frungillo, L.; de Oliveira, J.F.P.; Saviani, E.E.; Oliveira, H.C.; Martínez, M.C.; Salgado, I. Modulation of mitochondrial activity by *S*-nitrosoglutathione reductase in *Arabidopsis thaliana* transgenic cell lines. *Biochim. Biophys. Acta* **2013**, *1827*, 239–247. [CrossRef] [PubMed]
54. Tichá, T.; Lochman, J.; Činčalová, L.; Luhová, L.; Petřivalský, M. Redox regulation of plant *S*-nitrosoglutathione reductase activity through post-translational modifications of cysteine residues. *Biochem. Biophys. Res. Commun.* **2017**, *494*, 27–33. [CrossRef]
55. Rustérucci, C.; Espunya, M.C.; Díaz, M.; Chabannes, M.; Martínez, M.C. *S*-nitrosoglutathione reductase affords protection against pathogens in Arabidopsis, both locally and systemically. *Plant Physiol.* **2007**, *143*, 1282–1292. [CrossRef] [PubMed]
56. Lee, U.; Wie, C.; Fernandez, B.O.; Feelisch, M.; Vierling, E. Modulation of Nitrosative Stress by *S*-Nitrosoglutathione Reductase Is Critical for Thermotolerance and Plant Growth in Arabidopsis. *Plant Cell* **2008**, *20*, 786–802. [CrossRef] [PubMed]
57. Kwon, E.; Feechan, A.; Yun, B.W.; Hwang, B.H.; Pallas, J.; Kang, J.G.; Loake, G. AtGSNOR1 function is required for multiple developmental programs in Arabidopsis. *Planta* **2012**, *236*, 887–900. [CrossRef] [PubMed]
58. Shi, Y.F.; Wang, D.L.; Wang, C.; Culler, A.H.; Kreiser, M.A.; Suresh, J.; Cohen, J.D.; Pan, J.; Baker, B.; Liu, J.Z. Loss of GSNOR1 Function Leads to Compromised Auxin Signaling and Polar Auxin Transport. *Mol. Plant.* **2015**, *8*, 1350–1365. [CrossRef]
59. Wang, P.; Du, Y.; Hou, Y.J.; Zhao, Y.; Hsu, C.C.; Yuan, F.; Zhu, X.; Tao, W.A.; Song, C.P.; Zhu, J.K. Nitric oxide negatively regulates abscisic acid signaling in guard cells by *S*-nitrosylation of OST1. *Proc. Natl. Acad. Sci. USA* **2015**, *112*, 613–618. [CrossRef]

60. Airaki, M.; Leterrier, M.; Valderrama, R.; Chaki, M.; Begara-Morales, J.C.; Barroso, J.B.; del Río, L.A.; Palma, J.; Corpas, F.J. Spatial and temporal regulation of the metabolism of reactive oxygen and nitrogen species during the early development of pepper (*Capsicum annuum*) seedlings. *Ann. Bot.* **2015**, *116*, 679–693. [CrossRef]
61. Gong, B.; Wen, D.; Wang, X.; Wei, M.; Yang, F.; Li, Y.; Shi, Q. S-nitrosoglutathione reductase-modulated redox signaling controls sodic alkaline stress responses in *Solanum lycopersicum* L. *Plant Cell Physiol.* **2015**, *56*, 790–802. [CrossRef]
62. Salgado, I.; Martínez, M.C.; Oliveira, H.; Frungillo, L. Nitric oxide signaling and homeostasis in plants: A focus on nitrate reductase and S-nitrosoglutathione reductase in stress-related responses. *Braz. J. Bot.* **2013**, *36*, 89–98. [CrossRef]
63. Corpas, F.J.; Chaki, M.; Fernández-Ocaña, A.; Valderrama, R.; Palma, J.M.; Carreras, A.; Begara-Morales, J.C.; Airaki, M.; del Río, L.A.; Barroso, J.B. Metabolism of reactive nitrogen species in pea plants under abiotic stress conditions. *Plant Cell Physiol.* **2008**, *49*, 1711–1722. [CrossRef] [PubMed]
64. Chaki, M.; Valderrama, R.; Fernández-Ocaña, A.M.; Carreras, A.; Gómez-Rodríguez, M.V.; Pedrajas, J.R.; Begara-Morales, J.C.; Sánchez-Calvo, B.; Luque, F.; Leterrier, M.; et al. Mechanical wounding induces a nitrosative stress by downregulation of GSNO reductase and an increase in S-nitrosothiols in sunflower (*Helianthus annuus*) seedlings. *J. Exp. Bot.* **2011**, *62*, 1803–1813. [CrossRef] [PubMed]
65. Airaki, M.; Leterrier, M.; Mateos, R.M.; Valderrama, R.; Chaki, M.; Barroso, J.B.; del Río, L.A.; Palma, J.M.; Corpas, F.J. Metabolism of reactive oxygen species and reactive nitrogen species in pepper (*Capsicum annuum* L.) plants under low temperature stress. *Plant Cell Environ.* **2012**, *35*, 281–295. [CrossRef] [PubMed]
66. Ziogas, V.; Tanou, G.; Filippou, P.; Diamantidis, G.; Vasilakakis, M.; Fotopoulos, V.; Molassiotis, A. Nitrosative responses in citrus plants exposed to six abiotic stress conditions. *Plant Physiol. Biochem.* **2013**, *68*, 118–126. [CrossRef] [PubMed]
67. Kubienová, L.; Tichá, T.; Jahnová, J.; Luhová, L.; Mieslerová, B.; Petřivalský, M. Effect of abiotic stress stimuli on S-nitrosoglutathione reductase in plants. *Planta* **2014**, *239*, 139–146. [CrossRef]
68. Espunya, M.C.; De Michele, R.; Gómez-Cadenas, A.; Martínez, M.C. S-nitrosoglutathione is a component of wound- and salicylic acid-induced systemic responses in *Arabidopsis thaliana*. *J. Exp. Bot.* **2012**, *63*, 3219–3227. [CrossRef]
69. Díaz, M.; Achkor, H.; Titarenko, E.; Martínez, M.C. The gene encoding glutathione-dependent formaldehyde dehydrogenase/GSNO reductase is responsive to wounding, jasmonic acid and salicylic acid. *FEBS Lett.* **2003**, *543*, 136–139. [CrossRef]
70. Wünsche, H.; Baldwin, I.T.; Wu, J. S-Nitrosoglutathione reductase (GSNOR) mediates the biosynthesis of jasmonic acid and ethylene induced by feeding of the insect herbivore *Manduca sexta* and is important for jasmonate-elicited responses in *Nicotiana attenuate*. *J. Exp. Bot.* **2011**, *62*, 4605–4616. [CrossRef]
71. De Pinto, M.C.; Locato, V.; Sgobba, A.; Romero-Puertas, M.C.; Gadaleta, C.; Delledonne, M.; De Gara, L. S-Nitrosylation of Ascorbate Peroxidase Is Part of Programmed Cell Death Signaling in Tobacco Bright Yellow-2 Cells. *Plant Physiol.* **2013**, *163*, 1766–1775. [CrossRef]
72. Bai, X.G.; Chen, J.H.; Kong, X.X.; Todd, C.D.; Yang, Y.P.; Hu, X.Y.; Li, D.Z. Carbon monoxide enhances the chilling tolerance of recalcitrant *Baccaurea ramiflora* seeds via nitric oxide-mediated glutathione homeostasis. *Free Radic. Biol. Med.* **2012**, *53*, 710–720. [CrossRef] [PubMed]
73. Sehrawat, A.; Deswal, R. S-nitrosylation analysis in *Brassica juncea* apoplast highlights the importance of nitric oxide in cold-stress signaling. *J. Proteome Res.* **2014**, *13*, 2599–2619. [CrossRef] [PubMed]
74. Cheng, T.; Chen, J.; Ef, A.A.; Wang, P.; Wang, G.; Hu, X.; Shi, J. Quantitative proteomics analysis reveals that S-nitrosoglutathione reductase (GSNOR) and nitric oxide signaling enhance poplar defense against chilling stress. *Planta* **2015**, *242*, 1361–1390. [CrossRef] [PubMed]
75. Pető, A.; Lehotai, N.; Feigl, G.; Tugyi, N.; Ördög, A.; Gémes, K.; Tari, I.; Erdei, L.; Kolbert, Z. Nitric oxide contributes to copper tolerance by influencing ROS metabolism in Arabidopsis. *Plant Cell Rep.* **2013**, *32*, 1913–1923. [CrossRef] [PubMed]
76. Leterrier, M.; Airaki, M.; Palma, J.M.; Chaki, M.; Barroso, J.B.; Corpas, F.J. Arsenic triggers the nitric oxide (NO) and S-nitrosoglutathione (GSNO) metabolism in Arabidopsis. *Environ. Pollut.* **2012**, *166*, 136–143. [CrossRef]
77. Lehotai, N.; Kolbert, Z.; Pető, A.; Feigl, G.; Ördög, A.; Kumar, D.; Tari, I.; Erdei, L. Selenite-induced hormonal and signaling mechanisms during root growth of *Arabidopsis thaliana* L. *J. Exp. Bot.* **2012**, *63*, 5677–5687. [CrossRef] [PubMed]

78. Yang, L.; Tian, D.; Todd, C.D.; Luo, Y.; Hu, X. Comparative Proteome Analyses Reveal that Nitric Oxide Is an Important Signal Molecule in the Response of Rice to Aluminum Toxicity. *J. Proteome Res.* **2013**, *12*, 1316–1330. [CrossRef]
79. Arasimowicz-Jelonek, M.; Floryszak-Wieczorek, J.; Drzewiecka, K.; Chmielowska-Bąk, J.; Abramowski, D.; Izbiańska, K. Aluminum induces cross-resistance of potato to *Phytophthora infestans*. *Planta* **2014**, *239*, 679–694. [CrossRef]
80. Manai, J.; Gouia, H.; Corpas, F.J. Redox and nitric oxide homeostasis are affected in tomato (*Solanum lycopersicum*) roots under salinity-induced oxidative stress. *J. Plant Physiol.* **2014**, *171*, 1028–1035. [CrossRef]
81. Tanou, G.; Ziogas, V.; Belghazi, M.; Christou, A.; Filippou, P.; Job, D.; Fotopoulos, V.; Molassiotis, A. Polyamines reprogram oxidative and nitrosative status and the proteome of citrus plants exposed to salinity stress. *Plant Cell Environ.* **2014**, *37*, 864–885. [CrossRef]
82. Zhou, S.; Jia, L.; Chu, H.; Wu, D.; Peng, X.; Liu, X.; Zhang, J.; Zhao, J.; Chen, K.; Zhao, L. Arabidopsis CaM1 and CaM4 Promote Nitric Oxide Production and Salt Resistance by Inhibiting S-nitrosoglutathione Reductase via Direct Binding. *PLoS Genet.* **2016**, *12*, e1006255. [CrossRef] [PubMed]
83. Chen, X.; Tian, D.; Kong, X.; Chen, Q.; Abd-Allah, E.F.; Hu, X.; Jia, A. The role of nitric oxide signaling in response to salt stress in *Chlamydomonas reinhardtii*. *Planta* **2016**, *244*, 651–669. [CrossRef] [PubMed]
84. Signorelli, S.; Corpas, F.J.; Borsani, O.; Barroso, J.B.; Monza, J. Water stress induces a differential and spatially distributed nitro-oxidative stress response in roots and leaves of *Lotus japonicas*. *Plant Sci.* **2013**, *201–202*, 137–146. [CrossRef] [PubMed]
85. Ma, L.; Yang, L.; Zhao, J.; Wei, J.; Kong, X.; Wang, C.; Zhang, X.; Yang, Y.; Hu, X. Comparative proteomic analysis reveals the role of hydrogen sulfide in the adaptation of the alpine plant *Lamiophlomis rotata* to altitude gradient in the Northern Tibetan Plateau. *Planta* **2015**, *241*, 887–906. [CrossRef] [PubMed]
86. Kovacs, I.; Holzmeister, C.; Wirtz, M.; Geerlof, A.; Fröhlich, T.; Römling, G.; Kuruthukulangarakoola, G.T.; Linster, E.; Hell, R.; Arnold, G.J.; et al. ROS-Mediated Inhibition of S-nitrosoglutathione Reductase Contributes to the Activation of Anti-oxidative Mechanisms. *Front. Plant Sci.* **2016**, *7*, 1669. [CrossRef]
87. Linh, L.H.; Linh, T.H.; Xuan, T.D.; Ham, L.H.; Ismail, A.M.; Khanh, T.D. Molecular Breeding to Improve Salt Tolerance of Rice (*Oryza sativa* L.) in the Red River Delta of Vietnam. *Int. J. Plant Genom.* **2012**, *2012*, 949038. [CrossRef]
88. Bai, X.G.; Yang, L.; Tian, M.; Chen, J.; Shi, J.; Yang, Y.; Hu, X. Nitric Oxide Enhances Desiccation Tolerance of Recalcitrant *Antiaris toxicaria* Seeds via Protein S-Nitrosylation and Carbonylation. *PLoS ONE* **2011**, *6*, e20714. [CrossRef]
89. Malik, S.I.; Hussain, A.; Yun, B.W.; Spoel, S.H.; Loake, G.J. GSNOR-mediated de-nitrosylation in the plant defence response. *Plant Sci.* **2011**, *181*, 540–544. [CrossRef]
90. Janus, Ł.; Milczarek, G.; Arasimowicz-Jelonek, M.; Abramowski, D.; Billert, H.; Floryszak-Wieczorek, J. Normoergic NO-dependent changes, triggered by a SAR inducer in potato, create more potent defense responses to *Phytophthora infestans*. *Plant Sci.* **2013**, *211*, 23–34. [CrossRef]
91. Thalineau, E.; Truong, H.N.; Berger, A.; Fournier, C.; Boscari, A.; Wendehenne, D.; Jeandroz, S. Cross-Regulation between N Metabolism and Nitric Oxide (NO) Signaling during Plant Immunity. *Front. Plant Sci.* **2016**, *7*, 472. [CrossRef]
92. Kneeshaw, S.; Gelineau, S.; Tada, Y.; Loake, G.J.; Spoel, S.H. Selective protein denitrosylation activity of Thioredoxin-h5 modulates plant Immunity. *Mol. Cell* **2014**, *56*, 153–162. [CrossRef] [PubMed]
93. Ghanta, S.; Bhattacharyya, D.; Sinha, R.; Banerjee, A.; Chattopadhyay, S. *Nicotiana tabacum* overexpressing γ-ECS exhibits biotic stress tolerance likely through NPR1-dependent salicylic acid-mediated pathway. *Planta* **2011**, *233*, 895–910. [CrossRef] [PubMed]
94. Kovacs, I.; Durner, J.; Lindermayr, C. Crosstalk between nitric oxide and glutathione is required for NONEXPRESSOR OF PATHOGENESIS-RELATED GENES 1 (NPR1)-dependent defense signaling in *Arabidopsis thaliana*. *New Phytol.* **2015**, *208*, 860–872. [CrossRef] [PubMed]
95. Yun, B.W.; Skelly, M.J.; Yin, M.; Yu, M.; Mun, B.G.; Lee, S.U.; Hussain, A.; Spoel, S.H.; Loake, G.J. Nitric oxide and S-nitrosoglutathione function additively during plant immunity. *New Phytol.* **2016**, *211*, 516–526. [CrossRef] [PubMed]

96. Zheng, W.; Miao, K.; Liu, Y.; Zhao, Y.; Zhang, M.; Pan, S.; Dai, Y. Chemical diversity of biologically active metabolites in the sclerotia of *Inonotus obliquus* and submerged culture strategies for upregulating their production. *Appl. Microbiol. Biotechnol.* **2010**, *87*, 1237–1254. [CrossRef] [PubMed]
97. Zhao, Y.; He, M.; Ding, J.; Xi, Q.; Loake, G.J.; Zheng, W. Regulation of Anticancer Styrylpyrone Biosynthesis in the Medicinal Mushroom *Inonotus obliquus* Requires Thioredoxin Mediated Transnitrosylation of S-nitrosoglutathione Reductase. *Sci. Rep.* **2016**, *6*, 37601. [CrossRef] [PubMed]

© 2019 by the authors. Licensee MDPI, Basel, Switzerland. This article is an open access article distributed under the terms and conditions of the Creative Commons Attribution (CC BY) license (http://creativecommons.org/licenses/by/4.0/).

MDPI
St. Alban-Anlage 66
4052 Basel
Switzerland
Tel. +41 61 683 77 34
Fax +41 61 302 89 18
www.mdpi.com

Plants Editorial Office
E-mail: plants@mdpi.com
www.mdpi.com/journal/plants